CW00602878

Lysosomal Cysteine Proteases

SECOND EDITION

Series editor

Peter Sheterline, *Department of Human Anatomy and Cell Biology, University of Liverpool, Liverpool L69 3BX, UK*

Advisory Editors

Martin Humphries, *School of Biological Sciences, University of Manchester, Stopford Building, Manchester M13 9PT, UK*

Peter Downes, *Department of Biochemistry, Medical Sciences Institute, University of Dundee, Dundee DD1 4HN, Scotland*

Editorial Board

Alan Barrett, *Peptidase Laboratory, The Babraham Institute, Babraham Hall, Cambridge, CB2 4AT, UK.*

Julio E. Celis, *Aarhus University, Department of Medical Biochemistry, Ole Worms Alle, Bldg 170, University Park, DK-8000, Aarhus, Denmark.*

Benny Geiger, *Department of Chemical Immunology, Weizmann Institute of Science, Rehovot 76100, Israel.*

Keith Gull, *School of Biological Sciences, University of Manchester, Oxford Road, Manchester, UK.*

Tony Hunter, *The Salk Institute, 10010 North Torrey Pines Road, La Jolla, CA 92037, USA.*

Katsuhiko Mikoshiba, *Department of Molecular Neurobiology, Institute of Medical Sciences, University of Tokyo, 4-6-1 Shirokanedai, Minato-ku Tokyo 108, Japan.*

Thomas D. Pollard, *Department of Cell Biology/Anatomy, Johns Hopkins University School of Medicine, 725 North Wolfe St, Baltimore, MD 21205-2105, USA.*

Erkki Ruoslahti, *Cancer Research Center, La Jolla Cancer Research Foundation, 10901 North Torrey Pines Road, La Jolla, CA 92037-1062, USA.*

Urs Rutishauser, *Department of Genetics, Case Western Reserve University School of Medicine, 2119 Abington Road, Cleveland, OH 44106-2333, USA.*

Solomon H. Snyder, *Department of Neuroscience, Pharmacology and Molecular Sciences, Johns Hopkins University School of Medicine, 725 North Wolfe St, Baltimore, MD 21205, USA.*

Rupert Timpl, *Abteilung Proteinchemie, Max-Planck-Institut für Biochemie, D82152 Martinsried, Germany.*

Titles Published

Tyrosine Phosphoprotein Phosphatases (second edition). *Barry Goldstein*

Lysosomal Cysteine Proteinases (second edition). *Heidrun Kirschke, Alan J. Barrett and Neil D. Rawlings*

Helix–Loop–Helix Transcription Factors (third edition). *Trevor Littlewood and Gerard Evan*

Forthcoming Titles

Actins (fourth edition). *Peter Sheterline, Jon Clayton and John Sparrow*

Actin monomer-binding proteins. *Uno Lindberg and Clarence Schutt*

Intermediate filament proteins (third edition). *Roy Quinlan, Chris Hutchison and Birgit Lane*

Gelsolin family. *Horst Hinssen*

Kinesins (third edition). *Viki Allan and Rob Cross*

Myosins (second edition). *Jim Sellers and Holly Goodson*

Cadherins. *Sasha Bershadsky and Benny Geiger*

Integrins. *Martin Humphries*

IgCAMs (third edition). *Fritz Rathjen and Thomas Brümmendorf*

Matrix metalloproteinases. *Fred Woessner and Hideaki Nagase*

Network- and filament-forming collagens. *Michel Van der Rest and Bernard Dublet*

Nuclear receptors (second edition). *Hinrich Gronemeyer and Vincent Laudet*

EF-hand calcium-binding proteins (third edition). *Hiroshi Kawasaki and Bob Kretsinger*

Heterotrimeric G proteins (third edition). *Steve Pennington*

Serine phosphoprotein phosphatases. *Patricia Cohen and David Barford*

Signalling domains. *Bruce Mayer and Matthias Wilmann*

Proline isomerases. *Andrzej Galat*

Subtilisins. *Roland Siezen*

Thermolysins. *Rob Beynon and Ann Beaumont*

Lysosomal Cysteine Proteases

SECOND EDITION

Heidrun Kirschke

Institut für Physiologische Chemie, Medizinische Fakultät,
Martin-Luther-Universität Halle-Wittenberg, Halle (Saale), Germany

Alan J. Barrett and Neil D. Rawlings

MRC Peptidase Laboratory, Department of Immunology,
The Babraham Institute, Babraham, Cambridgeshire, United Kingdom

OXFORD   NEW YORK   TOKYO
OXFORD UNIVERSITY PRESS
1998

Oxford University Press, Great Clarendon Street, Oxford OX2 6DP
Oxford New York
Athens Auckland Bangkok Bogota Bombay
Buenos Aires Calcutta Cape Town Dar es Salaam
Delhi Florence Hong Kong Istanbul Karachi
Kuala Lumpur Madras Madrid Melbourne
Mexico City Nairobi Paris Singapore
Taipei Tokyo Toronto Warsaw
and associated companies in
Berlin Ibadan

Oxford is a trade mark of Oxford University Press

Published in the United States
by Oxford University Press, Inc., New York

© Oxford University Press, 1998

All rights reserved. No part of this publication may be
reproduced, stored in a retrieval system, or transmitted, in any
form or by any means, without the prior permission in writing of Oxford
University Press. Within the UK, exceptions are allowed in respect of any
fair dealing for the purpose of research or private study, or criticism or
review, as permitted under the Copyright, Designs and Patents Act, 1988, or
in the case of reprographic reproduction in accordance with the terms of
licences issued by the Copyright Licensing Agency. Enquiries concerning
reproduction outside those terms and in other countries should be sent to
the Rights Department, Oxford University Press, at the address above.

This book is sold subject to the condition that it shall not,
by way of trade or otherwise, be lent, re-sold, hired out, or otherwise
circulated without the publisher's prior consent in any form of binding
or cover other than that in which it is published and without a similar
condition including this condition being imposed
on the subsequent purchaser.

A catalogue record for this book is available from the British Library

Library of Congress Cataloging in Publication Data
(Data available)

ISBN 0 19 850249 4

Typeset and printed by the Alden Group, Oxford

Series preface

The Protein Profile *series has been developed from a recognition that individuals find it increasingly difficult to access readily the enormous amount of information accumulated by the international research community; information which is crucial for the efficiency and quality of their activities.*

The *Protein Profile* series aims to provide a practical, comprehensive and accessible information source on all major families of proteins. Each volume of *Protein Profile* is focused on a single family or sub-family of proteins, and contains tables and figures presenting as comprehensive an accumulation of structural, kinetic and biochemical information as is available on that particular protein group, coupled to an extensive bibliography. Every volume will be refined and updated to provide users with a practical up-to-date single source of information by the publication of new editions approximately every two years.

Content

The text provides a brief overview of the biological context of the function of the protein group followed by an overview of available information on:

- function
- kinetic and biochemical properties
- sequences, sequence relationships and sequence features
- domain structure
- mutations
- 3-D structure
- binding sites of protein
- ligand binding sites and interactions with drugs
- derivatization sites
- proteolytic cleavage sites

Each volume follows the same format but with different emphases depending on the protein family. The text is extensively supported by tables and figures listing key information gathered from the literature with comprehensive reference to primary sources.

Available references pertinent to the properties, structure and function of the proteins are listed in a full bibliography. References are numbered for access in the text, but are also arranged alphabetically under headings to allow browsing. Reviews are listed at the beginning and key papers are identified.

An **online version** of the Protein Profile Series will be available during 1998. For the latest information about the series, see our web page at
`http://www.oup.co.uk/protein_profile`

Protein Profile

online

Protein Profile is the most comprehensive resource available for information on protein families. **Protein Profile** is available as a series of books, but will also be published online from the middle of 1998 – a new service which will provide a focal point for information in the field of protein science.

A subscription to **Protein Profile Online** will give you:

> ➢ online access to all the **Protein Profile** volumes published by Oxford University Press

> ➢ frequent updates of the data

> ➢ links to abstracts or, if you subscribe to online journals, to complete papers

> ➢ links back to the original sequence and structural data

There is no better way of getting quickly to the information you need.

Protein Profile Online will be introduced during 1998 and there will be a free trial period for you to explore and provide us with feedback on the content and presentation. If you would like to be informed by e-mail or post of the latest news of **Protein Profile Online** please complete the online information request form which you can find at:

http://www.oup.co.uk/protein_profile/?b97

Alternatively, you can write to: Journals Marketing Department, Oxford University Press, Great Clarendon Street, Oxford OX2 6DP, UK. *Tel: +44 (0)1865 267907 Fax: +44 (0)1865 267835 E-mail: jnl.orders@oup.co.uk*

Contents

Cathepsin K (EC 3.4.22.38)

Dipeptidyl peptidase I (EC 3.4.14.1)

Legumain (EC 3.4.22.34)

Other lysosomal cysteine proteinases

Structures and evolution

Reaction with inhibitors

List of tables

Acknowledgement

The authors would like to thank Dr Ria Baumgrass (Potsdam-Rehbruecke, Germany) for drawing the structures of the genes and mRNA.

Abbreviations

Ac, Acetyl
BANA, Benzyloxycarbonyl-L-arginyl-2-naphthylamide
Boc, t-Butyloxycarbonyl
Bu, Butyl
Bu^t, t-Butyl
Bz, Benzoyl
Bzl, Benzyl
Cit, Citrulline
CM, Carboxymethyl
conA, Concanavalin A
CTLA-2α, Antigens from mouse cytotoxic lymphocytes
CTLA-2β, Antigens from mouse cytotoxic lymphocytes
DEAE, Diethylaminoethyl
DPP I, Dipeptidyl peptidase I
EDTA, Ethylene diamine tetra-acetate
ER, Endoplasmic reticulum
EST, Expressed sequence tags
FISH, Fluorescent *in-situ* hybridization
FPLC, Fast protein liquid chromatography
GlySc, Glycinal semicarbazone

Har, Homoarginine
IgG, Immunoglobulin G
IL, Interleukin
M6P, Mannose-6-phosphate
NHNap, 2-Naphthylamide
NHNapOMe, 2-(4-Methoxy)naphthylamide
NHMec, 7-(4-Methyl)coumarylamide
NHMecF, 7-(4-Fluoromethyl)coumarylamide
NHPhNO$_2$, 4-Nitroanilide
OMe, Methylester
OPhNO$_2$, (4-Nitro)phenylester
PAMs, Accepted point mutations
Ph, Phenyl
rcath, Recombinant cathepsin
SDS–PAGE, sodium dodecyl sulfate–polyacrylamide gel electrophoresis
Suc, Succinyl
TIMP-1, Tissue inhibitor of metallopeptidases
Z, Benzyloxycarbonyl

Introduction

Lysosomal cysteine proteinases are proteolytic enzymes that in their mature form are localized in lysosomes and the catalytic activity of which is based on a cysteine residue in the active site. They are synthesized as preproenzymes with molecular masses of 37 000–55 000 Da, and are processed to the mature, catalytically active enzymes with molecular masses of 23 000–30 000 Da.

Lysosomal cysteine proteinases catalyse the hydrolysis of peptide, amide, ester and thiol ester bonds. The mechanism of this hydrolysis is discussed in several reviews, e.g. [269]. The catalytic activity of these enzymes is dependent on pH values below 7 and the presence of reducing compounds.

The main *in vivo* function of the cysteine proteinases in lysosomes is the degradation of proteins that have been taken up by the cell or originate from other compartments of the same cell. The end-products of the acidic overall proteolysis by cysteine proteinases and all other lysosomal proteolytic enzymes are amino acids and dipeptides which diffuse through the lysosomal membrane and are available again for protein synthesis in the cell.

Four cysteine peptidases are ubiquitous in lysosomes of animals: cathepsins B, H, L and dipeptidyl peptidase I [8, 29, 33, 35]. Cathepsin L has endopeptidase activity, cathepsin B has endopeptidase and peptidyl dipeptidase activity, dipeptidyl peptidase I acts mainly as a dipeptidyl peptidase, and cathepsin H acts mainly as an aminopeptidase. In contrast to these enzymes with broad substrate specificity, another lysosomal cysteine endopeptidase, legumain, exhibits a strict specificity of hydrolysing only asparaginyl bonds [1783]. Cathepsin S and several recently discovered lysosomal cysteine peptidases, such as cathepsin K and others, seem to have a specific tissue distribution which implies tissue-specific functions.

Studies on mice deficient in cathepsin L [1790] and those deficient in cathepsin B (J. Deussing and C. Peters, personal communication) suggest that the enzymes can obviously substitute for each other, because the mice were viable and fertile.

However, obviously, lysosomal cysteine proteinases also have functions outside the lysosomes and the cells. Some cells such as macrophages, fibroblasts, transformed cells and others secrete the precursors of the enzymes. The released procathepsins can be activated – but proteolysis by cysteine proteinases outside the lysosomes is regulated by physiological pH values and natural inhibitors. An alteration of this balance in favour of the enzymes leads to uncontrolled proteolysis as seen in several disorders such as, for example, inflammation, arthritis and tumour growth.

Cathepsin B
(EC 3.4.22.1)

Introduction

In 1941 Fruton and co-workers [1377] proposed a classification of the proteolytic enzymes of animal tissues in which they included an enzyme that hydrolysed Bz-Arg-NH$_2$, a synthetic substrate of trypsin. At that time, this thiol-dependent enzyme was termed *cathepsin II*, but the nomenclature was further revised by Tallan *et al.* in 1952 [1401], who renamed cathepsin II as *cathepsin B*. Further characterization was done by Greenbaum and Fruton [1379]. Otto and co-workers, however, showed that two enzymes had contributed to the activity assigned to 'cathepsin B' in earlier work. Thus, gel filtration of enzymes from rat liver lysosomes separated components of 25 kDa and 52 kDa, both of which hydrolysed Bz-Arg-NH$_2$. The 25 kDa endopeptidase was briefly known as *cathepsin B'* and *cathepsin B$_1$*, but in reviewing this work in 1971, Otto [1272] termed the enzyme *cathepsin B1*. The second enzyme (cathepsin B2) was a carboxypeptidase, and when this was renamed 'lysosomal cysteine-type carboxypeptidase' (EC 3.4.18.1), the original name of cathepsin B could be used once more for the endopeptidase. In 1972, the nomenclature committee of International Union of Biochemistry (IUB) recommended the name *cathepsin B*, with the EC number 3.4.22.1.

Cathepsin B appears to be ubiquitous in mammals, having been detected in nearly all organs and tissues. Also enzymes with similar properties occur in many lower organisms.

Cathepsin B is released from normal cells, such as macrophages, fibroblasts, osteoclasts and others, and from malignantly transformed cells, together with the other lysosomal enzymes under normal culturing conditions or when the lysosomal system is stimulated. The secretion forms are mainly the proenzyme and several truncated forms, which differ from the mature, processed form of cathepsin B in being stable at neutral pH. Catalytically active forms seem to be secreted by special cells and cells cultured at pH values below 7.0 [1565].

The properties of cathepsin B vary very little between species.

The enzyme is catalytically active in the presence of SH-containing compounds at pH 5–6.

Gene structure and chromosomal location

The cathepsin B gene is located on chromosome 8p22–p23.1 [1148, 1210]. Gong and co-workers [1153] showed that there are at least 12 exons and 11 introns in the human gene, which is more than 20 kbp in length (Fig. 1).

The size of the most abundant form of cathepsin B mRNA from human, rat and mouse is 2.2-2.3 kb [605, 667, 674, 689, 691, 815, 984, 1153, 1560, 1570, 1576, 1709, 1719], from cattle 2.6 kb [786, 1802] and from birds 1.8 kb and 2.4 kb [1786]. (For accession numbers of the cDNA see Table 1.) Several splice variants have been detected, e.g. 4.0 kb and 5.0 kb transcripts in murine and human normal tissues and tumours [605, 667, 674, 689, 691, 1153, 1570, 1576, 1719]. The 2.3 kb and 4.0 kb (and 5.0 kb) mRNA forms differ in their 3' untranslated regions: the former is terminated in exon 11, and the latter results from processing at a cryptic intron donor site in exon 11 and splicing to exon 12 [1153], as shown in Fig. 1. Other splice variants show differences in their 5' untranslated regions; they lack exon 2 [1152, 1778, 1782], lack exons 1 and 2 [1778], or contain additional exons 2a and 2b [1778] (Fig. 1). All these splice variants encode the normal preprocathepsin B. However, mRNA variants from tumours lacking exons 2 and 3 [1153], or exons 1, 2 and 3 [1778], produce truncated procathepsin B without the signal peptide and more than half of the propeptide [1153]. This truncated

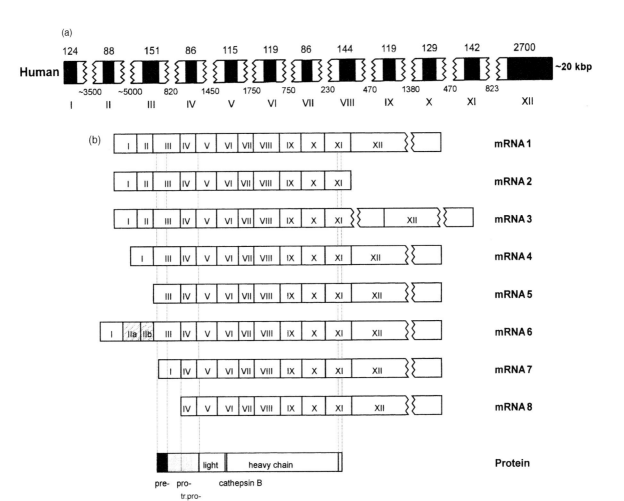

Figure 1
Schematic structures of the human cathepsin B gene, mRNA variants and protein. Exons are numbered with roman numerals from 5′ to 3′ in the direction of transcription. (a) Gene structure: black rectangles represent exons with sizes given above in base pairs; white, broken areas represent introns. (b) Structures of mRNA variants and protein [1244]: mRNA1, mRNA2, mRNA3 [1153], mRNA4 [1152, 1778, 1782], mRNA5, mRNA6 [1778], mRNA7 [1153], mRNA8 [1778]; pre- and propeptides, truncated propeptide (tr.pro-) and mature cathepsin B are indicated.

procathepsin B with M_r 32 000 Da has been expressed *in vitro* and shown to be catalytically active [1519]. This form is not expected to be transported into lysosomes *in vivo*, however.

The human [1153] and murine [1145] cathepsin B genes have been characterized as housekeeping genes with GC-rich regions around exon 1 which are known to bind the transcription factor Sp1. However, in addition to this constitutive expression, there seems to be a possibility of regulated transcription [1244].

Cathepsin B expression could be regulated post-transcriptionally through several mRNA splice variants that have different stabilities and translation rates.

A number of compounds, including cytokines, phorbol esters, dexamethasone and interferon-γ alter the levels of cathepsin B mRNA, activity and protein, but, as pointed out by Berquin and Sloane [1244], these agents may change the differentiation state of the cell rather than acting directly on the expression of cathepsin B.

Table 1 Accession numbers of the cathepsin B cDNA in the EMBL/GenBank database

Species	Accession No.
Homo sapiens	L16510
	L22569
	L38712
	M13230
	U44029
Rattus norvegicus	M11305
Mus musculus	M14222
	X54966
Bos taurus	L06075
	M64620
Gallus gallus	X73074
	U18083

Relative levels of cathepsin B mRNA have been determined in rat [1719] and murine tissues [689]. Comparison of the mRNA and protein [348] levels in rat organs, and mRNA and activity [689] levels in murine organs, has revealed that transcriptional and translational product levels correlate less well in the tissues studied.

Domain structure

Species: human; compiled from [365, 1134, 1146, 1192]

		Papain numbering
Signal peptide	1–17	p1–p65
N-Propeptide	18–79	p66–p133
Mature enzyme	80–333	1–212B
Light chain	80–126	1–43
Heavy chain	129–333	46–212B
C-Propeptide	334–339	212C–212H
Active site Cys	108	25
Active site His	278	159
Active site Asn	298	175
Disulfide bond	93, 122	14, 39
Disulfide bond	105, 150	22, 63
Disulfide bond	141, 207	56, 95
Disulfide bond	142, 146	57, 60A
Disulfide bond	179, 211	85, 99
Disulfide bond	187, 198	92A, 92L

		Papain numbering
Carbohydrate-binding site	192	92F
Molecular mass (calculated)	37 807 Da	
Number of residues	339 AA	

Other species: rat [1126, 1194, 1198]; bovine [786, 1131, 1175, 1176, 1178, 1181]; mouse [259, 1134, 1149, 1186]

Molecular masses determined by SDS–PAGE

Procathepsin B	42 000 Da
Single chain	30 000 Da
Heavy chain	25 000 Da
Light chain	~5 000 Da

Isoelectric point: 4.5–5.5 (mature cathepsin B) [306, 360, 427, 1198, 1236]

Precursors

After synthesis of preprocathepsin B by ribosomes bound to the endoplasmic reticulum, the signal peptide is removed (see p. 39).

Purification of latent procathepsin B has been described from a microsomal fraction of rat liver by chromatography on conA-Sepharose, Sepharose-Gly-Phe-GlySc and immobilized anti-cathepsin B IgG [1107].

The recombinant procathepsin B has been purified from culture media of different yeast strains [1110, 1126, 1758, 1766] and from inclusion bodies of *Escherichia coli* [1391].

Studies on the maturation of procathepsin B, mainly in experiments *in vitro*, have revealed that the propeptide can be cleaved autocatalytically [1111, 1126, 1766] as well as by cathepsin D [1107, 1121, 1126], pepsin [607, 671, 1391], serine peptidases [805, 1102] and metallopeptidases [1104]. Only auto-catalytic processing (at pH 4.5 for 10 min at 40°C), described as an intramolecular mechanism [1766], produced mature cathepsin B with the correct amino terminus, whereas the products of all the other enzymes were elongated by a few amino acid residues [1126]. Neutrophil elastase (EC 3.4.21.37) was shown to generate a 'truncated procathepsin B' with M_r 38 000 Da which was alkaline-stable and proteolytically active and which has been detected in purulent sputum [805, 806, 1632].

The propeptide has been shown to be a potent slow-binding inhibitor of cathepsin B at pH 6.0 [456]. The non-covalent complexes of mature cathepsin B and its propeptide formed during autocatalytic maturation of the proenzyme [1766] are stable at pH $\geq$6 and dissociate at pH $\leq$4.5 where the propeptide is degraded [1110].

The three-dimensional structure of procathepsin B has been reported [1758, 1827].

Purification

Cathepsin B has been purified from tissues of several species. The enzyme has been quantified in various rat tissues in concentrations of 92 ng of cathepsin B per 1 mg of protein of skeletal muscle to 1147 ng per 1 mg of protein of kidney [348].

The isolation of cathepsin B starts usually from tissue extracts, fractionated by ammonium sulfate or acetone, or from a lysosomal fraction.

The behaviour of cathepsin B in column chromatography is generally predictable from its physicochemical properties. Cathepsin B can be separated from cathepsin L by most cation exchange chromatography media, to which the latter shows very high affinity. Cathepsin B binds to DEAE–cellulose (DEAE–Sephacell) at low ionic strength and below pH 7, whereas cathepsins H and S are eluted under these conditions.

Covalent chromatography on organomercurial–Sepharose was effective in the purification of cathepsin B [306], but not conA–Sepharose, which has very low affinity for cathepsin B.

Cathepsin B has been purified from crude extracts or ammonium sulfate fractions by affinity chromatography using 2,2'-dipyridyl disulfide-activated thiol–Sepharose [293], immobilized peptidyl cystamine [323] and immobilized semicarbazone of Gly-Phe-glycinal [384].

The enzyme has been isolated from tumours [365, 677].

Cathepsin B has been crystallized [1226, 1228, 1236, 1821].

The enzyme obtained by most of the described methods is a mixture of single-chain and double-chain forms and is only partially active. In contrast, purification procedures, including activated thiol–Sepharose or other affinity chromatography media [293, 323, 384] reacting with the active site cysteine, yield preparations in which the cathepsin B is essentially 100% active.

Purification of cathepsin B has been described [115, 216, 293, 301, 306, 310, 315, 316, 318, 321–324, 329, 332, 334, 337, 358–360, 365, 367, 369–371, 373, 374, 380, 382, 384, 390, 394–397, 399, 404, 410, 411, 1107, 1236].

Activation and inhibition

Cathepsin B is catalytically active and stable in the pH range 5–6.5, but is irreversibly inactivated at higher pH values. Optimal activity is achieved in the presence of thiol compounds and EDTA [1294].

Like all cysteine peptidases, cathepsin B reacts with general thiol-blocking agents such as iodoacetate, iodoacetamide and N-ethylmaleimide.

Some covalent inhibitors such as peptidyl diazomethanes and epoxysuccinyl peptides are suitable active site titrants of cathepsin B, because they react rapidly and selectively with the enzyme and not with thiol compounds generally (see p. 48; Tables 24, 27).

Special epoxysuccinyl peptides such as CA074 have proved to be selective inhibitors of cathepsin B (see Table 27) [445, 566].

Cathepsin B is sensitive to leupeptin, a tight-binding, reversible inhibitor which also inhibits serine peptidases (see p. 53; Table 28).

Some members of the cystatin family such as cystatins A and D and the kininogens have a lower affinity for cathepsin B than for the other cysteine peptidases (see Table 31).

Hydrolysis of synthetic substrates

Cathepsin B has a broad substrate specificity and cleaves several synthetic substrates in the presence of SH-containing compounds at about pH 6.0 (Table 2). Cathepsin B exhibits a broad P1 specificity

Table 2 Kinetic constants for cathepsin B with synthetic substrates

Substrate	pH	k_{cat} (s^{-1})*	K_m (µM)	k_{cat}/K_m (mM^{-1}s^{-1})	Species	References
Z-R-R-NHMec	6.0	158.0	184	859	Human	81
	6.0	150.0	180	833	Rat	172
	6.0	23.0	230	100	Rat	279
Z-R-R-NHNap	6.0	157.0	170	924	Human	1167
	6.0	178.0	190	937	Human	173
Z-F-R-NHMec	6.0	364.0	252	1444	Human	81
	6.0	364.0	223	1632	Rat	93
			225		Porcine	396
Bz-F-V-R-NHMec	6.0	17.5	29	603	Rat	93
Z-V-V-R-NHMec	6.0	14.0	31	452	Rat	93
Boc-F-L-R-NHMec	6.0	46.2	107	432	Rat	93
Boc-F-F-R-NHMec	6.0	22.7	114	199	Rat	93
Bz-R-NHNap	6.0	25.0	4300	6	Human	8
	6.0	24.0	4000	6	Rat	172
	6.0	27.0	3500	8	Bovine	172

* k_{cat} values are based on molarities determined by titration [8]; temperature was 25–30°C.

[1297]. The enzyme has a hydrophobic S2 subsite allowing hydrolysis of substrates such as Z-Phe-Arg-NHMec with Phe in the P2 position. In addition, Glu205 located at the bottom of the S2 pocket (see Fig. 6) can interact with polar amino acid residues in the P2 position and thus cleave substrates such as Z-Arg-Arg-NHMec (see p. 37). Studies on the specifity of the S3 and S1′ subsites revealed a preference of cathepsin B for Tyr in the P3 position [1354] and a large hydrophobic residue in the P1′ position [235].

The specificity of cathepsin B for hydrolysis of dibasic substrates is unique among the lysosomal cysteine peptidases [173, 224]. However, these substrates are susceptible to hydrolysis by non-cysteine proteases that occur in tissues, so controls must be included to establish with certainty the identity of the enzyme that is being assayed [366].

The methylcoumarylamide substrates are suitable for both continuous and stopped assays [8, 308]. There are several methods of visualizing active cysteine proteinases in gel media and in tissue sections. If a 4-methoxynaphthylamide substrate is used, the reaction product can be detected by coupling with a diazonium salt after blocking of the thiol activator [306, 307, 366]. Naphthylamine released from a substrate can also be coupled with 5-nitrosalicylaldehyde [634].

The influence of pH on k_{cat}/K_m has been studied [147, 167, 168, 173, 291, 292].

Hydrolysis of polypeptides and proteins

Very little relationship has been found between the specificity of cathepsin B for cleavage of synthetic substrates and that for cleavage of peptide bonds in polypeptides.

The specificity of the endopeptidase activity of cathepsin B is unclear. Some proteins and polypeptides are degraded by the peptidyl- dipeptidase activity of cathepsin B (Table 3).

Data from the cleavage by cathepsin B of the oxidized insulin B chain do not allow one to recognize distinct specificities of the S1 and S2 subsites of cathepsin B; the amino acid in P1 varies greatly, and glycine and arginine as well as hydrophobic amino acids can be found as P2 residues of the cleavage site [232].

Inactivation, activation and degradation of several enzymes and proteins by cathepsin B have been described (Table 3).

Table 3 Degradation and modification of proteins by cathepsin B

Protein	Mode of action by cathepsin B	pH	References
Actin	Degradation	6.5	242, 334
		5.0	218, 262
Acetyl-CoA-carboxylase	Stepwise degradation; two polypeptides of 125 and 115 kDa following degraded to 50–100 kDa polypeptides	6.3	288
Albumin	Degradation	3.5–5.0	82
Antigens	Processing		220, 258, 259
Basement membranes	Degradation of glomerular basement membrane and basement membrane matrix	6.0	80, 81, 112, 140, 142, 205, 283
Class II-invariant chain	Limited proteolysis		256
Collagen	Hydrolysis in the telopeptide region, liberating mainly α chains; degradation of collagen in basement membranes and bone	3.5–4.5	94, 96, 125, 128, 206, 296, 360, 587
Enkephalin; Leu-, Met-	Degradation; peptidyl dipeptidase action	6.0	76, 207, 208, 210, 211, 334
Fibrinogen	Degradation	6.0	133
Fibronectin	Degradation		94, 154, 725
Fructose-1,6-bisphosphate aldolase	Inactivation; release of four dipeptides from the C terminus	6.0	88, 144, 163, 181, 244, 253
Glucagon	Degradation; peptidyl dipeptidase action: release of five dipeptides from the C terminus	3.0	74
Histones	Degradation		146, 263
IgG (heavy chain)	Identification of three major cleavages		121
Insulin A chain	Degradation; peptidyl dipeptidase action: release of six dipeptides from the C terminus	6.0	92
Insulin B chain	Various split positions		165, 232, 245
Laminin	Degradation	6.5–7.0	94, 140, 142, 191, 282, 725
Myelin proteins	Degradation	6.0	297
Myosin (heavy chain)	Degradation		218, 242, 262, 334
Osteocalcin	Degradation		247
Osteonectin	Degradation		247
Parathyroid hormone	Degradation		132
Proapolipoprotein AII	Conversion to the mature protein		139
Procathepsin D	Activation of the precursor 53 kDa → 48 kDa	4.0–7.0	266
Proelastase (pancreatic)	Activation of the precursor	3.8	143
Procollagenase	Activation of the precursor		120
Proinsulin	Conversion to insulin	5.0–6.0	79, 116, 828, 1291
Prorenin	Activation of the precursor 53 kDa → 50 kDa → 40 kDa	5.5	45, 153, 201, 257, 265, 274, 278, 286, 289, 937, 1014, 1339
Proplasminogen activator (urokinase type)	Activation of the precursor; cleavage at Lys158-Ile159	5.0–6.0	174, 645, 647, 703
Prostromelysin	Activation of the precursor		238
Proteoglycans	Multiple but distinct cleavage sites on link proteins		99, 129, 261, 959, 1179
Substance P	Degradation	6.0	208
Thyroglobulin	Limited proteolysis; three cleavage points	6.0	118
Troponin T	Degradation		218, 242
Trypsinogen	Activation of the zymogen	3.8, 5.0	738, 768

Cathepsin H
(EC 3.4.22.16)

Introduction

In 1976 Kirschke and co-workers [169] named a lysosomal peptidase *cathepsin H*. This enzyme was termed *L20C21* in 1972 by the same authors [340] and later *cathepsin B₃* [343], because of similarities to cathepsin B (then known as cathepsin B₁). Histochemical demonstration by Sylvén in 1968 [718] of a thiol-dependent lysosomal enzyme active against Leu-NHNap is now attributable to cathepsin H. In 1975 Davidson and Poole [1372] partially purified an enzyme hydrolysing Bz-Arg-NHNap, but different from cathepsin B, from rat liver lysosomes. This enzyme obviously was cathepsin H. A *BANA hydrolase* from rat skin described by Järvinen and Hopsu-Havu in 1975 [1385] was identified in 1985 by immunological methods as cathepsin H [1009]. A *benzoylarginine-β-naphthylamide hydrolase* isolated from rabbit lung by Singh and Kalnitsky in 1978 [392] was reported to degrade collagen [264] and in 1983 was named *cathepsin I* by Kalnitsky *et al.* [1261]. However, in later work by Kirschke and co-workers in 1986 [171], preparations of this enzyme from rabbit lung and rat lung did not show collagenolytic activity, but showed high cross-reactivity with an antibody to cathepsin H from rat liver. There was therefore no reason to retain the name cathepsin I for the enzyme from rabbit lung. The name *cathepsin H* (EC 3.4.22.16) was recommended by the nomenclature committee of IUB in 1981.

Cathepsin H belongs to a group of cysteine proteinases that are important for the overall degradation of proteins in lysosomes, where it may act mainly as an aminopeptidase with broad substrate specificity, since no other aminopeptidases have yet been identified in lysosomes.

Cathepsin H appears to be ubiquitous in mammalian cells and shows a close relationship not only to cathepsins L and S but also to proteinases of the cellular slime mould and of plants.

Cathepsin H shows only small variations between species.

Gene structure and chromosomal location

Work with human–mouse somatic cell hybrids has shown that the cathepsin H gene maps to chromosome 15q24–q25 [1210].

The cathepsin H gene of rat consists at least of 12 exons and spans in total more than 17.5 kbp [1158] (Fig. 2).

Cathepsin H mRNA was found to be 1.7 kb (human) [1151], 1.8 kb (rat) [1157] and 1.5 kb (mouse) [1678]. No splice variants have been detected.

In contrast to cathepsin B [1153], but similar to cathepsin S [1196], the promoter region of rat cathepsin H has low GC content, suggesting that cathepsin H is not a constitutive gene product. On the other hand, no TATA or CAAT boxes have been detected [1158].

Interferon-γ has been shown to upregulate the expression of cathepsin H mRNA [1678].

Overexpression of cathepsin H has also been detected in some diseases such as cancer and arthritis, but the mechanisms of regulation are not yet known. The accession numbers of the cathepsin H cDNA in the EMBL/GenBank database are given in Table 4.

Domain structure

Species: human; compiled from [1150, 1151, 1173, 1191]

		Papain numbering
Signal peptide	1–22	p1–p29
N-Propeptide	23–115	p36–p133
Mini chain	98–105	p113–p121
Mature enzyme	116–335	1–212A
Heavy chain	116–292	1–169A
Light chain	293–335	169F–212A

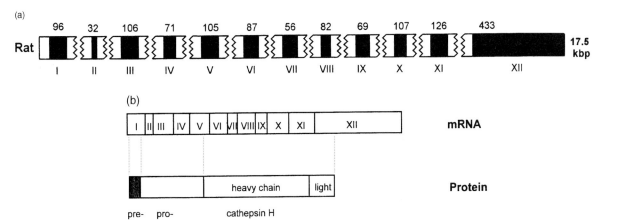

Figure 2
Schematic structures of the rat cathepsin H gene, mRNA and protein. Exons are numbered with roman numerals from 5′ to 3′ in the direction of transcription. (a) Gene structure: black rectangles represent exons, with sizes given above in base pairs; white, broken areas represent introns. (b) Structures of mRNA and protein [1244]: pre- and propeptides and mature cathepsin H are indicated [1158].

	Papain numbering	
Active site Cys	141	25
Active site His	281	159
Active site Asn	301	175
Disulfide bond	138, 181	22, 63
Disulfide bond	172, 214	56, 95
Disulfide bond	272, 322	153, 200
Carbohydrate-binding site	72	p86
Carbohydrate-binding site	101	p117
Carbohydrate-binding site	230	112
Molecular mass (calculated)	37 403 Da	
Number of residues	335 AA	

Other species: rat [995, 1157, 1158, 1198]; mouse [1678]

Molecular masses determined by SDS–PAGE

Procathepsin H	41 000 Da
Single chain	28 000 Da
Heavy chain	23 000 Da
Light chain	∼5 000 Da

Isoelectric point: 5.7–7.1 (mature cathepsin H) [302, 318, 344, 356, 369, 380, 388, 389, 1771]

Precursors

After synthesis of preprocathepsin H by ribosomes bound to the endoplasmic reticulum, the signal peptide is removed (see p. 39).

Table 4 Accession numbers of the cathepsin H cDNA in the EMBL/GenBank database

Species	Accession No.	Coding region	References
Homo sapiens	X16832	Prepro-minichain	1150
Homo sapiens	X07549	Minichain–mature	1151
Rattus norvegicus	Y00708	Complete	1157, 1158
Mus musculus	U06119	Complete	1678

Latent procathepsin H is reported to be converted to the mature enzyme by cathepsin D [1119] and metallopeptidases [1104].

Purification

Cathepsin H has been purified from tissues of several species and from tumour cells.

The enzyme has been quantified in various rat tissues in concentrations of 27 ng cathepsin H per 1 mg of protein of brain to 1429 ng per 1 mg of protein of kidney [348].

With some tissues that can be obtained immediately after death, it is feasible to take advantage of a 50-fold purification factor attainable by isolation of lysosomes [341, 342, 344]. Otherwise, working with an extract of the whole tissue one encounters the difficulty that cathepsin H is partially complexed with endogenous inhibitors. Separation of cathepsin H from the enzyme–inhibitor complexes requires autolysis at acidic pH values followed by fractionation with ammonium sulfate [392] or acetone [389].

The behaviour of cathepsin H in column chromatography is generally predictable from its physicochemical properties. Cathepsin H can be separated from cathepsin L by most cation exchange chromatography media, to which the latter shows a very high affinity. Separation from cathepsins B and S is also possible, but there are differences between species [295]. Cathepsins H and S are eluted from DEAE–cellulose (DEAE–Sephacell) at low ionic strength and below pH 7, whereas cathepsin B binds to anion exchange chromatography media under these conditions.

Chromatography on conA–Sepharose was effective in the purification of cathepsin H [389, 408].

Cathepsin H is reported to have been purified in a one-step procedure by affinity chromatography using an immobilized inhibitor from potato tubers [381].

Cathepsin H has been obtained by most of the described methods as a mixture of single-chain and double-chain forms containing about 50–60% active enzyme in the preparation. Purification by activated thiol–Sepharose (activated thiopropyl–Sepharose, etc.) gave completely active preparations, and the single-chain form of cathepsin H has been obtained using an immobilized inhibitor [381].

Purification of cathepsin H has been described [264, 275, 293, 300, 302, 315, 318, 337, 344, 355, 356, 358, 367, 369, 380–382, 388, 389, 392, 397, 410–412, 1771].

Activation and inhibition

Purified cathepsin H is stable in the pH range 5.0–6.5, but seems to be unstable in diluted tissue extracts after freezing. Optimal activity is achieved in the presence of thiol compounds and EDTA at pH 6.5–6.8. Cathepsin H is irreversibly inactivated above pH 7.0.

General thiol-blocking agents such as iodoacetate, 4-chloromercuribenzoate and N-ethylmaleimide react with cathepsin H.

The rate constant for the reaction of E-64 with cathepsin H was found to be relatively low (see Table 27). Therefore, the titration of cathepsin H with E-64 was less satisfactory. The incompleteness of the reaction at low concentrations of both cathepsin H and E-64 causes deviation from linearity of the titration curve below 30% activity, but above this, the inactivation followed a linear relationship [431].

Some other covalent inhibitors such as Phe-CH$_2$F and Ser(OBzl)-CHN$_2$ are selective inhibitors of cathepsin H, but the inactivation rates are also relatively low [419]. Phe-acyl-hydroxamate reacts faster with cathepsin H than the other affinity-labelling reagents (see Table 25) and may be a better active site titrant.

In contrast to cathepsins B and L, cathepsin H is inhibited relatively poorly by leupeptin (see Table 28).

Bestatin and puromycin, strong inhibitors of the cytosolic aminopeptidases, do not react with cathepsin H. In the presence of one of these inhibitors (20–100 μM), the aminopeptidase activity of cathepsin H can be determined in crude tissue extracts.

The affinity of most of the cystatins to cathepsin H is very high, displaying K_i values in the picomolar range (see Table 31).

α_2-Macroglobulin has been reported to inhibit both the aminopeptidase and endopeptidase activity of cathepsin H [512].

Table 5 Kinetic constants for cathepsin H with synthetic substrates

Substrate	pH	k_{cat} (s^{-1})*	K_m (μM)	k_{cat}/K_m (mM^{-1}s^{-1})	Species	References
R-NHNap	6.5	16.6	71	234	Rat	172
	6.5	13.8	100	138	Rabbit	264
	6.8		97		Human	389
R-NHMec	6.8	8.0	100	80	Rat	279
	6.8	4.8	97	49	Rat	1229
	6.8	3.1	40	77	Bovine	295
R-NHMecF	6.8	10.0	100	100	Rat	279
R-NHPhNO$_2$	6.8	13.0	200	65	Rat	279
Cit-NHMec	6.8	3.1	76	40	Rat	1229
K-NHNap	6.5	12.6	92	137	Rat	172
	6.5	11.5	120	96	Rabbit	264
Bz-F-V-R-NHMec	6.8	1.6	25	64	Bovine	295
Bz-R-NHNap	6.5	11.6	1200	10	Rabbit	264
	6.8	8.8	460	19	Porcine	397
Bz-R-NHMec	6.8	0.63	321	2	Rat	1229
Z-P-A-A-A-P-NH$_2$	6.5	7.7	3500	2	Rat	91
Z-P-A-A-A-P	6.5	5.3	7400	1	Rat	91
Suc-A-A-A-A-NHPhNO$_2$	6.5	0.9	700	1	Rat	91

* k_{cat} values are based on molarities determined by titration [8]; temperature was 30°C.

Hydrolysis of synthetic substrates

Cathepsin H acts as both an aminopeptidase and an endopeptidase (Table 5). The pH optimum of cathepsin H for hydrolysis of most substrates in the presence of SH-containing compounds is 6.5–6.8. Specific substrates for cathepsin H amongst the lysosomal cysteine proteinases are N-terminal unblocked amino acid

Table 6 Degradation and modification of proteins by cathepsin H

Protein	Mode of action by cathepsin H	pH	References
Albumin	Degradation	3.5–5.0	82
α_1-Proteinase inhibitor	Inactivation	Neutral	199
Angiotensin I, II	Degradation	6.0	108
Basement membrane	Degradation of glomerular basement membrane and basement membrane matrix	6.0	80, 112, 140, 142, 283
Bradykinin	Inactivation; cleavage at Gly4-Phe5, Phe5-Ser6	Neutral	198
		6.0	108
Complement (C5)	Degradation and generation of peptides with chemotactic activity		248
Fibrinogen	Degradation		725
Fibronectin	Degradation		725
Leukotriene D$_4$	Transformation to leukotriene E$_4$ by release of Gly		298
Prorenin	Activation of the precursor		201
Proteoglycans	Degradation		261, 282
Substance P	Degradation		108

derivatives such as Arg-NHMec, and Arg-NHNap. Among the 2-naphthylamide substrates, the most sensitive to cathepsin H are those of phenylalanine, tryptophan, arginine, leucine, lysine and alanine (approximately in this order) [275]. Cathepsin H hydrolyses endopeptidase substrates, such as Bz-Arg-NHNap, Bz-Phe-Val-Arg-NHMec, and acts on Pro-Gly-Phe and Pro-Arg-NHNap much like a dipeptidyl peptidase. Pro—X bonds in peptides or prolyl derivatives are not hydrolysed by cathepsin H. Substrates such as Arg-NHMec and Arg-NHNap are susceptible to hydrolysis by non-cysteine proteases that occur in tissues (e.g. arginine aminopeptidase, EC 3.4.11.6). Controls must be included to establish with certainty the identity of the enzyme that is being assayed.

The influence of pH on k_{cat}/K_m has been studied [292, 1229].

Hydrolysis of polypeptides and proteins

The peptide bond specificity of the endopeptidase action of cathepsin H has not yet been determined. Cathepsin H was shown to cleave several proteins (Table 6), but the endopeptidase activity of the enzyme is limited. Collagen and laminin, for instance, were not degraded by cathepsin H.

Cathepsin L
(EC 3.4.22.15)

Introduction

In 1971, Bohley and co-workers [1248] and, in 1972, Kirschke *et al.* [340] described a new enzyme from rat liver lysosomes which degraded proteins but not synthetic substrates. In 1974 the enzyme was named *cathepsin L* (in which 'L' stands for lysosomes) by the same authors [343, 1284]. In 1976 and 1977 cathepsin L was characterized in detail by Kirschke and colleagues [169, 341]. At about the same time, Towatari and co-workers [285, 401] isolated a new cathepsin from rat liver lysosomes which also proved to be cathepsin L. The first indication that this enzyme might be involved in the process of malignant tumour growth came from work in 1978 on the precursor of cathepsin L called *major excreted protein* (MEP) by Gottesman [630]. The name *cathepsin L* (EC 3.4.22.15) was recommended by the nomenclature committee of IUB in 1978.

Cathepsin L appears to be ubiquitous in vertebrate cells and shows almost no differences in properties between species [32, 170, 171, 212, 213].

Cathepsin L seems to have its main functions in the degradation of proteins in lysosomes and also after secretion as a precursor outside the cell. The function of procathepsin L or cathepsin L outside the cell is not quite clear yet. Procathepsin L is secreted in large amounts from normal cells such as macrophages, fibroblasts, osteoclasts and others, from cells treated with growth factors or tumour promoters, and from malignantly transformed cells. The involvement of the enzyme in several processes such as activation of proplasminogen activator (urokinase type) initiating a proteinase cascade [629], degradation of matrix proteins (see Table 9 below), inhibition of normal antigen processing [221] and promotion of proliferation processes [904] has been reported. Studies on cathepsin L-deficient mice has not yet confirmed an essential role of cathepsin L in intracellular protein catabolism nor in putative extracellular functions. Transgenic mice lacking cathepsin L mRNA and activity are viable, but they are partially devoid of hair and their mortality is enhanced [1790 and unpublished results by Roth *et al.*].

Gene structure and chromosomal location

The human cathepsin L gene maps to chromosome 9q21−q22 [1135, 1144].

The cathepsin L gene consists of eight exons and seven introns and spans ~5.1 kbp (human), ~7.4 kbp (mouse) [1135] and 8.5 kbp (rat) [1159]. The exons of different species have been shown to have the same size with the exceptions of exons 1 and 8 (Fig. 3), but the introns vary widely in their lengths [1135, 1159].

The size of cathepsin L mRNA from human, rat and mouse is 1.5−1.8 kb [599, 656, 904, 1152, 1162, 1165, 1203, 1709, 1788], but variants with sizes of 2.3 kb [1709] and ~4.0 kb [1165] have also been detected. Some of the accession numbers of the cathepsin L cDNA in the EMBL/GenBank database are given in Table 7.

Four splice variants of human cathepsin L mRNA that have been analysed differ in their 5$'$ untranslated regions (Fig. 3). Exon 1 encompasses 278 nucleotides [1135, 1152], 251 and 188 nucleotides, whereas the two latter forms lack 27 and 90 nucleotides, respectively, from the 3$'$ end of exon 1 [1807]. Another variant lacks exon 1 and contains exon 1a, derived from the 3$'$ end of intron 1 [1135, 1165].

All these splice variants encode the normal preprocathepsin L, but differ in their stability and translational activity.

The synthesis of cathepsin L is induced by oncogenes, growth factors and tumour promoters [19], the latter having been shown to stimulate

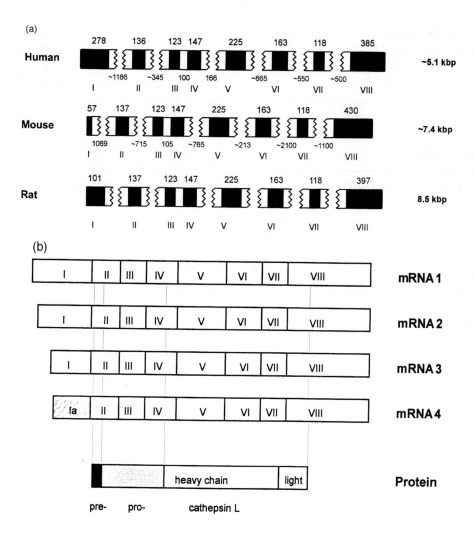

Figure 3

Schematic structures of the cathepsin L genes, mRNA and protein. Exons are numbered with roman numerals from 5′ to 3′ in the direction of transcription. (a) Structures of the genes of the human, mouse and rat [1135, 1159]: black rectangles represent exons, with sizes given above in base pairs; white, broken areas represent introns, with sizes given below in base pairs. (b) Structures of the human cathepsin L mRNA variants and of protein [1244]; mRNA1 and mRNA4 [1135], mRNA2 and mRNA3 [1807]; pre- and propeptides and mature cathepsin L are indicated.

Table 7 Accession numbers of the cathepsin L cDNA in the EMBL/GenBank database

Species	Accession No.	Coding region	References
Homo sapiens	L06426	Complete	1135
Rattus norvegicus	Y00697	Complete	1162
Mus musculus	J02583	Complete	1182
Mus musculus	Y00314	Complete	1205
Sarcophaga peregrina	D16533	Complete	1659

expression at the transcriptional level, requiring protein synthesis [1776]. Other compounds such as interferon-γ [1679] and cytokines [904] seem to change the differentiation of the cell rather than acting directly on the expression of cathepsin L.

Domain structure

Species: human; compiled from [1152, 1164, 1165, 1174, 1191]

		Papain numbering
Signal peptide	1–17	p1–p22
N-Propeptide	18–113	p37–p133
Mature enzyme	114-333	1–211
Heavy chain	114–288	1–169B
Light chain	292–333	169C–211
Active site Cys	138	25
Active site His	276	159
Active site Asn	300	175
Disulfide bond	135, 178	22, 63
Disulfide bond	169, 211	56, 95
Disulfide bond	269, 322	153, 200
Carbohydrate-binding site	221	106
Molecular mass (calculated)	37 564 Da	
Number of residues	333 AA	

Other species: rat [1159, 1162, 1202]; mouse [715, 1165, 1182, 1205]; chicken [1143, 1208, 1209]

Molecular masses determined by SDS–PAGE

Procathepsin L	38 000 Da
Single chain	28 000 Da
Heavy chain	24 000 Da
Light chain	∼4 000 Da

Isoelectric point: 5.0–6.3 (mature cathepsin L) [341, 360, 362, 363, 409, 1141].

Precursors

Cathepsin L is synthesized as a preproenzyme and the signal peptide is cotranslationally removed in the endoplasmic reticulum (see p. 39). The carbohydrate chain of the remaining procathepsin L is processed in the Golgi apparatus and the proenzyme is targeted by the mannose-6-phosphate marker to the lysosomal–endosomal compartment. In some cells, and after overexpression, procathepsin L is predominantly secreted as a result of an intrinsically low affinity for the mannose-6-phosphate receptor [613, 1103].

Lack of glycosylation leads to secretion of stable procathepsin L [1106], whereas the propeptide has been shown to be indispensable for proper folding and for its secondary effects, such as stabilization of the proenzyme molecule, trafficking out of the endoplasmic reticulum and mannose phosphorylation [1773]. In addition, a six-residue sequence within the propeptide mediates binding of procathepsin L to an integral microsomal membrane protein (M_r 43 000 Da) and thus may be important for the proper targeting of this lysosomal enzyme [1115, 1769].

Purification of the latent procathepsin L has been described from the culture medium of normal and transformed cells [135, 328, 336, 904] and from spermatozoa [1393].

The isoelectric points of mouse procathepsin L are 7.3, 7.0 and 6.6 [904].

Studies on the activation of procathepsin L have revealed that limited proteolysis of the propeptide can occur autocatalytically to produce mature enzyme [1127, 1393] at pH 3.0–3.5 [214, 1114], at negatively charged surfaces at pH 5.0–6.0 [1113, 1312], by cathepsin D [1120, 1121, 1129] and by metallopeptidases [1104]. After autocatalysis at 4°C, an intermediate form appears with M_r 31 000 Da and a six amino acid extension at the amino terminus possessing almost no activity [1764]. Similar intermediate forms, with M_r 31 000–33 000 Da have been detected in a non-secreting transformed lymphocyte cell line [1537].

The propeptide of cathepsin L has been identified as an inhibitor [1428]. The ability of the propeptide to act as an inhibitor of mature cathepsin L may also be revealed by studies on the crystallographic structure of procathepsin L [1818, 1819, 1824].

Recent studies suggest that secreted procathepsin L, not the mature enzyme, may be involved in special physiological processes such as cell proliferation and differentiation [1258, 1593, 1659] and steroidogenesis, the latter in a complex with TIMP-1 [1630].

Purification

Cathepsin L has been purified from tissues of several species. The enzyme has been quantified in various rat tissues in concentrations of 39 ng of cathepsin L per 1 mg of protein of skeletal muscle to 606 ng per 1 mg of protein of kidney [304].

Cathepsin L was first detected in a lysosomal fraction from rat liver [169, 285, 340, 341, 343, 401], taking advantage of a 50-fold purification factor and a separation from cytosolic inhibitors attainable by isolation of lysosomes. Otherwise, working with an extract of the whole tissue, one encounters the difficulty that cathepsin L is partially complexed with endogenous inhibitors. Separation of cathepsin L from the enzyme–inhibitor complex requires autolysis at acidic pH values followed by fractionation with ammonium sulfate [362, 363]. Three-phase partitioning in t-butanol–water–ammonium sulfate has also been used [378]. Cathepsin L can be separated from the other lysosomal cysteine proteinases by most cation exchange chromatography media, to which

the enzyme shows an anomalously high affinity [285, 340, 341, 362, 363, 372].

Cathepsin L binds only partially to conA–Sepharose.

Purification of cathepsin L and procathepsin L has been described [135, 169, 216, 285, 302, 304, 315, 316, 328, 336, 337, 340, 341, 356, 358, 360, 362, 363, 369, 372, 378, 382, 401, 409, 411, 1141, 1370, 1375].

Activation and inhibition

Mature cathepsin L is stable in the pH range 4.5–6.0. The enzyme is catalytically active in the presence of thiol compounds and EDTA at pH 3.5–7.0, but the stability is decreased below pH 4.5 and above pH 6.0. The stability is also affected by ionic strength and reductive agents [1293, 1294].

Like all cysteine peptidases, cathepsin L reacts with general thiol-blocking reagents such as 4-chloromercuribenzoate, iodoacetate and N-ethylmaleimide.

Some covalent inhibitors such peptidyl diazomethanes and epoxysuccinyl peptides are suitable active site titrants of cathepsin L because they react rapidly and selectively with the enzyme and not with thiol compounds generally (see p. 48; Tables 24, 27).

Table 8 Kinetic constants for cathepsin L with synthetic substrates

Substrate	pH	k_{cat} (s^{-1})*	K_m (µM)	k_{cat}/K_m (mM^{-1} s^{-1})	Species	References
Z-F-R-NHMec	5.5	17.0	2.4	7083	Human	362
	6.0	25.8	2.2	11727	Human	81
	5.5	17.6	2.8	6286	Rat	93
	5.5	23.0	2.0	11500	Rat	172
	6.0	20.0	1.8	11111	Rabbit	362
	5.5	18.0	3.0	6000	Bovine	172
	5.5	10.0	2.2	4545	Bovine	212, 213
	5.5	30.0	1.8	16667	Sheep	212, 213
	5.6	15.8	1.68	9405	Chum salmon	409
Z-F-R-NHMecF	6.0	28.0	42.0	667	Rat	279
Boc-F-F-R-NHMec	5.5	7.5	1.4	5357	Rat	93
	5.5	7.9	1.5	5267	Rat	93
Z-V-V-R-NHMec	5.5	8.5	4.8	1771	Rat	93
Bz-F-V-R-NHMec	5.5	1.1	1.8	611	Rat	93
Z-Lys-OPhNO$_2$	5.5	19.8	10.0	1980	Rat	8

* k_{cat} values are based on molarities determined by titration [8]; temperature was 25–30°C.

Table 9 Degradation and modification of proteins by cathepsin L

Protein	Mode of action by cathepsin L	pH	References
Actin	Degradation	4.0	219, 372
Albumin	Degradation; also ^{14}C labelled	3.5–5.0	82, 231, 346
α_1-Proteinase inhibitor	Inactivation; cleavage at Glu354-Ala355 and Met358-Ser359	5.0–5.5	159, 199
Angiotensin I	Degradation	Neutral	198
Angiotensin II	Inactivation; formation of two tetrapeptides	Neutral	198
Antigens	Uptake of procathepsin L by cells interferes with antigen processing by total degradation of antigens	<7.0	221, 222
Azocasein	Degradation; substrate used to quantify enzyme activity, also in the presence of 3 M urea	6.0 / 5.5	346, 407 / 231
Basement membrane	Degradation of glomerular basement membrane and basement membrane matrix	3.5–6.0	80, 81, 112, 140, 142, 205, 283
Bradykinin	Inactivation; cleavage at Gly4-Phe5	Neutral	198
C3, third component of complement	Degradation	7–7.5	1527, 1526
Collagen, soluble and insoluble	Hydrolysis in the telopeptide region, conversion of γ and β chains to α chains; degradation of collagen in basement membranes and bone	3.5 / 3.3–3.6 / 5–7 / 5.9	135, 1167 / 142, 282 / 105, 128 / 160, 206, 296, 587
Cytosol proteins	Degradation of short- and long-lived cytosol proteins	3.5	85
Epidermal growth factor receptor	Degradation		149, 150
Enkephalin; Leu-, Met-	Degradation; cleavage at Gly2-Gly3		207
Elastin	Degradation of the insoluble substrate	5.5	215
Fibronectin	Degradation		135, 140, 142, 725
Fructose-1,6-bisphosphate aldolase	Inactivation		163, 231
Glucagon	Degradation; major cleavages at Thr7-Ser8, Asp15-Ser16, Met27-Asn28	3.0	162
Haemoglobin	Degradation; used as substrate, also ^{14}C labelled	5.0	231, 346
Glucose-6-phosphate dehydrogenase	Inactivation	6.0	285
Insulin B chain	Split positions		134, 161
Intrinsic factor (gastric)	[125]I labelled; cleavage products: 33 000, 31 000 Da	–	1302
Laminin	Degradation	3.5–5.5	135, 140, 142, 282, 725
Myosin, heavy chain	Degradation	4.2	219, 372
Osteocalcin	Degradation		247
Osteonectin	Degradation		247
Proplasminogen activator (urokinase type)	Activation of the precursor; cleavage at Lys158-Ile159	5–7	629, 703
Proteoglycans	Multiple but distinct cleavage sites		99, 261, 959
Thyroglobulin	Limited proteolysis; four cleavage points	3.5–5.0	118, 119
Troponin T, troponin I	Degradation	4.0	219

Z-Phe-Phe(4-NO$_2$)-CHN$_2$ and especially Z-Phe-Tyr(OBu^t)-CHN$_2$ have proved to be selective inhibitors, because they react several (10^3) times faster with cathepsin L than with cathepsins B and S (see Table 24).

Cathepsin L is sensitive to leupeptin, a tight-binding, reversible inhibitor which also inhibits serine peptidases (see p. 53; Table 28).

Natural protein inhibitors, the cystatins, are very potent inhibitors of cathepsin L (see Table 31).

Hydrolysis of synthetic substrates

Cathepsin L has a restricted specificity in the hydrolysis of synthetic substrates. The S1 subsite of the enzyme has a broad specificity. Cathepsin L clearly favours aromatic residues in the P2 [1219] and hydrophobic residues in the P3 position [134, 161, 162] and prefers amino acids with small or long but non-branched side chains in the P1$'$ position [235]. Table 8 shows some substrates cleaved by cathepsin L. However, these substrates are not specific amongst the lysosomal cysteine proteinases, also being hydrolysed by cathepsins B and S [93, 362]. Z-Phe-Arg-NHMec (Table 8) at low concentrations ($K_m \sim 2\,\mu$M) has proved to be a suitable substrate, and is widely used for the determination of cathepsin L [8, 362, 489, 1167].

The specificity for cathepsin L can be achieved by use of Z-Phe-Phe-CHN$_2$, an irreversible inhibitor specific at low concentrations ($0.5\,\mu$M) for cathepsin L [362, 489]. At higher concentrations or longer incubation times than 10 min, Z-Phe-Phe-CHN$_2$ reacts also with cathepsin B and cathepsin S [170, 554].

Using Z-Phe-Arg-NHNapOMe as a substrate, cathepsin L activity has been determined at pH 3.5, where cathepsins B and S have mainly lost their activities [708]. Under these conditions the precursor of cathepsin L is activated, and so the activity of mature cathepsin L and the latent activity of procathepsin L are determined.

The dipeptidyl rhodamine diamide substrates such as (Z-Phe-Arg)$_2$- and (Z-Arg-Arg)$_2$- proved to be more selective for cathepsin L than for cathepsin B [75], and have been used to determine the activity of cathepsin L in viable cells [385].

Hydrolysis of polypeptides and proteins

Cathepsin L is a strong endopeptidase degrading several proteins and inactivating many enzymes (Table 9).

The peptide bonds cleaved by cathepsin L in polypeptides or proteins are predictable from studies on the specificity of the enzyme hydrolysing insulin B chain [134, 161] and glucagon [162]. Cathepsin L has a requirement for hydrophobic amino acid residues in subsites S2 and S3. The specificities of cathepsin L for cleavage of synthetic substrates and for cleavage of peptide bonds in polypeptides are the same.

Cathepsin S
(EC 3.4.22.27)

Introduction

In 1975 the name *cathepsin S* was given to a cysteine peptidase purified from bovine lymph nodes by Turnsek and co-workers [402] and in 1981 from spleen by Locnikar *et al.* [355]. The enzyme showed similarities to cathepsin L, but differed in a number of respects, including pI, M_r, pH stability, activity against synthetic substrates and sensitivity to inhibitors, as shown by Kirschke and Brömme and co-workers, 1984–1989 [93, 170, 172, 346]. In 1991 studies on the amino acid sequences by Ritonja *et al.* and Wiederanders and co-workers [1190, 1213] confirmed that cathepsins S and L are different enzymes. The name *cathepsin S* (EC 3.4.22.27) was recommended by the nomenclature committee of International Union of Biochemistry and Molecular Biology (IUBMB) in 1992.

Cathepsin S is one of the lysosomal cysteine proteinases known so far to show a restricted tissue distribution. The highest levels have been detected in spleen, sessile lung macrophages and heart [34, 172, 295, 346, 1141, 1195], but the occurrence of the enzyme has also been described in ileum, brain, thyroid and ovary [1180].

Cathepsin S has been shown to be preferentially expressed in cells of mononuclear-phagocytic origin [1710] and overexpressed in cortical neurones and glia in samples of brain from patients with Alzheimer's disease [1684]. Cells doubly transfected with both β-amyloid precursor protein and cathepsin S, but not with the other cathepsins, secreted high levels of amyloid β-peptides, indicating a function of cathepsin S in the pathogenesis of Alzheimer's disease [1337]. Cathepsin S was also shown to be essential in B cells for effective proteolysis of the invariant chain necessary to render class II molecules competent for binding peptides [1345].

In contrast to cathepsins L, B and H, cathepsin S is both stable and catalytically active in the presence of SH-containing compounds at pH 7.0. As a result of this, cathepsin S may be a new potential participant in the physiological protein breakdown, pathological degradative and invasive processes and in antigen presentation.

Gene structure and chromosomal location

The human cathepsin S gene maps to chromosome 1q21 [1196]. This is the region where the cathepsin K gene is located [1793], suggesting that the two enzymes may be evolved from the same ancestral gene by gene duplication (see p. 24).

The human cathepsin S gene contains six exons and six introns and the total gene spans about 17–18 kbp [1196] (Fig. 4).

The size of the most abundant cathepsin S mRNA from rat is 1.4 kb [1180], from human 1.9 kb [1083] and from bovine 1.7 kb [1213]. Additionally, a 6.0 kb transcript has been detected in rat lung [1180]. Some of the accession numbers of the cathepsin S cDNA in the EMBL/GenBank database are given in Table 10.

Sequencing of the 5′-flanking region of the human cathepsin S gene has revealed no classical TATA and CAAT boxes – but also no GC-rich regions. Shi *et al.* [1196] suggested a specific regulation of the cathepsin S gene, because of the presence of several AP1 transcription factor binding sites and microsatellites. No signals for specific regulation are yet known, however.

Domain structure

Species: human; compiled from [1083, 1195, 1196]

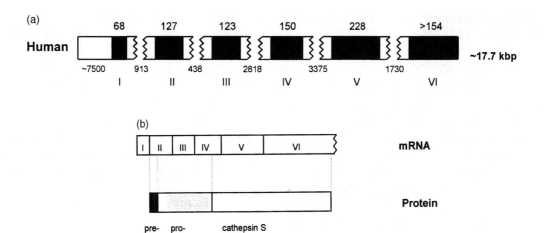

Figure 4

Schematic structures of the human cathepsin S gene, mRNA and protein. Exons are numbered with roman numerals from 5' to 3' in the direction of transcription. (a) Structure of the human cathepsin S gene [1196]: black rectangles represent exons, with sizes shown above in base pairs; white, broken areas represent introns, with sizes given below in base pairs. (b) Structures of the mRNA and protein [1244]; pre- and propeptides and mature cathepsin S are indicated [1196].

		Papain numbering
Signal peptide	1–16	p1–p22
N-Propeptide	17–114	p37–p133
Mature enzyme	115–331	1–211
Active site Cys	139	25
Active site His	278	159
Active site Asn	298	175
Disulfide bond	136, 180	22, 63
Disulfide bond	170, 213	56, 95
Disulfide bond	272, 320	153, 200
Carbohydrate-binding site	104	p119
Molecular mass (calculated)	37 479 DA	
Number of residues	331 AA	

Other species: rat [1180]; bovine [1141, 1190, 1213]

Molecular masses determined by SDS–PAGE

Procathepsin S	37 000 Da
Cathepsin S	24 000 Da

Isoelectric point: 6.3–7.0 (mature cathepsin S) [346, 355, 1141]

Precursors

Cathepsin S is synthesized as a preproenzyme and the signal peptide is cotranslationally removed in the endoplasmic reticulum (see p. 39). The carbohydrate chain bound to the propeptide is processed in the Golgi apparatus and the latent precursor is targeted by the mannose-6-phosphate signal to the lysosomal

Table 10 Accession numbers of the cathepsin S cDNA in the EMBL/GenBank database

Species	Accession No.	Coding region	References
Homo sapiens	M90696	Complete	1083
Rattus norvegicus	L03201	Complete	1180
Bos taurus	X62001	Complete	1213

endosomal compartment, where maturation by limited proteolysis takes place.

Human procathepsin S has been cloned and expressed in Sf9 cells using a baculovirus system, where the latent precursor was activated in the lysed cells and the mature enzyme subsequently purified with a final yield of 45–50 mg from 1 l of cell extract [1368]. A high-yield expression of procathepsin S has also been described from *E. coli* using a T7 RNA polymerase expression system. The recombinant procathepsin S was purified from inclusion bodies and activated by autocatalysis. The yield was 2 mg of cathepsin S from 1 l of bacterial culture [1390].

Maturation of procathepsin S ($M_r \sim 37\,000$ Da) can be achieved by treatment with subtilisin at pH 8.0, 25°C for 1 h or by autoactivation at pH 4.5, 40°C for 3 h in the presence of dithiothreitol and EDTA [311].

The propeptide has proved to be a selective inhibitor of cathepsin S [1768].

Purification

Cathepsin S has been purified from beef spleen [172, 295, 345, 346, 355, 1141] and lymph nodes [355, 402].

In tissue extracts, cathepsin S is partially complexed with endogenous inhibitors, to which the enzyme has a high affinity [440]. Separation of cathepsin S from the enzyme–inhibitor complex requires autolysis at acidic pH values followed by fractionation with ammonium sulfate.

Separation from cathepsins H and L can be achieved by chromatography on CM–Sephadex C-50, to which the latter shows a very high affinity. Separation of cathepsin H from cathepsin S together with cathepsin B is important in this step of purification: with a gradient of increasing NaCl concentration, cathepsins S and B are eluted before cathepsin H (bovine) or after cathepsin H (human and rat) [295, 345].

Cathepsins S and B have been separated by chromatofocusing, beginning at about pH 8. Between pH 8 and 6, cathepsin S was eluted, and then cathepsin B using acetate buffer pH 4.5 [346]. Bound to the gel, cathepsin B was stable at pH 8.

Recombinant cathepsin S has been purified after expression in transfected baby hamster kidney cells [1083], yeast [311] and as a fusion protein in COS cells [1195] and bacterial cells [1196]. Recombinant rat cathepsin S has been purified as a fusion protein in bacterial cells [1180].

Cathepsin S has been obtained by the described methods as a mature enzyme with a molecular mass of 24 000 and about 40% activity as revealed by titration with E-64. By inclusion of an affinity chromatography step in the procedure (e.g. activated thiol– or thiopropyl–Sepharose) the inactive molecules are removed and the resulting cathepsin S preparation is 100% catalytically active.

Activation and inhibition

Cathepsin S is stable in a pH range of 5.0–7.5. The enzyme has the unique property among the lysosomal cysteine peptidases such as cathepsins B, H, L and K to be stable at and above pH 7.0. Optimal activity is achieved in the presence of thiol compounds and EDTA at pH 6.5.

Cathepsin S reacts with general thiol-blocking agents such as iodoacetate, iodoacetamide and N-ethylmaleimide.

E-64 and some peptidyl diazomethanes (see Tables 24, 27) have proved to be suitable active site titrants, because they react rapidly and selectively with the enzyme and not with thiol compounds generally.

Cystatins have a high affinity for cathepsin S, displaying K_i values in the picomolar range (see Table 31).

Hydrolysis of synthetic substrates

Cathepsin S is active in the presence of SH-containing compounds against synthetic substrates from pH 5.0 to 8.0 with optimal activity at pH 6.5 (Table 11).

Tests with a series of synthetic endopeptidase substrates revealed that all compounds were more or less sensitive to cathepsins S, L and B, and no peptide derivative has been described so far that is specific for cathepsin S at acidic pH values. However, changing

Table 11 Kinetic constants for cathepsin S with synthetic substrates

Substrate	pH	k_{cat} (s^{-1})*	K_m (µM)	k_{cat}/K_m $(mM^{-1}s^{-1})$	Species	References
Z-V-V-R-NHMec	6.5	40.5	17.5	2314	Bovine	93
	6.5	15.0	18.1	830	Human (rcath)	311
Bz-F-V-R-NHMec	6.5	13.0	8.1	1605	Bovine	93
	7.5	1.6	4.6	348	Bovine	295
Z-F-V-R-NHMec	6.5	2.9	12.1	240	Human (rcath)	311
Boc-F-L-R-NHMec	6.5	6.1	7.3	836	Bovine	93
Z-F-R-NHMec	6.5	4.7	14.7	320	Bovine	93
	7.5	2.0	15.1	132	Bovine	295
	6.5	1.9	22.4	85	Human (rcath)	311
Boc-F-F-R-NHMec	6.5	8.6	37.5	229	Bovine	93
	6.5	3.3	48.0	69	Human (rcath)	311

* k_{cat} values are based on molarities determined by titration [8]; temperature was 25°C for incubation at pH 6.5 and 30°C at pH 7.5. rcath, recombinant cathepsin.

the assay conditions to pH 7.5 and using the most susceptible substrates, such as Bz-Phe-Val-Arg-NHMec or Z-Val-Val-Arg-NHMec, allows the specific determination of cathepsin S. During a preincubation period of up to 1 h at pH 7.5, the activities of cathepsins L, B and H are completely abolished, whereas cathepsin S retains 60–70% of its activity [345, 346]. Controls must be included to establish with certainty the identity of the enzyme that is being assayed.

Although cathepsins S and L show similarities in their substrate specificities, detailed studies with a variety of substrates have revealed significant differences in the S2 subsite specificity of the two enzymes.

Cathepsin S favours branched hydrophobic residues, whereas cathepsin L prefers bulky aromatic residues in the P2 position [93, 311, 1219].

Several amino acid residues have been replaced in cathepsin S by site-directed mutagenesis in order to investigate the residues determining the S2 substrate specificity. The replacement of Gly133 (papain numbering) in cathepsin S by an alanine residue, as in cathepsins B and L, changed the specificity of cathepsin S toward that of the latter enzymes, i.e. to prefer bulky aromatic residues in the P2 position. A mutation of Phe205 → Glu in cathepsin S resulted in a change of specificity toward that of cathepsin B, with a preference to hydrolyse dibasic substrates.

Table 12 Degradation and modification of proteins by cathepsin S

Protein	Mode of action by cathepsin S	pH	References
Albumin	^{14}C labelled; degradation	5.0, 7.5	346
Azocasein	Degradation, also in the presence of 3 M urea	5.5, 6.0, 7.5	170, 172, 346
Collagen	Insoluble and soluble; degradation	3.5	346, 360
Elastin	Insoluble, ^{3}H labelled; degradation	4.0–7.0	295, 1195
Fibronectin	Degradation	6.0, 7.5	1710
Haemoglobin	Native, acid denatured and ^{14}C labelled; degradation	5.0, 7.5	346, 355
Insulin B chain	Oxidized; degradation; major cleavages at Glu13-Ala14, Leu17-Val18, Phe25-Tyr26	7.9	93
Invariant chain (Ii) (MHC class II)	Digestion of Ii; generation of α, β-CLIP complexes	–	1345
Laminin	Degradation	6.0, 7.5	1710
Proteoglycans	Degradation	6.0, 7.5	1710

This specificity was improved in the double mutants having Gly133 → Ala and Phe205 → Glu of cathepsin S [1219]. The S1′ subsite specificity of cathepsin S is broad, like that of papain [235].

Hydrolysis of polypeptides and proteins

Cathepsin S is capable of hydrolysing protein substrates as fast as other lysosomal endopeptidases, such as cathepsin L [346]. It shows collagenolytic [346] and elastinolytic [295, 1195] activities (Table 12). But cathepsin S has the unique property among the lysosomal cysteine proteinases of degrading soluble proteins as well as insoluble elastin at acidic and neutral pH values [346, 295].

The cleavage specificity of cathepsin S seems to be very similar to that of cathepsin L, as revealed by the cleavage sites in the insulin B chain [93]: hydrophobic, less bulky residues are preferred in the P2 and P3 positions of the substrate.

Cathepsin K (EC 3.4.22.38)

Introduction

In 1994 Tezuka and co-workers [1201] identified, in an osteoclast cDNA library from rabbit, clones encoding a novel cysteine peptidase named *OC-2*. In the same year, human equivalents of the rabbit gene were mentioned independently in two abstracts by Drake *et al.* [1787] and Li *et al.* [1797] and the sequence was published in 1995 (in order of date of submission) by Li *et al.* [1798], Inaoka *et al.* [1792], Shi *et al.* [1811] and Brömme and Okamoto [1780]. The cDNA of cathepsin K from mouse [1642, 1806] and chicken [1520] have also been cloned. Names such as *cathepsin X* [1798], *cathepsin K* [1792], *cathepsin O* [1811, 1760, 1643] and *cathepsin O2* [1780] were given to the novel human osteoclast cysteine peptidase – but the name eventually recommended by the nomenclature committee of the IUBMB was *cathepsin K*.

Cathepsin K was the temporary name of a high molecular weight enzyme which has only partially been purified, and no detailed description of its properties has been given [353].

Unlike cathepsins B, L and H, cathepsin K shows a restricted tissue distribution, being found predominantly in osteoclasts [1201, 1288, 1645, 1780, 1792, 1798]. The enzyme has also been detected by northern blot analysis in ovary [1780] and in low levels in some human tissues such as lung, brain, small intestine, colon and pancreas [1780, 1792], as well as monocyte-derived macrophages [1811]. The predominant expression of cathepsin K in osteoclasts leads to the suggestion that the enzyme may play a major role in osteoclast-mediated bone resorption. Studies on pycnodysostosis, an autosomal recessive osteochondrodysplasia, supported this suggestion. Pycnodysostosis results from gene defects in cathepsin K. Osteoclasts of such patients lack cathepsin K immunoreactive protein; they function normally in demineralizing bone, but do not adequately degrade the matrix proteins [1652].

Gene structure and chromosomal location

Genomic DNA of cathepsin K has been isolated from a human placental genomic library [1793]. Eight exons and seven introns have been analysed of a 12.5 kbp gene [1793].

The human cathepsin K gene maps to chromosome 1q21 [1793], the same location as the cathepsin S gene [1196], so cathepsins S and K may well have evolved from a single ancestral gene by gene duplication. The accession numbers of the cathepsin K cDNA in the EMBL/GenBank database are shown in Table 13.

The size of human cathepsin K mRNA has been determined to be ~2.0 kb [1645, 1787, 1792], ~1.9 kb [1798], ~1.8 kb [1780], 1.7 kb [1811] and

Table 13 Accession numbers of the cathepsin K cDNA in the EMBL/GenBank database

Species	Accession No.	Coding region	References
Homo sapiens	U13665	Complete	1811
Homo sapiens	X82153	Complete	1792
Homo sapiens	U20280	Complete	1798
Mus musculus	X94444	Complete	1806
Oryctolagus cuniculis	D14036	Complete	1201
Gallus gallus	U37691	Complete	1520

1.6 kb [1797]; the mRNA of rabbit cathepsin K is ~2.0 kb [1201].

All-*trans*-retinoic acid has been reported to regulate the expression of cathepsin K in rabbit osteoclasts at the transcriptional level [1717].

Domain structure

Species: human; compiled from [1798, 1811, 1792, 1780]

		Papain nomenclature
Signal peptide	1–15	p1–p20
N-Propeptide	16–114	p38–p133
Mature enzyme	115–329	1–211
Active site Cys	139	25
Active site His	276	159
Active site Asn	296	175
Disulfide bonds	136, 178	22, 63
Disulfide bonds	171, 210	56, 95
Disulfide bonds	269, 318	153, 200
Carbohydrate-binding site	103	p120
(Carbohydrate-binding site	161*	47*)
Molecular mass (calculated)	36 869 Da	
Number of residues	329 AA	

* Due to the subsequent Pro164 (50) the probability of glycosylation is 50%

Other species: rabbit [1201]

Molecular masses determined by SDS–PAGE

Procathepsin K	38 000 Da [1645], 43 000 Da [1288]
Mature cathepsin K	27 000–29 000 Da [1288, 1286, 1384]

Precursor

Procathepsin K has proved to be catalytically inactive and is processed to the mature active enzyme by treatment with pepsin at pH 4.0 [1288], or by autoactivation at pH 4.0, 60°C for 5 min or at 4°C for 24 h in the presence of preactivated cathepsin K [1286].

Purification

Purification of recombinant cathepsin K has been described in detail at the level of the precursor [1286] and the mature enzyme [1288]. The natural, mature cathepsin K has been isolated only from a giant cell tumour, according to a preliminary report [1384]. Procathepsin K has been purified (at pH 8.0 throughout) by ion exchange chromatography (Mono S), blue dye chromatography and gel filtration (Superdex 75) and subsequently activated [1286].

Purification of mature cathepsin K at pH 5.5 started with the activation of procathepsin K in the medium, followed by hydrophobic chromatography (butyl–Sepharose) and ion exchange chromatography (Mono S) [1288].

Mature cathepsin K has been isolated from an extract of a human giant cell tumour by ammonium sulfate precipitation and ion exchange chromatography (DEAE–cellulose) [1384].

Hydrolysis of synthetic substrates

Cathepsin K is catalytically active against synthetic substrates in the presence of SH-containing compounds. The pH optimum for activity is between 6.0 and 6.5 [1288]. Recombinant human cathepsin K displays a bell-shaped pH profile with flanking pK values of 4.0 and 8.13 [1288].

The kinetic constants (Table 14) for different preparations of recombinant human cathepsin K show wide variations with the same substrate. Therefore, the catalytic efficiency (given as k_{cat}/K_m) in Table 14 does not characterize the S2P2 subsite specificity of cathepsin K. However, data for the same enzyme preparation show a preference for leucine over phenylalanine in dipeptidyl substrates in the P2 position. Cathepsin L prefers phenylalanine over leucine

Table 14 Kinetic constants for cathepsin K with synthetic substrates

Substrate	pH	k_{cat} (s^{-1})*	K_m (µM)	k_{cat}/K_m (mM^{-1} s^{-1})	Species	References
Z-F-R-NHMec	6.5	0.9	7.5	120	Human (rcath)	1288
	5.5	3.5	58	60	Human (rcath)	1286
	6.5	7.6	14.1	539	Human (rcath)	1310
	?	960†	17.5	54 857	Human	1384
Z-L-R-NHMec	6.5	0.98	3.8	258	Human (rcath)	1288
	5.5	3.4	8.3	410	Human (rcath)	1286
Z-V-R-NHMec	6.5	1.1	13.1	81	Human (rcath)	1288
	5.5	2.9	33	86	Human (rcath)	1286
Z-L-L-R-NHMec	6.5	0.02	0.4	50	Human (rcath)	1288
	5.5	0.08	13	6	Human (rcath)	1286
Boc-V-L-R-NHMec	5.5	0.52	30	17	Human (rcath)	1286
Tos-G-P-R-NHMec	5.5	2.2	1100	1.9	Human (rcath)	1286
Bz-F-V-R-NHMec	5.5	0.4	340	1.2	Human (rcath)	1286
Z-V-V-R-NHMec	6.5	0.01	18.5	0.5	Human (rcath)	1288
	5.5	0.02	44	0.5	Human (rcath)	1286

* k_{cat} values are based on molarities determined by titration [8].
† Titration was not indicated.

in the P2 position, whereas the specificity of cathepsin S resembles that of cathepsin K [1288].

Substrates containing two basic amino acid residues, normally specific for cathepsin B, are not hydrolysed by cathepsin K, because the glutamate residue in the S2 subsite pocket of cathepsin B is replaced in cathepsin K by leucine [1288]. Compounds such as Z-Leu-Leu-NHMec or Z-Leu-Leu-Leu-NHMec have proved to be inhibitors and not substrates of the enzyme, whereas Z-Leu-Leu-Arg-NHMec, Z-Leu-Arg-NHMec and Z-Phe-Arg-NHMec are inhibitors only at concentrations greater than K_m [1286].

No specific substrate for cathepsin K has yet been discovered, and it is therefore not possible to determine the activity of cathepsin K in the presence of other lysosomal cysteine peptidases.

Hydrolysis of polypeptides and proteins

Cathepsin K has been characterized as an endopeptidase lacking additional aminopeptidase or dipeptidyl peptidase activity.

Table 15 Degradation of proteins by cathepsin K

Protein	Mode of action by cathepsin K	pH	References
Collagen, type I	Primary cleavage in the telopeptide region; conversion of γ and β chains to α chains; but also degradation of the monomers	5–6	1288, 1311, 1286, 1384
Fibrinogen	[125]I labelled fibrinogen; degradation of α, β and γ chains	4.5	1288, 1286
Elastin	Degradation of insoluble ^{3}H labelled elastin	5.5 (4.5, 7.0)	1288
Denatured collagen, type I	Total degradation	5.5–7.0	1288
Osteonectin	Degradation products: 34, 14 and 10 kDa; cleavage at Gln3-Gln4 and . . . Lys-Leu-Arg ↓ Val-Lys-Lys . . .	5.5	1286

Table 15 shows some proteins that have been used as substrates of cathepsin K.

Cathepsin K displays higher elastinolytic activity than cathepsins S or L or pancreatic elastase at pH 4.5, 5.5 and 7.0, respectively [1288]. The gelatinase activity of cathepsin K is also very high and exceeds that of cathepsins L and S [1288].

The cleavage specificity of cathepsin K has been revealed by the cleavage sites in osteonectin: Ala-Pro-Gln ↓ Gln-Glu-Ala-Leu... and ...Gln-Lys-Leu-Arg ↓ Val-Lys-Lys-Ile... [1286], where the internal cleavage (↓) shows the same requirement of the enzyme for the amino acids in subsites S1 and S2 as in synthetic peptide substrates (Table 14).

Dipeptidyl peptidase I (EC 3.4.14.1)

Introduction

A peptidase that was discovered in 1948 in pig kidney by Gutman and Fruton [1380] and described as having chymotrypsin-like specificity was termed, in 1952, *cathepsin C* by Tallan and co-workers [1401]. The enzyme was also known as *glucagon-degrading enzyme* [1316] and *dipeptidyl transferase* [364]. In 1966 McDonald and co-workers [1332] renamed cathepsin C *dipeptidyl arylamidase I* and, in 1969 [228], *dipeptidyl aminopeptidase I,* in recognition of its exopeptidase activity. In semi-systematic enzyme nomenclature, the term *dipeptidyl peptidase* describes the reaction in which an N-terminal dipeptide is hydrolysed from a polypeptide. Consistent with this, the name *dipeptidyl peptidase I* (here abbreviated *DPP I*) was recommended by the nomenclature committee of the IUB in 1981. A high molecular mass, thiol-dependent enzyme that hydrolysed Z-Phe-Arg-NHMec was termed *cathepsin J* by Liao and Lenney in 1984 [353], but was later shown to be identical to DPP I by Nikawa and co-workers [368].

DPP I is a lysosomal enzyme with the major catalytic activity of hydrolysing dipeptides from the N terminus of polypeptides and proteins at pH 5–6. At pH 7–8 the enzyme exhibits dipeptidyl transferase activity resulting in the formation of polymer dipeptide derivatives [228, 1240, 1301]. Since DPP I behaves as a strict exopeptidase with most substrates, it came as surprise when Nikawa and co-workers [368] showed in 1992 that an enzyme previously termed *cathepsin J*, which hydrolysed Z-Phe-Arg-NHMec, is identical to DPP I. The action of DPP I on this N-blocked artificial substrate may be quite exceptional, however, since none of 18 other peptidyl aminomethylcoumarins was hydrolysed, and many comparable nitroanilides and 2-naphthylamides had given negative results in earlier work. The action of DPP I on Z-Phe-Arg-NHMec was further characterized by Kuribayashi and co-workers [190], who also showed evidence of true endopeptidase activity on a chemically modified derivative of lysozyme. However, the significance of this finding is not yet clear, since no endopeptidase action of DPP I has been detected in many earlier studies of the specificity of the enzyme [228], or in work in which DPP I was used for peptide sequencing. There has been no report of endopeptidase action of DPP I on any natural protein.

Unique among the lysosomal cysteine proteinases is the requirement of halide ions (Cl^-, Br^-, I^-) in addition to an SH-containing compound for the catalytic activity of this enzyme [225].

Several metabolic functions have been attributed to DPP I in addition to its main function in protein degradation in lysosomes. These include activation of neuraminidase and of serine proteinases in granules of cytotoxic lymphocytes, mast cells and myeloic cells [230] and activation of platelet factor XIII [202]. Other biological roles that have been described are in Duchenne muscular dystrophy [775, 776, 831, 854], in cell growth [831] and as a marker for cells with cytolytic potential [229].

The DPP I gene has not been sequenced. The sizes of the rat mRNA are 2.1 and 2.7 kb [1160]. Some of the accession numbers of the DPP I cDNA in the EMBL/GenBank database are given in Table 16.

Domain structure

Species: human; compiled from [1804]

		Papain numbering
Signal peptide	1–24	
N-Propeptide	25–230	
Mature enzyme	231–463	1–212C
Heavy chain	231–394	1–154
Light chain	395–463	155–212C

Species	Accession No.	Coding region	References
Homo sapiens	X87212	Complete	1804
Rattus norvegicus	D90404	Complete	1160

		Papain numbering
Active site Cys	258	25
Active site His	405	159
Active site Asn	427	175
Disulfide bond	255, 298	22, 63
Disulfide bond	291, 331	56, 95
(Disulfide bond	321, 337	85, 101)
(Disulfide bond	355, 448	114, 200)
Carbohydrate-binding site	29	
Carbohydrate-binding site	53	
Carbohydrate-binding site	119	
Carbohydrate-binding site	276	43
Molecular mass (calculated)	51 848 Da	
Number of residues	463 AA	

Other species: rat [368, 1160, 1170]

The propeptide of DPP I is extremely long and does not show identities to propapain (see Fig. 13).

Molecular masses determined by SDS–PAGE

Pro-DPP I	55 000–56 000 Da
Mature enzyme	21 000–25 000 Da
Heavy chain	18 000 Da
Light chain	~7 000 Da

Additional molecular masses of DPP I have been determined mainly by gel filtration and by sedimentation equilibrium, to be 135 000–200 000 Da, 91 000–92 000 Da and 73 000 Da, because of oligomerization of pro-DPP I and of the mature forms (probably to a tetramer)

Isoelectric point: 5.3-5.9 [299, 305, 358, 376]

Purification

DPP I is widely distributed in mammalian tissues, but the expression pattern is different to that of cathepsins B, L and H.

DPP I has been purified from several organs and tissues of many species. The main sources for the purification of DPP I are spleen, liver and kidney. Usually the preparation starts with an acidic tissue extract [364] (although a mitochondrial–lysosomal fraction has also been used [335, 368]) and proceeds with ammonium sulfate fractionation. Separation from the other lysosomal proteinases can be achieved by gel filtration, in which DPP I exhibits a higher molecular mass than the other cathepsins, as a result of its oligomerization. Usually the purification procedures include anion and cation exchange chromatography [225, 227, 305, 309, 318, 358, 364, 368, 376].

Activation and inhibition

DPP I is stable in the presence of 1% NaCl at pH 5.0–6.0. Maximal activity is achieved in the presence of thiol compounds, EDTA and NaCl or NaBr at pH 5.0–6.0 [225, 1332]. The dipeptidyl transferase activity of the enzyme is maximal at pH 7.7 and obviously also requires thiol compounds and halide ions.

DPP I reacts with general thiol-blocking reagents such as iodoacetate and 4-chloromercuriphenyl sulfonate.

E-64 inactivates DPP I only slowly, but other covalent inhibitors such as Gly-Phe-diazomethane and Phe-Ala-dimethylsulfonium salt (see p. 49) cause rapid inactivation [419]. The latter may be used as an active site titrant.

The cystatins, reversible tight-binding inhibitors, react with DPP I.

Hydrolysis of synthetic substrates

DPP I acts as a specific dipeptidyl peptidase on almost all substrates (Table 17). The dipeptidyl peptidase

Table 17 Kinetic constants for DPP I with synthetic substrates

Substrate	pH	k_{cat} (s^{-1})	K_m (mM)	k_{cat}/K_m (mM^{-1} s^{-1})	Species	References
A-A-NHNap	6.0	248	0.19	1300	Rat	228
A-A-NH$_2$	6.0	1160	51.0	23	Rat	228
A-A-OMe	5.0	178	1.4	127	Rat	228
G-R-NHNap	6.0	1300	0.1	13 000	Rat	228, 860
G-R-NH$_2$	6.0	1160	30.0	39	Rat	228
G-K-OMe	5.0	1125	1.6	703	Rat	228
G-F-NHNap	6.0	79	0.17	465	Rat	228
G-F-NH$_2$	6.0	263	18.0	15	Rat	228
G-F-OMe	5.0	282	0.43	656	Rat	228
G-F-PhNO$_2$	5.0		2.51		Bovine	967
H-S-NHNap	6.0	116	0.02	5800	Rat	223
S-M-NHNap	6.0	510	0.17	3000	Rat	228
S-M-OMe	5.0	268	1.1	244	Rat	228
Z-F-R-NHMec	5.0		0.23		Bovine	190

activity is broad, but the enzyme does not hydrolyse substrates that contain a basic amino acid (arginine or lysine) in the first (N terminal) position, or proline in the second or third positions [197, 228, 229].

Derivatives of dipeptides such as Gly-Arg-, Pro-Arg-, Gly-Phe- and Ser-Tyr- have been used for the determination of DPP I activity. Gly-Arg- and Pro-Arg-NHNapOMe served as substrates for histochemical studies [860]. None of the substrates is entirely selective for DPP I, because aminopeptidases and cathepsin H can degrade them, as well. DPP I is very much more active on Gly-Arg- substrates than

on the more commonly used Gly-Phe- compounds [228].

Hydrolysis of polypeptides and proteins

By virtue of its main activity of catalysing the consecutive release of dipeptides from the N terminus of polypeptides and proteins (Table 18), DPP I has been used as a sequencing enzyme [101, 102, 104, 109, 197, 223, 226–228, 287] and for trimming of polypeptide chains [194, 230].

Table 18 Degradation and modification of proteins by DPP I

Protein	Mode of action by DPP I	pH	References
Angiotensin II amide	Dipeptidyl peptidase action; release of two dipeptides	6.0	226, 227
β corticotropin	Dipeptidyl peptidase action; release of five dipeptides	5.0–5.5	228
Glucagon	Dipeptidyl peptidase action; release of eight dipeptides	5.0	223, 228
Insulin A chain	Oxidized; dipeptidyl peptidase action; release of 10 dipeptides	5.0	103
Insulin B chain	Oxidized; dipeptidyl peptidase action; release of 13 dipeptides	5.0	102
Insulin-like growth factor II	Dipeptidyl-peptidase action: release of one dipeptide without change of biological activity	–	1321
Lysozyme, modified	Degradation by endopeptidase action; N-terminal amino acid is Lys		190
Prorganzyme A	Dipeptidyl peptidase action; activation	–	1324
Secretin	Dipeptidyl peptidase action; release of nine dipeptides	5.0	223

Legumain (EC 3.4.22.34)

Introduction

It has been known since the initial discovery of Csoma and Polgár in 1984 [1292] that the seeds of several legumes and other plants contain an endopeptidase specific for the hydrolysis of asparaginyl bonds. The enzyme has been termed *asparaginyl endopeptidase* [1259] and *vacuolar processing enzyme* [1305], but the name recommended by the nomenclature committee of the IUBMB is *legumain*. Kembhavi *et al.* [1389] in 1993 isolated legumain from moth bean (*Vigna aconitifolia*) and described a convenient fluorimetric assay in which the substrate was Z-Ala-Ala-Asn-NHMec. The cloning and sequencing of legumain from castor bean [1381] revealed some sequence identities with the 'haemoglobinase', from the blood fluke, *Schistosoma mansoni*. The parasite enzyme has now been shown also to be an asparaginyl endopeptidase and termed *Schistosoma* legumain [1638].

In reviewing the cysteine peptidases, Rawlings and Barrett [1187] assigned the number C13 to the peptidase family containing the legume asparaginyl endopeptidase and the *Schistosoma* endopeptidase. This family shows no relationship to the papain family, C1, to which all the other lysosomal cysteine proteinases belong. During 1995, the publication of human EST sequences made it clear that the human genome contains a form of legumain. This prompted Barrett and co-workers in 1997 [1783] to clone and sequence the cDNA for human legumain and to isolate and characterize the enzyme from pig kidney, using the assay that they had introduced earlier for the moth bean enzyme. The properties of the mammalian enzyme were so similar to those of the plant and *Schistosoma* forms that it seems appropriate to retain the same name, *legumain,* for the mammalian enzyme. Quite independently of Chen and co-workers [1783], in 1996 Tanaka *et al.* [1814] cloned and sequenced a human cDNA for a putative cysteine proteinase, *PRSC1*, that we can now identify as legumain.

Mammalian legumain shows maximal activity on the synthetic substrate at pH 5.8, and is very unstable above pH 6 [1783]. The mature protein contains about 10% carbohydrate and the cDNA encodes a putative signal peptide, so there were grounds for suggesting that this could well be a lysosomal enzyme. Subsequent work (P. M. Dando, unpublished results) has confirmed that legumain is indeed lysosomal in rat kidney.

Since there have been only two publications concerning mammalian legumain at the time of writing, the present account must necessarily be brief, but the novelty of a lysosomal cysteine proteinase with striking specificity, but being unrelated to the papain family containing the other lysosomal cysteine proteinases, seems to demand a mention.

Chromosomal location

Tanaka and co-workers [1814] determined the chromosomal location of the putative peptidase, *PRSC1*, or legumain, as 14q32.1 by FISH. The gene has not been sequenced.

Domain structure

The cDNA of human legumain seems to encode a pre-proprotein of 433 amino acid residues and 49 kDa. The prepro sequence is estimated to be quite short, about 25 residues, yet the mature (and deglycosylated) enzyme runs as a protein of 31 kDa in SDS polyacrylamide gel electrophoresis. This clearly indicates that the mammalian legumain is C-terminally processed, as are the plant and *Schistosoma* forms. Pig legumain is *N*-glycosylated, and can be deglycosylated by treatment with *N*-glycosidase F.

Purification

The purification of legumain from pig kidney [1783] required a somewhat unconventional procedure, because the enzyme is stable only in the range pH 3-6 and at low salt concentrations it tends to be adsorbed to any solid material present. The tissue was homogenized and centrifuged and the supernatant was fractionated with ammonium sulfate. When the active fraction was dialysed at pH 5, a heavy precipitate formed, to which the legumain was adsorbed. The enzyme was eluted by raising the salt concentration at pH 6.0 and was run on SP–Sepharose at pH 5.5, requiring 0.4 M NaCl for elution. The enzyme was then run on Mono S in FPLC and bound to thiopropyl–Sepharose activated with 2-pyridyl disulfide. The purified enzyme was stable to storage at pH 5.8, 4°C, in the presence of a thiol compound, showing little loss of activity over several months.

Activation and inhibition

Pig legumain is stable in the pH range 4–6, but is irreversibly denatured at higher pH values. This is strongly reminiscent of the behaviour of most of the lysosomal cysteine peptidases of the papain family (e.g. cathepsins B, H and L).

Typically for a cysteine peptidase, legumain is inhibited by general thiol-blocking agents such as iodoacetate and iodoacetamide (reacting only slowly), as well as by N-ethylmaleimide. N-Phenylmaleimide was the most potent irreversible inhibitor [1783]. E-64 and leupeptin do not inhibit pig kidney legumain, sharply distinguishing the enzyme from most members of the papain family, C1.

Human cystatin C and cystatin from chicken egg white inhibit pig legumain with low nanomolar K_i values and it is possible to determine the molar concentration of legumain active sites by titration of the enzyme with cystatin that has itself been standardized with papain and E-64 [1783].

Legumain differs from the majority of mammalian cysteine endopeptidases, which belong to peptidase families C1 (papain) and C2 (calpain), in being unaffected by E-64, but inhibited by cystatins, as well as in its selectivity for the hydrolysis of asparaginyl bonds. Interestingly, the cytosolic cysteine endopeptidases of the caspase family show an equally strict specificity for hydrolysis of aspartyl bonds [1364].

Hydrolysis of synthetic substrates

The assay substrate for mammalian legumain [1783] was Z-Ala-Ala-Asn-NHMec, originally described for the legume enzyme [1389], but Bz-Asn-OPhNO$_2$ was also hydrolysed. The purified enzyme showed a requirement for activation by a thiol compound and was maximally active at pH 5.8, in a sodium citrate buffer.

Hydrolysis of polypeptides and proteins

Pig kidney legumain acts on polypeptides as a strict asparaginyl endopeptidase. The asparaginyl bonds of neurotensin and vasoactive intestinal peptide are hydrolysed selectively. The three bonds cleaved in a model protein, the 500-amino acid recombinant C-fragment of tetanus toxoid, were asparaginyl bonds, but 44 other asparaginyl bonds in the protein were unaffected, showing that there are additional determinants of specificity. For plant seed legumain, Hara-Nishimura and co-workers [1305] have pointed out that sites of cleavage are typically in markedly hydrophilic parts of the substrate protein and this could account for the selection of bonds cleaved in the tetanus toxoid fragment as well.

A number of lysosomal proteins such as pig cathepsin D [1830] and human β-N-acetylhexosaminidase [1767] are known to be processed by cleavage of asparaginyl bonds, and it is likely that this is an action of legumain [1783]. The hydrolysis of asparaginyl bonds is prominent in the post-translational processing of lysosomal hydrolases. Examples are the cleavages that generate the two-chain forms of cathepsins B and H (see Table 22).

The conservation of the strict asparaginyl endopeptidase specificity of legumain, at least since the divergence of plants and animals perhaps 1000 million years ago, suggests that there is a biological need in the eukaryotic cell for an enzyme with this specificity, but if so, the natural substrates have yet to be identified. As a lysosomal enzyme, legumain is likely to be secreted from cells under some conditions, and like cathepsin L [1294] may be active in the pericellular environment.

Other lysosomal cysteine proteinases

Some lysosomal cysteine proteinases that differ in their physical and catalytic properties from cathepsins B, H, L, K, S and DPP I have been described. These enzymes have been named cathepsins T, N, O and fructose-1,6-bisphosphatase-converting enzyme.

Amino acid sequence analysis revealed that cathepsin J [353] is identical to DPP I [368]. Cathepsin M [251, 252, 379, 541] seems to be identical to cathepsin B by their similarity in size, carbohydrate and amino acid composition, pattern of peptides formed on digestion by trypsin [321] and catalytic properties [122].

The primary structures of cathepsin T and cathepsin N are not known yet.

Cathepsin T (EC 3.4.22.24)

The enzyme from rat liver and kidney has a molecular mass of about 35 000 Da, and has the unique property among the lysosomal cysteine proteinases of converting the 52.5 kDa form of tyrosine aminotransferase to the 48 kDa form [49, 137, 138, 145, 327, 792]. But tyrosine aminotransferase from species other than the rat and mouse, such as the guinea-pig, rabbit, bovine and sheep, is resistant to this conversion by cathepsin T [826].

Cathepsin N

The enzyme ($M_r \sim 35\,000$ Da) has been purified from spleen of cattle [319] and rabbit [360] and human placenta [126]. The latter may be identical with the 'placental cathepsin', the amino acid sequence of which has been identified. The main properties of the enzyme shared with cathepsins L and S were its high collagenolytic activity at pH 3.5 and negligible degradation of synthetic substrates. The pattern of cleavage sites in the insulin B chain by cathepsin N from placenta [127] is different from those by cathepsins L and S. Comparisons of the properties and distribution of cathepsin N with other lysosomal cathepsins have been described [360, 411, 838].

Cathepsin O

A cDNA has been cloned from a human breast cancer cDNA library encoding a cysteine peptidase different from cathepsins B, H, K, L, S and DPP I. The enzyme was named cathepsin O.

Procathepsin O is most closely related to procathepsin H, showing 34.7% amino acid sequence identity. Cathepsin O mRNA (2.5 kb, 5.0 kb) is detectable in a wide variety of human tissues with the highest levels in kidney, liver, ovary and placenta. The translational product has not yet been identified in human tissues. Procathepsin O (M_r 33 000 Da) is expressed in *E. coli*. The activated recombinant enzyme shows very little activity at pH 6.0 against Z-Phe-Arg-NHMec and Z-Arg-Arg-NHMec [1207].

Converting enzyme

The fructose-1,6-bisphosphatase-converting enzyme has been described as a 70 kDa cysteine proteinase bound to the lysosomal membrane and exhibiting the property of converting the catalytically active 36 kDa form of fructose-1,6-bisphosphatase to its active 29 kDa form [251, 540, 541, 988].

Structures and evolution

Structures

The lysosomal cysteine peptidases considered here belong to the papain family of peptidases, termed family C1 by Rawlings and Barrett [1187]. (At the time of writing, it is just emerging that legumain, an endopeptidase specific for the hydrolysis of asparaginyl bonds, is probably an additional lysosomal cysteine proteinase, of a separate family, C13 [1783]. It is too early to say much more about this, at this time.) Among the amino acid sequences known for the papain family, numbering more than 250, are those of enzymes from bacteria, fungi, protozoa and plants, as well as animals. Most of the proteins in the family are endopeptidases, but there are also some exopeptidases and proteins without known catalytic activity. The available amino acid sequences of vertebrate lysosomal cysteine peptidases (Table 19) provide a basis for dividing them into three groups, as is illustrated by a dendrogram for the human enzymes in Fig. 14 (below). One group contains enzymes that are closest in structure to papain: cathepsins H, K, L and S, whereas cathepsin B and dipeptidyl peptidase I represent separate groups, remote from the cathepsin L set and from each other.

Three-dimensional structures have been determined crystallographically for several members of the papain family. Among the lysosomal peptidases significant structures include those of human cathepsin B, procathepsin B and procathepsin L [1228, 1758, 1818, 1821, 1827, 1828]. Figures 5 and 6 depict the structures of papain and human cathepsin B, which show many similarities. The proteins share a two-domain structure in which the defined structural elements, α helices and β strands, are similar in number and location. The L domain, which is shown on the left in the figures, is composed of almost all of the N-terminal half of the protein and contains only helices (H1, H2 and H3) as structural

elements. The right-hand R domain comprises the extreme N-terminal residues and the C-terminal half of the molecule. It is this domain that contains β strands (S1, S3, S4, S5 and S6) in the form of a β barrel, together with two helices (H4 and H5) that close the ends of the barrel.

The active site takes the form of a deep cleft between the domains, below which two antiparallel β strands (S2 and S7) link the two domains. The active site Cys25 is within a long helix (H1) that traverses the molecule along the edge of the L domain; His159 is contained in the long, twisted strand S4. (In the structure of papain, strand S4 is treated as two consecutive strands because the angle of the Ala164-Gly165 link cannot be accommodated in a single strand [1224].) Asn175, which orientates the imidazole ring of His159, forms the third element of the active site and is found at the end of strand S5, running antiparallel to S4. The most obvious difference between the structures of papain and cathepsin B is the large 'occluding loop' that blocks the top (in the orientation of Fig. 6) of the active site cleft and contains residues His92C and His92D. This is undoubtedly of functional importance (see below).

In view of the similarity of the primary structures, it is safe to assume that the other lysosomal cysteine peptidases of family C1 retain the papain fold [1223]. It can be seen from the alignment of the amino acid sequences (Fig. 7) that there is marked conservation of structure in those parts of the molecules that contain residues that are directly implicated in catalytic activity: Cys25, His159 and Asn175 (Table 20). The sequence Gln19 to Phe28 is identical between papain and cathepsin B. The region of well-conserved residues around His159 includes Gly165, which permits the unusual twist of strand S4. The region from Asn175 to Gly185 is also maintained, and helps form one rim of the active site cleft. Other regions

Table 19 Amino acid sequence data for vertebrate lysosomal cysteine peptidases

Classification	Sequence ID	References	Comments
Osteichthyes			
Cyprinus carpio (Carp)			
Cysteine endopeptidase	CCCYPR	Tsai and Huang (unpublished)	
Aves			
Gallus gallus (Chicken)			
Cathepsin B-like protein	GDCBLEMA	84	Fragment
Cathepsin L	CATL_CHICK	1208, 1143	
		1209	Identification of heavy and light chains
Cathepsin B	CATB_CHICK	1786	
Cathepsin K	GG37691	1520	
Mammalia			
Rattus norvegicus (Brown rat)			
Cathepsin B	CATB_RAT	1198	
Dipeptidyl peptidase I	CATC_RAT	1170	
Cathepsin H	CATH_RAT	1198	
Cathepsin L	CATL_RAT	1202	
		1159	Gene structure
Cathepsin S	CATS_RAT	1180	
Placental cathepsin	RNCATLR	1784	
Mus musculus (Mouse)			
Cathepsin B	CATB_MOUSE	1149, 1134	
		1189	Multiple leader sequences
Cathepsin H	CATH_MOUSE	1678	
Cathepsin L	CATL_MOUSE	1165	
Cathepsin K	MMPPCATHK	1806	
Oryctolagus cuniculis (Rabbit)			
Cathepsin K	CATK_RABIT	1201	
Felis catus (Cat)			
Cathepsin L	CATL_FELCA	1163	Fragment
Bos taurus (Cattle)			
Cathepsin B	CATB_BOVIN	1175, 1178	
Cathepsin L	CATL_BOVIN	1190	Fragment
Cathepsin S	CATS_BOVIN	1190, 1213	
Cathepsin 'X'	CATX_BOVIN	1193	Fragment
Homo sapiens (Human)		1173	Review
Cathepsin B	CATB_HUMAN	1192, 1134	
Cathepsin H	CATH_HUMAN	1150, 1191	
Cathepsin L	CATL_HUMAN	1152, 1165, 1191	
		1174	
Cathepsin S	CATS_HUMAN	1195, 1083	Identification of heavy and light chains
Cathepsin O	CATO_HUMAN	1207	
Cathepsin K	CATK_HUMAN	1792	
Dipeptidyl peptidase I	HSCATHCGE	1804	

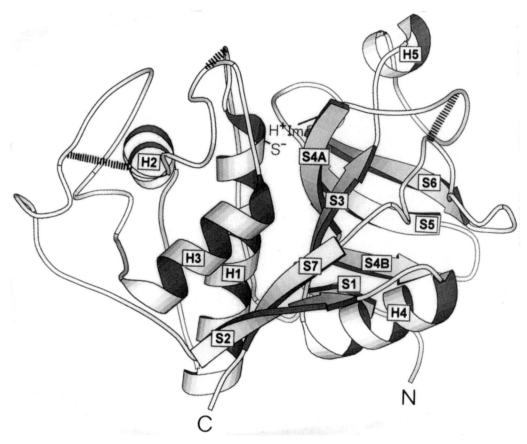

Figure 5

Three-dimensional structure of papain. The figure was plotted from data in the Brookhaven Database entry 9PAP by use of the MOLSCRIPT program [1225], with which α helices (H1–H5) are depicted as coils, and β strands (S1–S7) as arrows. The parts of Cys25 and His159 that comprise the catalytic thiolate–imidazolium pair have also been included. Disulfide bridges are shown as banded lines.

in which sequence is conserved include Gly62 to Gly66, which forms part of the active site wall, and Tyr86 to Tyr88. Tyr86 has hydrogen bonds with residues adjacent to the Cys56/Cys95 disulfide.

Examining the sequences of the lysosomal enzymes in the light of the structure of papain, one can see that most of the insertions and deletions of residues occur in the loops and turns between the defined structural elements. These are depicted schematically for cathepsins H, K, L and S and dipeptidyl peptidase I in Figs 8–12.

The pattern of disulfide bonds and free cysteine residues differs between the peptidases. Papain has only three disulfide bonds: Cys22–Cys63, Cys56–Cys95

and Cys153–Cys200. The same three bonds are probably present in cathepsins H, L and S, as well. Mouse and rat cathepsins L may have an additional disulfide, Cys12–Cys33. Cathepsin B retains the first two disulfides of papain, but has four additional bonds, two of them stabilizing the structure of the loop that forms the cap to the active site (Cys92A–Cys92L, Cys85–Cys99); one connecting helix H1 to an N-terminal loop (Cys14–Cys39) and one across a loop that follows helix H2 (Cys57–Cys60A). Dipeptidyl peptidase I almost certainly also retains the first two of the bridges present in papain (Cys22–Cys63, Cys56–Cys95) and the sequence suggests that there may be an extra bridge (Cys85–Cys101) in the L

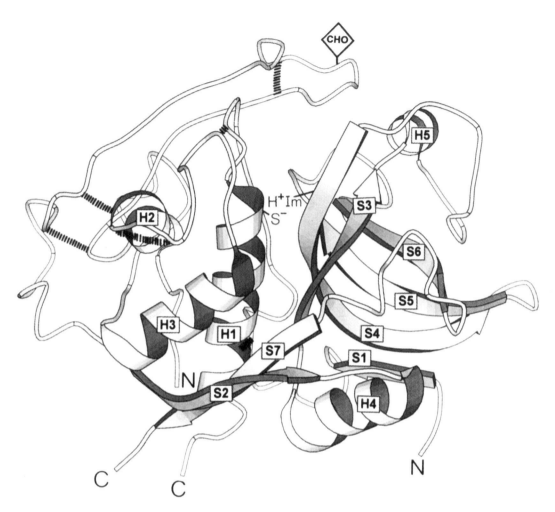

Figure 6

Three-dimensional structure of cathepsin B. The figure was plotted as described for papain (Fig. 5), from data in the Brookhaven Database entry 1HUC.

domain and another (Cys114–Cys200) in the R domain.

Dipeptidyl peptidase I is found as a protein of about 200 kDa, which may well be an octamer of the minimal subunit [1227]; the other lysosomal cysteine peptidases are monomeric.

Structural basis of specificity

The three-dimensional structure of cathepsin B has given some insights into the reasons for its distinctive properties. Cathepsin B is unusual amongst the papain-like endopeptidases in that it readily accepts Arg in the S2 specificity subsite, normally reserved for hydrophobic residues such as Phe, and because of this, Z-Arg-Arg-NHMec is an excellent, selective substrate [8]. The proposal that this is due to the substitution of Ser205 (papain) by Glu in the bottom of the S2 pocket, made originally in 1984 [1130], has recently been confirmed by site-directed mutagenesis [1219].

Cathepsin B is also unusual in showing peptidyl dipeptidase activity (removing C-terminal dipeptides)

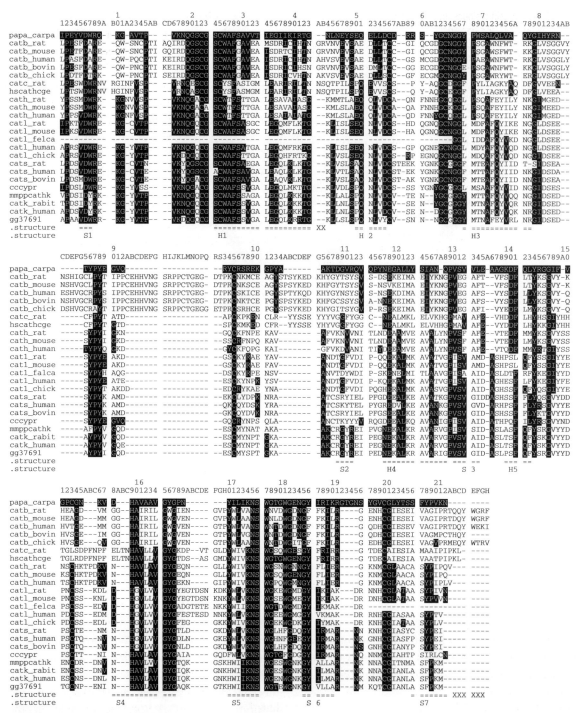

Figure 7

Alignment of amino acid sequences of the lysosomal cysteine peptidases and papain (papa_carpa). An alignment prepared by use of the PILEUP program [1140] was optimized by hand in the light of the three-dimensional structure of cathepsin B [1228]. The numbering system is that of mature papain, with insertions marked 'A, B . . .'. Residues in the other sequences that are identical to residues in papain are printed in white on black. H1–H5 and S1–S7 indicate the positions of the helices and strands known to be present in the structures of papain and cathepsin B. Full names of the sequences can be obtained from Table 19.

Table 20 Amino acid residues invariant between the lysosomal cysteine peptidases and papain

Residue	Location	Function
Pro2		
Asp6		
Arg8		
Gln19		
Cys22		Conserved disulfide bond (to Cys63)
Cys25	Helix 1	Active site
Phe28	Helix 1	Part of active site cleft
Cys56	Helix 2	Conserved disulfide bond (to Cys95)
Gly62		Part of active site wall
Cys63		Conserved disulfide bond (to Cys22)
Gly65		Part of active site wall
Gly66		Part of active site wall
Tyr88		Forms stabilizing H-bond to Gln19
Cys95		Conserved disulfide bond (to Cys56)
Pro129		
Tyr144		Interacts with propeptide
Gly147		
His159	Strand 4	Active site
Gly165	Strand 4	
Gly167	Strand 4	
Asn175	Strand 5	Orientates imidazolium ring of His159
Ser176		
Trp177		Shields His159 and Asn175 from solvent
Gly182		
Gly185	Strand 6	
Cys200		Disulfide bond (not cathepsin B)

Data for possible functions of the residues are taken from [1224, 1223].

in addition to its endopeptidase activity [74, 88]. This is well accounted for by the presence of the occluding loop that blocks the top of the active site cleft (Fig. 6). The basic residues His92C and His92D are located in this loop in such a way as to be well placed to interact with the free C-terminal carboxylate of a polypeptide chain, directing it into position for hydrolysis of the C-terminal dipeptide. These residues also explain the selectivity of the epoxide inhibitor CA074 for cathepsin B [445, 463] (see Table 27).

Table 21 summarizes the residues thought to be important for the subsite specificity of cathepsin B, with the known substitutions at these sites in the other lysosomal homologues. In papain, residues 67 and 157 have also been implicated in the S2 subsite,

but substitution in cathepsin S of Val157 for Leu failed to alter the specificity appreciably [1219]. The importance of the S2 subsite in determining specificity is especially apparent for residue 205 and substitution of Phe205 for Glu in cathepsin S alone is enough to confer a cathepsin B-like specificity, although an accompanying mutation of Gly133 to Ala increases the rate of hydrolysis of the synthetic substrate [1219]. Residues 133 and 160 are almost always Ala or Gly. The S1 subsite makes little contribution to the specificity of the lysosomal proteinases of the papain family, and this is emphasized by the lack of variation between the lysosomal cysteine peptidases. There is considerable variability in the S1' subsite, especially for residues 137 and 142; residue 157 is always hydrophobic and residue 177 is always Trp. The amino acid side chains forming the S2' subsite are invariable and, again, this site contributes little to specificity, with the important exception of His92C in cathepsin B, forming part of the occluding loop.

Biosynthesis and post-translational processing

The lysosomal cysteine peptidases follow the path of the other lysosomal hydrolases in their biosynthesis and trafficking to the lysosomes. Thus, the newly synthesized enzymes pass into the lumen of the ER and are then transported through the Golgi apparatus. They are carried from the *trans* Golgi network to the late endosomes by means of transport vesicles. The lysosomal hydrolases, which contain N-linked oligosaccharides, are recognized by a unique system in the *cis* Golgi network which specifically phosphorylates the mannose residues of the lysosomal proteins. The resulting mannose-6-phosphate (M6P) groups are, in turn, recognized by an M6P receptor protein in the *trans* Golgi network which segregates the hydrolases and helps to package them into budding transport vesicles, which deliver their contents to late endosomes and thence to lysosomes. The transport vesicles shuttle the M6P receptors back and forth between the *trans* Golgi network and late endosomes, but the low pH in the late endosome dissociates the lysosomal hydrolases from the

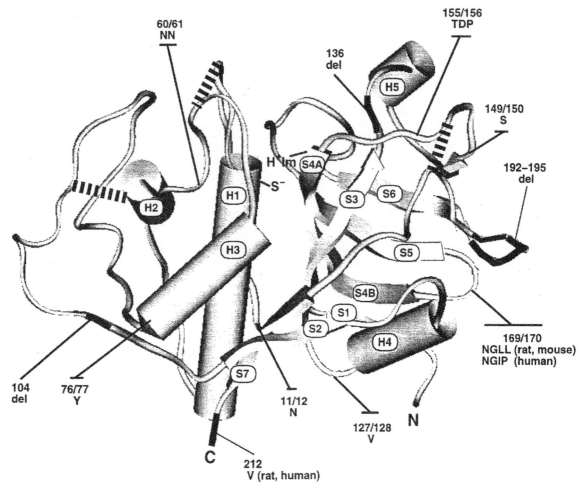

Figure 8

Cathepsin H structure. The positions of insertions and deletions relative to papain are shown on a plot of the papain molecule. The parts of the polypeptides believed to form these defined structural elements are indicated in Fig. 7. An insertion is indicated by a heavy line proportional to the length of insertion and the inserted sequence is given over or under the line. A deletion is indicated by a black segment.

receptor there. The result is that the receptor shuttles back empty and the hydrolases are unidirectionally transported to the endosomes and lysosomes [1238].

During and after their transfer to the lysosomes, the structures of the lysosomal cysteine peptidases are modified by several steps of limited proteolysis [1761]. As the molecules pass into the lumen of the endoplasmic reticulum, the signal peptides are removed. In late endosomes or in the lysosomes, the propeptides are

cleaved, giving rise to the active enzymes. Further proteolysis then generates the two-chain forms of cathepsins B, H and L, and there may be C-terminal trimming.

Since a carbohydrate moiety bearing M6P is of key importance in the recognition of the lysosomal enzymes for transport to the lysosomes, one might ask whether the structurally very similar cysteine peptidases are glycosylated at similar locations. The answer is no. The crystal structure of mature cathepsin

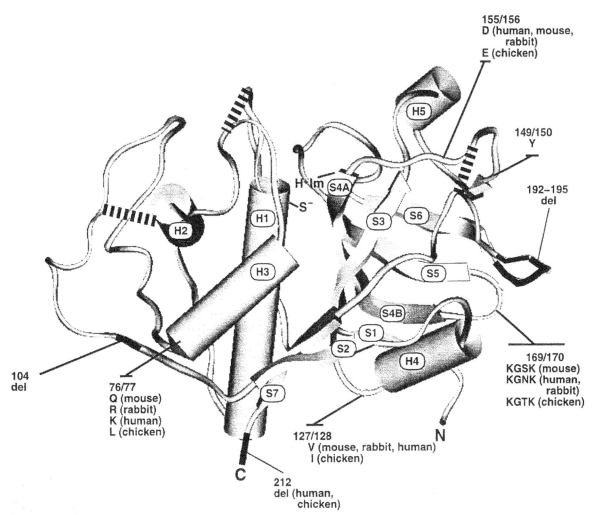

Figure 9
Cathepsin K structure. See Fig. 8 for explanation.

B shows carbohydrate attached to Asn92F [1228]. Cathepsin S, like papain, has no potential glycosylation sites in the mature enzyme, only in the propeptide, but in contrast, cathepsin L has potential glycosylation sites only in the mature protein. Clearly, targeting to the lysosome is not dependent on where in the molecule the carbohydrate moiety is attached.

All of the lysosomal cysteine peptidases are synthesized as inactive precursors with N-terminal extensions. The propeptides range from 62 residues in cathepsin B to 201 in dipeptidyl peptidase I (Fig. 13). There are similarities between the sequences of the propeptides of cathepsins H, K, L and S and papain, and these are related to antigens from mouse cytotoxic lymphocytes known as CTLA-2α and CTLA-2β [1139], and mouse testicular proteins known as testins [1220]. CTLA-2β has been shown to be a competitive inhibitor of papain ($K_i = 25$ nM), cathepsin L ($K_i = 24$ nM) and cathepsin H, but not cathepsin B [1137]; the testins are not known to be inhibitory [1220]. The crystallographic structure of procathepsin L has revealed the structural basis of inhibition of the enzyme in the proenzyme [1818]. The propeptide runs through the active-site cleft in

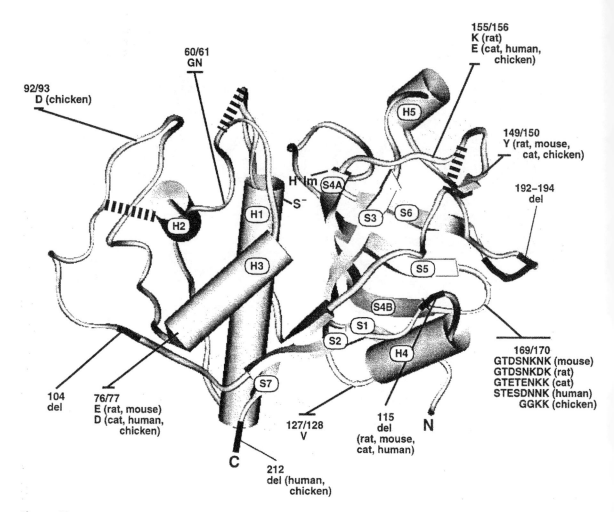

Figure 10
Cathepsin L structure. See Fig. 8 for explanation.

the opposite orientation to a substrate, but in the same direction as the inhibitor, E-64. Side chains of the propeptide occupy specificity subsites on both S and S′ sides of the catalytic groups.

The propeptide of cathepsin B has been shown to be an effective inhibitor of the mature endopeptidase, with a K_i of 0.4 nM [456, 1754, 1755]. It appears that during maturation of cathepsin B, a non-covalent enzyme–propeptide complex forms, which can be secreted [1110]. The propeptides of cathepsin B and dipeptidyl peptidase I show no relationship in amino acid sequence to those of other members of the

papain family, or to each other, but the crystal structure of procathepsin B shows that the propeptide runs through the substrate-binding cleft in very much the same way as that of cathepsin L [1758]. This suggests that there is a common mode of inhibition of activity in the proenzymes of the papain family, despite differences in length and sequence of the propeptides [1818]. In procathepsin B, the propeptide is in contact with two regions of the mature enzyme in addition to the substrate-binding cleft: a large surface loop, known as the prosegment-binding loop (residues 137–154), and the crevice between the occluding

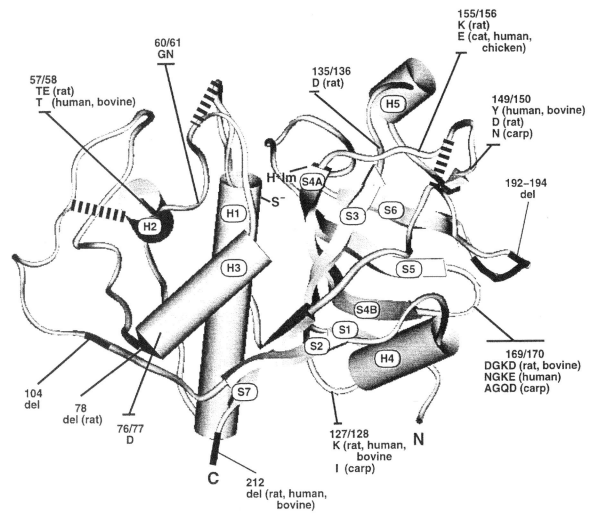

Figure 11

Cathepsin S structure. See Fig. 8 for explanation.

loop and the prosegment-binding loop. Activation of procathepsin B involves a small conformational change, mainly in the occluding loop, in which His111 moves 1.4 nm closer to the active site [1758].

Cathepsin B is unique among the lysosomal cysteine peptidases in that processing also occurs at the C terminus. In human and mouse cathepsins B the cleavage is at Xaa–Gln212C and six residues are removed [1134]. In bovine cathepsin B the cleavage occurs at Thr212A–His and only three residues are lost [786]. In the crystal structure of procathepsin B [1758],

this C-terminal extension forms a distorted two-turn α helix.

Mature cathepsin B is a two-chain molecule, as a result of limited proteolysis that removes a dipeptide between residues Asn43 and Val44 (Table 22) and produces an N-terminal light chain [1134]. This cleavage occurs below the antiparallel strands at the base of the active site, so that the L domain has three termini. Cathepsins H and L are also two-chain molecules, but in these, cleavage towards the C terminus produces C-terminal light chains. In

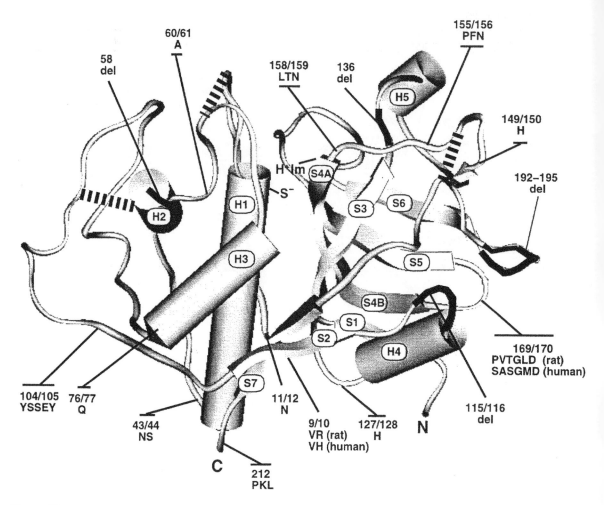

Figure 12
Dipeptidyl peptidase I structure. See Fig. 8 for explanation.

cathepsin H, the cleavage is at Asn169A [1151], which is contained in an extended loop with respect to papain (Fig. 8). Cathepsin L is processed in the same loop, at Thr169B [1152]. Chicken cathepsin L is reported to have a much shorter insertion at this position and processing occurs at Gly169A [1208]. In both cathepsins H and L the disulfide bridge Cys153–Cys200 links the two chains. Again, dipeptidyl peptidase I is processed to a two-chain molecule, with a cleavage at Ser(or Arg)154 [1170]. In contrast, both cathepsins K and S resemble papain in being single-chain molecules [1286, 1083]. Production of the

two-chain form of cathepsin B is inhibited by leupeptin, leading to the accumulation of a 33 kDa single-chain form, in a 54 h pulse-chase experiment in cultured human skin fibroblasts [876].

It will be noted from Table 22 that processing at asparaginyl bonds is prominent in the generation of the two-chain forms of cathepsins B and H. Such processing affects other lysosomal enzymes as well. Thus, the cleavage of cathepsin D at asparaginyl bonds has been discussed by Yonezawa et al. [1830] and a similar observation has been made for N-acetyl-hexosaminidase [1767].

Table 21 Specificity subsites in lysosomal cysteine peptidases

Subsite	Residue	Cathepsin B	Cathepsin H	Cathepsin L	Cathepsin K	Cathepsin S	DPP I
S2	68	Pro	Pro	Met, Ile	Met	Met	Pro
	133	Ala	Ala	Ala	Ala, Gly, Ser	Gly, Ala	Ala
	160	Ala	Ala	Gly	Ala	Gly, Ala	Ala
	205	Glu	Cys	Ala	Leu, Met	Tyr, Phe, Thr	Ile
S1	19	Gln	Gln	Gln	Gln	Gln	Gln
	25	Cys	Cys	Cys	Cys	Cys	Cys
	66	Gly	Gly	Gly	Gly	Gly	Gly
S1'	137	Val	Val	Ser, Gly	Ser	Ser, Thr, Arg	Val
	142	Leu	Met, Leu	Gln, Arg, Leu	Gln	Phe, Ile	Leu
	157	Met, Val	Val	Leu, Val, Met	Val, Leu, Ile	Val, Ile, Met	Phe
	177	Trp	Trp	Trp	Trp	Trp	Trp
S2'	19	Gln	Gln	Gln	Gln	Gln	Gln
	92C	His	–	–	–	–	–

Residues thought to be involved in substrate binding have been identified in cathepsin B only [1828]. Equivalent residues in other lysosomal cysteine peptidases are taken from the alignment (Fig. 7). Residues are numbered according to papain.

Evolution of the lysosomal cysteine peptidases

In addition to the lysosomal enzymes of vertebrate animals that are our primary topic here, the papain family contains numerous peptidases of invertebrate animals, plants, fungi, protozoa and eubacteria [1187]. The relationships amongst these can be investigated by the construction of phylogenetic trees [1779]. The lysosomal enzymes fall into three groups: one comprises the enzymes most related to cathepsin B, the second is a small group containing little more than dipeptidyl peptidase I and the largest group can be described broadly as 'cathepsin L-like endopeptidases'. This latter group is composed of cathepsins K, L and S, connected more remotely to cathepsin H. (The sequence of cathepsin O places it between cathepsin H and dipeptidyl peptidase I, but little is yet known of the properties of the protein.)

The three groups of cysteine peptidases that contain the lysosomal enzymes diverged during the evolution of the protozoa; all could potentially be represented in any modern multicellular organism. They can be discerned in a simple dendrogram for the human lysosomal cysteine peptidases (Fig. 14).

In addition to the vertebrate enzymes, the cathepsin B group also contains peptidases from protozoa, nematodes and platyhelminths, and one from wheat [1187]. An endopeptidase from the blood fluke, *Schistosoma,* has been described as cathepsin B, and is reported to have similar specificity to the mammalian enzyme [1154].

Among the common characteristics of the endopeptidases of the cathepsin K/L/S/H group is their pattern of disulfide bridges, but these enzymes also have clearly homologous propeptides, whereas it will be remembered that the propeptides of cathepsin B and dipeptidyl peptidase I are very different. Included within the immediate group of cathepsin L are a number of peptidases from arthropods, including crustaceans and insects [1197].

The cathepsin H group contains plant enzymes such as aleurain from barley and oryzain γ from rice, the first of which has also been shown to have similar specificity to cathepsin H [1155, 1230].

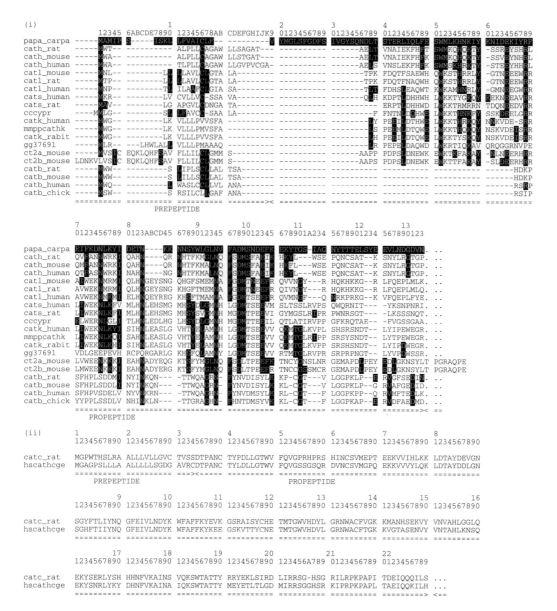

Figure 13

Sequences of the propeptides of the lysosomal cysteine peptidases. (i) Prepropeptides of cathepsins B, H, K, L and S aligned with that of papain. The presumed cleavage points between the signal (pre-) peptides and the propeptides, and between the propeptides and the mature proteins, are marked ' > < '. (ii) The prepropeptide of dipeptidyl peptidase I. Full names of the sequences can be obtained from Table 19.

Table 22 Processing of lysosomal cysteine endopeptidases in two chain forms

		43	
Cathepsin B	(Chicken)	Arg-Ile-Cys-Val-His-Thr-**Asn**┼Ala-Lys┼Val-Ser-Val-Glu-Val-Ser	(a)
	(Rat)	Arg-Ile-Cys-Ile-His-Thr-**Asn**┼Gly-Arg┼Val-Asn-Val-Glu-Val-Ser	(b)
	(Mouse)	Arg-Thr-Cys-Ile-His-Thr-**Asn**┼Gly-Arg┼Val-Asn-Val-Glu-Val-Ser	(c)
	(Cattle)	Arg-Ile-Cys-Ile-His-Ser-**Asn**┼Gly-Arg┼Val-Asn-Val-Glu-Val-Ser	(d)
	(Human)	Arg-Ile-Cys-Ile-His-Thr-**Asn**┼Ala-His┼Val-Ser-Val-Glu-Val-Ser	(e)

		169A	
Cathepsin H	(Rat)	Val-Gly-Tyr-Gly-Glu-Gln-**Asn**┼Gly-Leu-Leu-Tyr-Trp-Ile	(b)
	(Mouse)	Val-Gly-Tyr-Gly-Glu-Gln-**Asn**┼Gly-Leu-Leu-Tyr-Trp-Ile	(b)
	(Human)	Val-Gly-Tyr-Gly-Glu-Lys-**Asn**┼Gly-Ile-Pro-Tyr-Trp-Ile	(f)

		169B	
Cathepsin L	(Rat)	Gly-Tyr-Gly-Tyr-Glu-Gly-Thr┼Asp-Ser┼Asn-Lys-Asp-Lys-Tyr-Trp	(g)
	(Mouse)	Gly-Tyr-Gly-Tyr-Glu-Gly-Thr┼Asp-Ser┼Asn-Lys-Asn-Lys-Tyr-Trp	(h)
	(Cattle)	Gly-Tyr-Gly-Phe-Glu-Gly-Thr┼Asp-Ser┼Asn-Asn-Asn-Lys-Phe-Trp	(i)
	(Sheep)	Gly-Tyr-Gly-Phe-Glu-Gly-Thr┼**Asn**┼Asn-Lys-Phe-Trp-Ile-Val	(j)
	(Human)	Gly-Tyr-Gly-Phe-Glu-Ser-Thr┼Glu-Ser-Asp┼Asn-Asn-Lys-Trp-Leu	(f)

		154	
DPP I	(Rat)	Tyr-His-His-Thr-Gly-Leu-Ser┼Asp-Pro-Phe-Asn-Pro-Phe-Glu-Leu	(k)
	(Human)	Tyr-His-His-Thr-Gly-Leu-Arg┼Asp-Pro-Phe-Asn-Pro-Phe-Glu-Leu	(l)

It can be seen that in cathepsin B two residues are excised; in cathepsin L, one, two or three amino acids may be removed. Asparagine residues in the S1 subsites relative to the site of processing are shown in bold. Cathepsins K and S are single-chain molecules. Amino acid numbers are relative to papain (see Fig. 7). ┼: cleaved peptide bonds. References: (a) [84], (b) [1198], (c) [1134], (d) [1175], (e) [1192], (f) [1191], (g) [1202], (h) [715], (i) [1190], (j) [1808], (k) [1160], (l) [1374].

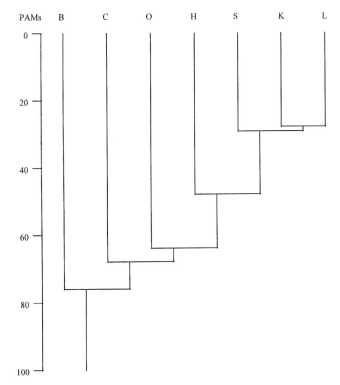

Figure 14

Phylogenetic tree for human lysosomal cysteine peptidases. The amino acid sequences were aligned essentially as in Fig. 7 and the phylogenetic tree was obtained using the KITSCH algorithm and the PHYLIP package [1187]. The identities of the human enzymes are: B, cathepsin B; C, dipeptidyl peptidase I; O, cathepsin O; H, cathepsin H; S, cathepsin S; K, cathepsin K; and L, cathepsin L. The left hand axis is accepted point mutations (PAMs).

Reaction with inhibitors

Introduction

Cysteine proteinases contain a highly reactive cysteine residue in their active centre. They react readily with many thiol-blocking reagents, such as Hg^{2+} derivates, disulfides, heavy metal ions, iodoacetate and N-ethyl-maleimide.

To achieve a more selective reaction with the enzymes rather than with SH-containing compounds generally, several peptidyl derivatives can be used. Compounds of this type can show selectivity amongst the cysteine proteinases. As will be shown in this section, nearly all of the synthesized cysteine proteinase inhibitors react (in some cases very weakly) with serine proteinases, with the exception of the epoxysuccinyl peptides. Selectivity amongst the individual cysteine proteinases has been obtained by variation of the peptidyl portion of the inhibitors, and in some cases (peptidyl-O-acyl-hydroxamates and peptidyl-(acyloxy)methanes) also by modification of the leaving group.

Synthetic inhibitors are indispensable in the study of the functions of cysteine proteinases *in vivo*, to determine their activity *in vitro*, to characterize the binding sites, catalytic functional groups and transition state geometries. The main aim is to develop inhibitors for use as therapeutic agents in diseases where involvement of lysosomal cysteine proteinases may lead to uncontrolled proteolysis or to induction or secretion of other enzymes.

Natural inhibitors belonging to the cystatin superfamily are selective for cysteine proteinases, whereas α_2-macroglobulin and related proteins react with all endopeptidases.

Covalent inhibitors

Peptidyl chloromethanes

These peptide derivatives (e.g. Z-Phe-Ala-CH$_2$Cl)

inhibit cysteine proteinases irreversibly. Two mechanisms have been proposed for the alkylation of the thiol group of the active site cysteine [50].

Peptidyl-chloromethanes (chloromethyl ketones) react very rapidly with cysteine proteinases (Table 23), but can be destroyed by dithiothreitol [543], and also inhibit serine proteinases [420, 543, 544, 1231].

Peptidyl fluoromethanes

Peptidyl fluoromethanes (fluoromethyl ketones, e.g. Z-Phe-Ala-CH$_2$F) react more slowly with cysteine proteinases than peptidyl chloromethanes [543, 544], but also alkylate the active site cysteine. The fluoromethanes are stable to high concentrations of dithiothreitol.

Fluoromethanes also react with serine proteinases, but, for example, Z-Phe-CH$_2$F inactivates chymotrypsin with a rate constant that is only 2% of that observed with Z-Phe-CH$_2$Cl [544].

Second-order rate constants for inhibition are given in Table 23.

Peptidyl fluoromethyl ketones have been used in *in vitro* and *in vivo* experiments [93, 416, 418–420, 454, 472, 543, 544, 559, 571, 733, 1053].

Peptidyl diazomethanes

The intrinsic reactivity of the diazomethyl ketone group (e.g. Z-Phe-Ala-CHN$_2$) to thiols is lower than that of fluoromethyl ketones, but the substrate-binding site of the enzyme accelerates the reaction with an inhibitor containing a suitable peptide sequence. Therefore, the peptidyl diazomethanes show some degree of selectivity between the cysteine proteinases [54, 448, 490, 555] (Table 24).

The inhibition of serine proteinases by peptidyl-diazomethanes is restricted to prolyl oligopeptidase

Table 23 Inactivation of cysteine proteinases by peptidyl chloromethanes and fluoromethanes

Inhibitor	k_{2app} (M^{-1} s^{-1})*			
	Cathepsin B	Cathepsin L	Cathepsin S	References
Pro-Phe-Har-CH$_2$Cl	500 000			1231
Pro-Phe-Arg-CH$_2$Cl	5 000 000	30 000 000		431
Pro-Gly-Arg-CH$_2$Cl	1670			1231
Pro-Phe-Arg(NO$_2$)-CH$_2$Cl	139 000			1231
Pro-Phe-D-Arg-CH$_2$Cl	10 500			1231
D-Phe-Phe-Arg-CH$_2$Cl	400 000			1231
Phe-Phe-Arg-CH$_2$Cl	327 000			1231
Phe-Ala-Lys-CH$_2$Cl	318 000			1231
Phe-Ala-Arg(NO$_2$)-CH$_2$Cl	250 000			1231
Phe-Ala-Arg-CH$_2$Cl	247 000			1231
D-Phe-Pro-Arg-CH$_2$Cl	3800			1231
Val-Val-Arg-CH$_2$Cl	348 000			1231
Gly-Val-Arg-CH$_2$Cl	180 000			1231
Ala-Lys-Arg-CH$_2$Cl	22 000			1231
Z-Leu-Leu-Phe-CH$_2$Cl	190 000	21 500 000		448
Tos-Lys-CH$_2$Cl	517	783		431
Z-Phe-Ala-CH$_2$F	18 000			543
	8200	88 000	2 778 000	93

* The second-order rate constant was $k_{2app} = k_{obs}/[I]$ [491].

by alkylation of the active site serine [561] and to members of the subtilisin family such as subtilisin Carlsberg and thermitase, where the active site histidine was alkylated [453].

The peptidyl diazomethanes have been widely used as inhibitors for lysosomal cysteine proteinases in *in vitro* and *in vivo* experiments [419, 421, 448, 465, 468, 469, 489, 490, 504, 513, 543, 553, 555, 575, 578, 580, 664, 733].

Peptidyl methyl sulfonium salts

These inhibitors are derived from the reaction of peptidyl chloromethanes with dimethylsulfide (e.g. Z-Phe-Ala-CH$_2$Cl + (CH$_3$)$_2$S → Z-Phe-Ala-CH$_2$S$^+$(CH$_3$)$_2$ + Cl$^-$). Irreversible inactivation is the result of transfer of the peptidyl group and not the methyl groups. The ylide structure formed by release of a proton may be the active affinity labelling form, because the rate of inactivation of cathepsin B increases with pH [551].

Reactions with peptidyl methyl sulfonium salts have been studied [419, 539, 551, 584].

N-peptidyl-O-acyl hydroxamates

The inhibitors have the structure Z-Phe-Ala-NHO-CO-R, and were developed as mechanism-based inhibitors of serine proteinases, but their reactivity with cysteine proteinases was much higher than with serine proteinases [441, 557].

Variation of the peptidyl residue as the affinity labelling group and the O-acyl residue as the leaving group allows one to increase the selectivity and the reactivity of these inhibitors (Table 25). Peptidyl-O-acyl hydroxamates as inhibitors of cysteine proteinases have been studied in several experiments [437–439, 441, 451, 557].

Peptidyl (acyloxy)methanes

These irreversible inhibitors have the structure Z-Phe-Ala-CH$_2$-O-CO-R [500]. They have in common with the peptidyl-O-acyl hydroxamates that the rate of inactivation depends on both the peptidyl portion and the leaving group R-COO$^-$ [501] (Table 26).

Table 24 Inactivation of cysteine proteinases by peptidyl diazomethanes

Inhibitor	k_{2app} (M^{-1} s^{-1})			
	Cathepsin B	Cathepsin L	Cathepsin S	References
Z-Phe-Thr-(OBzl)-CHN$_2$	300 000			555
Z-Phe-Cys(SBzl)-CHN$_2$	30 000	390 000		490, 555
Z-Phe-Ser(OBzl)-CHN$_2$	8800	1 000 000		490, 555
Z-Phe-Cit-CHN$_2$	6700	1 800 000		490, 555
Z-Phe-Ala-CHN$_2$	1200			555
		160 000		362
Z-Phe(I)-Ala-CHN$_2$	980	130 000		448
Z-Phe-Phe-CHN$_2$	350			555
		140 000		362
			910	554
Z-Phe-Thr(OBu^t)-CHN$_2$	160	60 000		490, 555
Z-Phe-Trp-CHN$_2$	200	140 000		490, 555
Z-Phe-Tyr(OCH$_3$)-CHN$_2$	30	180 000		490, 555
Z-Phe-Phe(4-NO$_2$)-CHN$_2$	50	190 000		490, 555
Z-Phe-Tyr(OBu^t)-CHN$_2$	10	200 000		490
			30	554
Z-Phe-Gly-CHN$_2$	700			578
Z-Phe-CHN$_2$	0.3			578
Z-Ala-Ala-CHN$_2$	140			465
Z-Ala-Phe-Ala-CHN$_2$	1 175			578
Z-Gly-Gly-Met-CHN$_2$	6			465
Z-Gly-Gly-Phe-CHN$_2$	5			465
Z-Leu-Met-CHN$_2$	4060	212 000		448
Z-Leu-Trp-CHN$_2$	<50	12 000		448
			22 000	554
Z-Leu-Tyr-CHN$_2$	<50	15 000		448
Z-Leu-Val-Gly-CHN$_2$	10 640	118 000		415
Z-Leu-Leu-Tyr-CHN$_2$	1300	1 500 000		448
			133 000	295
			2 100 000	554
Z-Leu-Leu-Nle-CHN$_2$			9 200 000	554
Boc-Lys-Leu-Tyr-CHN$_2$	570	5800		448
			40 000	554
Z-Tyr-Ala-CHN$_2$	1180	177 000		448
	1800	120 000	1740	99
Z-Tyr(I)-Ala-CHN$_2$	27 800	1 128 000		448
Z-Trp-Met-CHN$_2$		140 000	380 000	554
Z-Val-Val-Nle-CHN$_2$		12 000	4 600 000	554

Table 24 Continued

Inhibitor	k_{2app} (M^{-1} s^{-1})			
	Cathepsin B	Cathepsin L	Cathepsin S	References
Z-Val-Val-Tyr-CHN$_2$			340 000	554
Z-Val-Met-CHN$_2$			33 000	554
Z-Val-Val-CHN$_2$			2500	554

Some of the peptidyl (acyloxy)methyl ketones bind to serine proteinases, but possibly in a reversible manner, because no time-dependent inactivation of these enzymes has been observed [500].

Epoxysuccinyl peptides

Although the first member of this group of compounds, E-64 (which is [N-(L-3-*trans*-carboxyoxirane-2-carbonyl)-L-leucyl]-amido(4-guanido)-butane or epoxysuccinyl leucyl agmatine) was a natural inhibitor [471] secreted from *Aspergillus japonicus*, the epoxysuccinyl peptides are described in the section on synthetic inhibitors, because several analogues of E-64 have been synthesized (Table 27). The compounds N-(L-3-*trans*-propylcarbamoyloxirane-2-carbonyl)-L-isoleucyl-L-proline (CA074 [522]) and N-(L-3-*trans*-ethoxycarbonyloxirane-2-carbonyl)-L-isoleucyl-L-proline (CA030 [522]) proved to be selective in *in vitro* experiments for the inhibition of cathepsin B [522, 566], but in *in vivo* studies only compound CA074 selectively inhibited cathepsin B [566].

No inhibition of metallo-, serine or aspartic proteinases by E-64 and related epoxide compounds has been described. The epoxysuccinyl peptides are unreactive with low molecular mass, SH-containing compounds present in *in vitro* assay systems to guarantee complete catalytic activity of cysteine proteinases. Therefore, these inhibitors can be used as active-site titrants [8, 431].

The epoxysuccinyl peptides have been widely used in *in vitro* and *in vivo* experiments to study the functions of lysosomal cysteine proteinases [431, 432, 445, 463, 473, 474, 522, 523, 564, 566, 567, 938, 956].

Table 25 Inactivation of cysteine proteinases by peptidyl-*O*-acyl hydroxamates

Inhibitor	k_{2app} (M^{-1} s^{-1})*				
	Cathepsin B	Cathepsin L	Cathepsin S	Cathepsin H	References
Z-Phe-Ala-NHO-CO-2,4,6-(CH$_3$)$_3$-Ph	640 000				557
Boc-Phe-Ala-NHO-CO-4-NO$_2$-Ph	14 000	437 000	42 000	21	441
Z-Phe-Lys-NHO-CO-4-NO$_2$-Ph	35 000	3 538 000	471 000	760	439
Z-Phe-Phe-NHO-CO-C(CH$_3$)=CH$_2$	2800	1 222 000	21 000	19	441
Boc-Ala-Phe-Leu-NHO-CO-4-NO$_2$-Ph	12 000	696 000	229 000	32	441
Boc-Ala-Phe-NHO-CO-4-NO$_2$-Ph		600	30		441
Z-Lys-Lys-NHO-CO-4-NO$_2$-Ph	30 000	47 000	5600	230	439
Z-Val-Val-Lys-NHO-CO-4-NO$_2$-Ph	15 000	443 000	606 000	650	439
Phe-NHO-CO-4-NO$_2$-Ph				31 000	439
Boc-Phe-Gly-NHO-CO-Ph	2600	136 000	19 000		441
Boc-Gly-Phe-Phe-NHO-CO-4-NO$_2$-Ph	8300	800 000	267 000		441

* The second-order rate constant was $k_{2app} = k_2/K_i$ (M^{-1} s^{-1}) [491].

Table 26 Inactivation of cysteine proteinases by peptidyl (acyloxy)methanes

Inhibitor	k_{2app} $(M^{-1} s^{-1})^*$			
	Cathepsin B	Cathepsin L	Cathepsin S	References
Z-Phe-Ala-CH$_2$O-CO-2,6-(CF$_3$)$_2$-Ph	1 600 000	332 000	364 000	500
Z-Phe-Ala-CH$_2$O-CO-2,6-Cl$_2$-Ph	690 000	143 000	686 000	500
Z-Phe-Ala-CH$_2$O-CO-2,5-(CF$_3$)$_2$-Ph	38 000	2400	4700	500
Z-Phe-Ala-CH$_2$O-CO-2,6-(CH$_3$)$_2$-4-COOCH$_3$-Ph	58 000	3600	42 000	500
Z-Phe-Ala-CH$_2$O-CO-4-NO$_2$-Ph	610	44 000	3300	500
Z-Phe-Ala-CH$_2$O-CO-2,4,6-(CH$_3$)$_3$-Ph	14 000	4200	500	500
Z-Phe-Ala-CH$_2$O-CO-4-CN-Ph	280	1000	200	500
Z-Phe-Ala-CH$_2$O-CO-CH$_2$OCH$_3$	240	4400	500	500
Z-Phe-Ala-CH$_2$O-CO-4-CF$_3$-Ph	80	700	600	500
Z-Phe-Ala-CH$_2$O-CO-4-OCH$_3$-Ph		1000	100	500
Z-Phe-Ala-CH$_2$O-CO-CH(CH$_2$CH$_3$)$_2$	70	900	200	500
Z-Phe-Ala-CH$_2$O-CO-C(CH$_3$)$_3$	330	7800	2700	500
Z-Phe-Lys-CH$_2$O-CO-2,4,6-(CH$_3$)$_3$-Ph	230 000	71 000	120 000	500
Z-Phe-Cys(SBzl)-CH$_2$O-CO-2,6-(CF$_3$)$_2$-Ph	2 900 000	10 700 000	1 550 000	500
Z-Phe-Ser(OBzl)-CH$_2$O-CO-2,6-(CF$_3$)$_2$-Ph	2 600 000	4 290 000	52 000	500
Z-Phe-Phe-CH$_2$O-CO-2,4,6-(CH$_3$)$_3$-Ph	9900	100	400	500

* The second-order rate constant was $k_{2app} = k_{inact}/K_i$ $(M^{-1} s^{-1})$.

Reversible inhibitors

Peptide aldehydes

Peptide aldehydes are transition state analogue inhibitors of cysteine proteinases. The aldehyde group of the inhibitors reacts with the SH group of the active site cysteine to form the hemithioacetal. The reaction of the peptide aldehydes with cysteine proteinases occurs in two steps [427], the second of which occurs slowly [550].

The peptide aldehydes bind very tightly to cysteine proteinases, reaching K_i values in the range of 10^{-9}–10^{-10} M, but nevertheless the reaction is reversible.

Naturally occurring peptide aldehydes have been found in microbial culture filtrates by Umezawa [570]: antipain, leupeptin, chymostatin and others.

Peptide aldehydes also inhibit serine proteinases. The amino acid residues in the P1 and P2 positions are most important for determining enzyme selectivity. Cathepsins L and B are sensitive to these tight-binding inhibitors, but cathepsin H is inhibited relatively poorly (Table 28).

Peptide aldehydes have been used as inhibitors for lysosomal cysteine proteinases in several *in vitro* and *in vivo* experiments [427, 467, 492, 514, 523, 527, 550, 558, 735, 956].

Peptide-aldehyde semicarbazones

Peptide semicarbazones are potent inhibitors of cysteine proteinases depending on the length of the peptide chain (Table 29). The mechanism of inhibition is thought to be closely related to that of the peptide aldehydes [50]. Immobilized peptide semicarbazones have been used for affinity chromatographic purification of cathepsin B [384].

Peptidyl methanes

Peptidyl methanes (methyl ketones) with the structure Z-Phe-Ala-CH$_3$ represent a new class of reversible, competitive cysteine proteinase inhibitors. The affinity of the inhibitors to cysteine proteinases is dependent on the peptide chain length and on side-chain effects (Table 30).

Table 27 Inactivation of cysteine proteinases by epoxysuccinyl peptides

Inhibitor	k_{2app} (M^{-1} s^{-1})*				
	Cathepsin B	Cathepsin L	Cathepsin S	Cathepsin H	References
E-64 (L)	89 400	96 000		4000	431
	27 000				1444
		218 000	99 000		1443
E-64 (D)	1900	2700		65	431
Ep-459	69 500	27 500		3080	431
Ep-479	339 000	142 600		2070	431
Ep-460	175 000	231 000		778	431
Ep-174	388				431
Ep-429	64 500				431
Ep-475 (L,L)	298 000	206 000		2080	431
Ep-475 (L,D))	3790			28	431
CA074	112 000				445
OH-Eps-Phe-OBzl	910				1444
		434 000	61 000		1443
OH-Eps-Leu-OBzl	24 700				1444
		791 000	171 000		1443
OH-Eps-Arg-OBzl	8300				1444
		73 000	820		1443
OH-Eps-Phe-NHBzl	9000				1444
		27 765 000	501 000		1443
OH-Eps-Leu-NHBzl	37 600				1444
		4 224 000	542 000		1443
OH-Eps-Leu-Pro-OBzl	8700				463
OH-Eps-Leu-Pro-OH	2200				463
EtO-Eps-Leu-Pro-OBzl	30				463
EtO-Eps-Leu-Pro-OH	13 800				463
BuNH-Eps-Leu-Pro-OBzl	200				463
BuNH-Eps-Leu-Pro-OH	52 000				463

* The second-order rate constant was $k_{2app} = k_2/[\mathrm{I}]$ (M^{-1} s^{-1}) or $k_{2app} = k_{inact}/K_i$ (M^{-1} s^{-1}).

E-64 (L)   N-(L-3-trans-carboxyoxirane-2-carbonyl)-L-leucyl-4-guanidinobutylamide
Ep-459   N-(L−3-trans-carboxyoxirane-2-carbonyl)-L-leucyl-aminobutylamide
Ep-479   N-(L−3-trans-carboxyoxirane-2-carbonyl)-L-leucyl-aminoheptylamide
Ep-460   N-(L−3-trans-carboxyoxirane-2-carbonyl)-L-leucyl-carbobenzoxy-aminobutylamide
Ep-174   N-(L−3-trans-carboxyoxirane-2-carbonyl)-L-leucine
Ep-429   N-(L−3-trans-carboxyoxirane-2-carbonyl)-L-leucyl-L-leucine
Ep-475   N-(L−3-trans-carboxyoxirane-2-carbonyl)-L-leucyl-3-methyl-butylamide
CA074   N-(L−3-trans-propyl-carbamoyloxirane-2-carbonyl)-L-isoleucyl-L-proline
OH-Eps-Leu-Pro-OBzl   N-[(2S,3S-3-trans-(hydroxycarbonyl)oxirane-2-yl)carbonyl]-L-leucyl-L-proline benzyl ester
OH-Eps-Leu-Pro-OH   N-[(2S,3S-3-trans-(hydroxycarbonyl)oxirane-2-yl)carbonyl]-L-leucyl-L-proline
EtO-Eps-Leu-Pro-OBzl   N-[(2S,3S-3-trans-(ethoxycarbonyl)oxirane-2-yl)carbonyl]-L-leucyl-proline benzyl ester
EtO-Eps-Leu-Pro-OH   N-[(2S,3S-3-trans-(ethoxycarbonyl)oxirane-2-yl)carbonyl]-L-leucyl-proline
BuNH-Eps-Leu-Pro-OBzl   N-[(2S,3S-3-trans-(isobutylamino-carbonyl)oxirane-2-yl) carbonyl]-L-leucyl-L-proline benzyl ester
BuNH-Eps-Leu-Pro-OH   N-[(2S,3S-3-trans-(isobutylamino-carbonyl)oxirane-2-yl) carbonyl]-L-leucyl-L-proline

Table 28 Inhibition of cysteine proteinases by peptide aldehydes

Inhibitor	K_i (nM)			References
	Cathepsin B	Cathepsin L	Cathepsin H	
Ac-Leu-Leu-Arg-H	7			173
(leupeptin)	5			427
		<1		362
			5000	169
			6900	389
Ac-Gly-Phe-Nle-H	13			50
Z-Gly-Phe-Gly-H	510			50
Z-Arg-Ile-Phe-H	11			50
Z-Ile-Phe-H	130			50
Z-Phe-Gly-H	1250			50

Displaying K_i values in the micromolar range, these inhibitors could be used as ligands for affinity purification of cysteine proteinases [93, 436].

Natural protein inhibitors

Cystatins

The cystatin superfamily contains three families of proteins that are related functionally as cysteine proteinase inhibitors and evolutionarily by their amino acid sequence identity [1, 9, 1206]. These inhibitors occur

Table 29 Inhibition of cysteine proteinases by peptide-aldehyde semicarbazones

Inhibitor	Cathepsin B K_i (µM)	References
Z-Arg-Ile-Phe-H semicarbazone	0.5	50
Z-Ile-Phe-H semicarbazone	1.2	50
Z-Phe-Gly-H semicarbazone	20	50
Z-Gly-Phe-Gly-H semicarbazone	3.4	384
Gly-Phe-Gly-H semicarbazone	63	384

in all cells and body fluids of mammals and many lower organisms. Their function is suggested to be the protection of proteins outside the lysosomes from degradation by cysteine proteinases, but also to protect active cysteine proteinases outside the lysosomes from inactivation by the physiological neutral pH. The reaction of catalytically active cathepsin L with cystatin B was faster than the inactivation of cathepsin L at pH 7.4. At acidic pH values the inhibitor complex with cysteine proteinases dissociates, releasing catalytically active enzymes [568].

Gene defects leading to mutants or which prevent synthesis cause severe diseases such as amyloidosis syndrome of an Icelandic type, where the mutated cystatin C deposits selectively in the walls of cerebral blood vessels [1445] and one form of progressive myoclonus epilepsy, where cystatin B synthesis is disturbed [1471].

The interaction of cystatins with cysteine peptidases [1217] is a reversible and tight-binding one at the active site, but without formation and cleavage of covalents bonds. Cystatins also react with their inactive carboxymethylated target enzymes.

The affinity of the cystatins to the lysosomal cysteine proteinases is very high, displaying K_i values in the nanomolar and picomolar range (Table 31). They do not react with serine or other types of proteinases.

The cystatins are classified into three families as follows.

Table 30 Inhibition of cysteine proteinases by peptidyl methanes

Inhibitor	K_i (µM)			
	Cathepsin B	Cathepsin L	Cathepsin H	References
Z-Ala-Ala-CH₃	800	80	>1000	436
Z-Ala-Ala-Ala-CH₃	220	240	>1000	436
Z-Ala-Ala-Ala-Ala-CH₃	12	58	>500	436
Z-Phe-Phe-CH₃	58	10	>30	436
Z-Phe-Lys(Z)-CH₃	26	3		436
Z-Ala-Phe-Ala-CH₃	610	45		436
Z-Ala-Ala-Phe-Ala-CH₃	370	400	>200	436
Z-Ala-Ala-Phe-Leu-CH₃	160	18		436
Z-Ala-Ala-Phe-Phe-CH₃	16	1	>20	436
Z-Ala-Ala-Phe-Lys(Z)-CH₃	33	29		436
Ac-pAB-Ala-Phe-Ala-CH₃	9	3		436
Leu-CH₃	>1000	>1000	67	436
Phe-CH₃	>1000	>1000	45	436
Lys(Z)-CH₃	>1000	160	13	436

- *Family 1:* cystatin A, cystatin B (other names: stefin A, stefin B; ACPI, NCPI). Synthesized without signal peptides; $M_r = 11\,000–12\,000$ Da; contain no disulfide bonds; occur intracellularly in the cytosol.

- *Family 2:* cystatins C, D, S, SN, SA. Synthesized with signal peptides, $M_r = 13\,000–14\,000$ Da, contain disulfide bonds; are secreted and present in body fluids.

Table 31 Inhibition of cysteine proteinases by cystatins

Inhibitor	K_i (nM)					
	Cathepsin B	Cathepsin L	Cathepsin H	Cathepsin S	DPP I	References
Cystatin A	8.2	1.3	0.31	–	33	9
	1.9	0.03	0.4	–	–	1486
	–	–	–	0.05	–	440
Cystatin B	73	0.23	0.58	–	0.23	9
	130	0.035	–	–	–	1456
	–	–	–	0.07	–	440
	–	–	–	~0.008	–	1485
Cystatin C	0.25	0.005	0.28	–	3.5	9
	–	–	–	0.008	–	440
Cystatin D	–	25	8.5	0.24	–	1
	>1000	–	–	–	–	457
Cystatin SN	19	–	–	–	–	414
Chicken cystatin	1.7	0.019	0.064	–	0.35	9
H kininogen	400	0.019	1.1	–	–	9
L kininogen	600	0.017	1.2	–	>130	9
	–	–	0.72	–	–	1173
Corn cystatin I	290	17	5.7	–	–	1405

- *Family 3:* kininogens (other names: α-cysteine proteinase inhibitor or α-CPI). Exist in several forms (L kininogen, H kininogen); $M_r = 60\,000$–$120\,000$ Da; are glycoproteins; contain three cystatin domains, two of which are functional; occur mainly in blood plasma and in synovial and amniotic fluids.

There are a number of papers dealing with reactions of the cystatins with lysosomal cysteine proteinases [380, 413–415, 417, 422–426, 428–430, 434, 440, 442, 444, 455, 457–459, 462, 464, 470, 475–478, 484–486, 488, 493–497, 499, 506, 508, 509, 511, 515, 516, 518, 519, 521, 524–526, 529, 530–533, 535, 538, 540, 542, 545–547, 560, 562, 563, 568, 569, 573, 574, 577, 641, 658, 681, 724, 733, 1011, 1218].

Ha-*ras* oncogene products

Ha-*ras* oncogene proteins with $M_r \sim 21\,000$ Da have been expressed in *Escherichia coli*. These oncogene products were found to be potent inhibitors of cysteine proteinases. They seem to be related to the cystatins, because of amino acid sequence identities [433, 479–482, 541, 885].

Some mutants of the Ha-*ras* oncogene products which were shown to have lost their transforming activity toward NIH3T3 mouse fibroblasts have also lost, to a great extent, their inhibitory activity toward cathepsin L [549]. It is suggested that the inhibitory activity of the *ras* protein may be important in the early stage of carcinogenesis [1450].

Physiological functions and role in diseases

The locations of the cysteine proteinases described here are the lysosomes. At the acidic pH values in lysosomes these enzymes degrade proteins that have been taken up by the cells or that originate from the same cell, e.g. from cytosol. No disease is known yet that can be attributed to deficiency or inactivity of one of the lysosomal cysteine proteinases. A protein storage disease such as Batten's disease has been shown not to be caused by the failure of lysosomal proteinases but possibly by the abnormal post-translational modification of the stored proteolipid subunit c of the mitochondrial ATP synthase complex [841, 985].

The function of lysosomal cysteine proteinases in various diseases seems to be restricted to their proteolytic action outside the lysosomes after secretion from the cell or after translocation in different intracellular granules.

The secretion of the precursors of lysosomal cysteine proteinases by normal cells, such as macrophages, fibroblasts, osteoclasts and others, is accompanied by a regulated balance of catalytically active proteinases and their natural inhibitors. An alteration of this balance mainly in favour of the enzymes, e.g. by an overexpression of lysosomal cysteine proteinases and/or decreased synthesis of the cystatins, leads to uncontrolled proteolysis. This process, which we can observe in inflammation or tumour growth, is extremely complex, and several other proteinases such as metallo- and serine proteinases are involved, as well. The results of innumerable experiments prove that lysosomal cysteine proteinases play a role in several disorders – but their individual contributions are not yet clear.

Cancer and metastasis [585–736, 1495–1597]

Pancreatitis [737–771, 1598–1615]

Alzheimer's disease [271, 272, 808, 893, 1066, 1251a, 1648, 1657, 1671, 1748]

Arthritis [326, 416, 429, 814, 840, 845, 924, 935, 981, 1010, 1044, 1045, 1053, 1621, 1622, 1655, 1669, 1690]

Atherosclerosis [819, 883, 946]

Batten's disease (Lipofuscinosis) [822–824, 841, 936, 947, 985, 1253a, 1274a, 1731]

Diabetes [973, 997, 1030, 1627, 1636, 1722]

Gingivitis, periodontitis [314, 816, 833, 835, 919, 920, 1650, 1661, 1668, 1692, 1737, 1750]

Infection [881, 990, 991, 1639, 1737]

Inflammation [248, 314, 326, 422, 531, 777, 791, 804, 806, 868–873, 920, 1013, 1017, 1078, 1664]

Liver disorders [187, 296, 751, 752, 792, 884, 1644, 1662, 1670, 1691, 1749]

Lung disorders [774, 804, 810, 811, 853, 881, 925, 926, 978, 983, 996, 1001, 1002, 1021, 1037, 1070, 1632]
Effect of smoking [810, 811, 853, 926, 1002, 1037, 1070]

Lysosomal disorders [911, 1652]

Macrophage-concerning disorders [811, 853, 881, 883, 925, 955, 990, 991, 983, 1001, 1002, 1013, 1037, 1050, 1070, 1678, 1679]

Muscular disorders [182, 277, 775, 776, 815, 852, 854, 889, 890, 892, 896–898, 901, 905, 909, 910, 965, 1018, 1029, 1030, 1623, 1635, 1666, 1675, 1689, 1701, 1738, 1740, 1741, 1743, 1744]

Myocardial disorders [184, 1062, 1071, 1076]

Renal disorders [783–785, 887, 970, 973–976, 997, 1019, 1043, 1648, 1649, 1680, 1686–1688, 1705, 1706, 1720–1723, 1726, 1733]

Trauma [422, 777, 957, 1078, 1664, 1729]

Bibliography

References for first edition[†]

Reviews

1. **Abrahamson, M.** 1994. Cystatins. *Methods Enzymol.* **244**: 685–700.

2. **Aoyagi, T. and H. Umezawa.** 1975. Structures and activities of protease inhibitors of microbial origin. In *Proteases and Biological Control*, E. Reich, D. B. Rifkin and E. Shaw (eds), pp. 429–454. Cold Spring Harbor Laboratory Press, Cold Spring Harbor, NY.

3. **Barnhart, M. I., C. Quintana, H. L. Lenon, G. B. Bluhm, and J. M. Riddle.** 1968. Proteases in inflammation. *Ann. N. Y. Acad. Sci.* **146**: 527–539.

4. **Barrett, A.J.** 1977. Cathepsin B and other thiol proteinases. In *Proteinases in Mammalian Cells and Tissues*, A. J. Barrett (ed.), pp. 181–248. Elsevier/North-Holland Biomedical Press, Amsterdam.

5. **Barrett, A. J.** 1979. Protein degradation in health and disease. Introduction: the classification of proteinases. *Ciba Found. Symp.* 1–13.

6. **Barrett, A. J.** 1980. The many forms and functions of cellular proteinases. *Fed. Proc.* **39**: 9–14.

7. **Barrett, A. J.** 1985. The cystatins: small protein inhibitors of cysteine proteinases. *Prog. Clin. Biol. Res.* **180**: 105–116.

8. **Barrett, A. J. and H. Kirschke.** 1981. Cathepsin B, cathepsin H, and cathepsin L. *Methods Enzymol.* **80**(Pt C): 535–561.

9. **Barrett, A. J., N. D. Rawlings, M. E. Davies, W. Machleidt, G. Salvesen, and V. Turk.** 1986. Cysteine proteinase inhibitors of the cystatin superfamily. In *Proteinase Inhibitors*, A. J. Barrett and G. Salvesen (eds), pp. 515–569. Elsevier, Amsterdam.

10. **Beynon, R. J. and J. S. Bond.** 1986. Catabolism of intracellular protein: molecular aspects. *Am. J. Physiol.* **251**: C141-C152.

11. **Bohley, P., C. Hieke, H. Kirschke, and S. Schaper.** 1985. Protein degradation in rat liver cells. *Prog. Clin. Biol. Res.* **180**: 447–455.

12. **Chambers, A. F. and A. B. Tuck.** 1993. Ras-responsive genes and tumor metastasis. *Crit. Rev. Oncog.* **4**: 95–114.

13. **Chapman, H. A. J.** 1991. Role of enzyme receptors and inhibitors in regulating proteolytic activities of macrophages. *Ann. N. Y. Acad. Sci.* **624**: 87–96.

14. **Dean, R. T.** 1979. Lysosomes and protein degradation. *Ciba Found. Symp.* 139–149.

15. **Dean, R. T.** 1980. Protein degradation in cell cultures: general considerations on mechanisms and regulation. *Fed. Proc.* **39**: 15–19.

16. **Dice, J. F. and C. D. Walker.** 1979. Protein degradation in metabolic and nutritional disorders. *Ciba Found. Symp.* 331–350.

17. **Eijan, A. M. and E. Bal de Kier Joffe.** 1990. [Cathepsins: their role in tumor invasiveness]. *Medicina. B. Aires.* **50**: 171–174.

18. **Etherington, D. J.** 1979. Proteinases in connective tissue breakdown. *Ciba Found. Symp.* 87–103.

19. **Gottesman, M.M.** 1993. Cathepsin L and cancer. In *Proteolysis and Protein Turnover*, J. S. Bond and A. J. Barrett (eds), pp. 247–251. Portland Press, London.

20. **Horecker, B. L., S. Erickson Viitanen, E. Melloni, and S. Pontremoli.** 1985. Inactivation of rabbit liver and muscle aldolases by limited proteolysis by lysosomal cathepsin M. *Curr. Top. Cell Regul.* **25**: 77–89.

21. **Kane, S. E. and M. M. Gottesman.** 1990. The role of cathepsin L in malignant transformation. *Semin. Cancer Biol.* **1**: 127–136.

22. **Katunuma, N.** 1983. [Structure, function and regulation of lysosomal thiol proteinases]. *Seikagaku* **55**: 77–89.

23. **Katunuma, N.** 1989. Mechanisms and regulation of lysosomal proteolysis. *Revis. Biol. Celular.* **20**: 35–61.

24. **Katunuma, N.** 1989. Possible regulatory mechanisms of intracellular protein catabolism through lysosome. In *Intracellular Proteolysis. Mechanisms and Regulations*, N. Katunuma and E. Kominami (eds), pp. 3–23. Japan Scientific Societies Press, Tokyo.

25. **Katunuma, N.** 1990. New biological functions of intracellular proteases and their endogenous inhibitors as bioreactants. *Adv. Enzyme Regul.* **30**: 377–392.

26. **Katunuma, N., H. Kakegawa, Y. Matsunaga, T. Nikawa, and E. Kominami.** 1993. Different functional share of individual lysosomal cathepsins in normal and pathological conditions. *Agents Actions Suppl.* **42**: 195–210.

27. **Katunuma, N. and E. Kominami.** 1983. Structures and functions of lysosomal thiol proteinases and their endogenous inhibitor. *Curr. Top. Cell Regul.* **22**: 71–101.

28. **Katunuma, N. and E. Kominami.** 1985. Lysosomal sequestration of cytosolic enzymes and lysosomal thiol cathepsins. *Adv. Enzyme Regul.* **23**: 159–168.

[†] See p. 105 for new references for this edition

29. **Katunuma, N. and E. Kominami.** 1985. Lysosomal thiol cathepsins and their endogenous inhibitors. Distribution and localization. *Prog. Clin. Biol. Res.* **180**: 71–79.

30. **Kay, J.** 1978. Intracellular protein degradation. *Biochem. Soc. Trans.* **6**: 789–797.

31. **Kirschke, H.** 1981. On the substrate specificity of cathepsin L. *Acta Biol. Med. Ger.* **40**: 1427–1431.

32. **Kirschke, H. and A. J. Barrett.** 1985. Cathepsin L – a lysosomal cysteine proteinase. *Prog. Clin. Biol. Res.* **180**: 61–69.

33. **Kirschke, H. and A.J. Barrett.** 1987. Chemistry of lysosomal proteases. In *Lysosomes: Their Role in Protein Breakdown*, H. Glaumann and F. J. Ballard (eds), pp. 193–238. Academic Press, London.

34. **Kirschke, H., D. Bromme, and B. Wiederanders.** 1993. Cathepsin S, a lysosomal cysteine proteinase. In *Proteolysis and Protein Turnover*, J. S. Bond and A. J. Barrett (eds), pp. 33–37. Portland Press, London.

35. **Kirschke, H., J. Langner, S. Riemann, B. Wiederanders, S. Ansorge, and P. Bohley.** 1979. Lysosomal cysteine proteinases. *Ciba Found. Symp.* 15–35.

36. **Kirschke, H. and B. Wiederanders.** 1987. Lysosomal proteinases. *Acta Histochem.* **82**: 2–4.

37. **Kominami, E., N. Katunuma, and Y. Uchiyama.** 1989. Biosyntheses, processings and localizations of lysosomal cysteine proteinases. In *Intracellular Proteolysis. Mechanisms and Regulations*, N. Katunuma and E. Kominami (eds), pp. 52–60. Japan Scientific Societies Press, Tokyo.

38. **Kominami, E. and Y. Uchiyama.** 1993. Lysosomal cysteine proteinases as processing proteases: Localization in secretory granules. In *Proteolysis and Protein Turnover*, J. S. Bond and A. J. Barrett (eds), pp. 39–44. Portland Press, London.

39. **Kopecek, J.** 1984. Controlled biodegradability of polymers–a key to drug delivery systems. *Biomaterials* **5**: 19–25.

40. **Krutzsch, H. C.** 1983. Polypeptide sequencing with dipeptidyl peptidases. *Methods Enzymol.* **91**: 511–524.

41. **Lokshina, L. A. and E. A. Dilakian.** 1986. [Thiol peptide hydrolases from animal tissues, their structure and function]. *Mol. Biol. Mosk.* **20**: 1157–1175.

42. **Mason, R. W. and A. J. Barrett.** 1985. Properties of human liver cathepsin L. *Prog. Clin. Biol. Res.* **180**: 217–219.

43. **McDonald, J. K.** 1985. An overview of protease specificity and catalytic mechanisms: aspects related to nomenclature and classification. *Histochem. J.* **17**: 773–785.

44. **Monsky, W. L. and W. T. Chen.** 1993. Proteases of cell adhesion proteins in cancer. *Semin. Cancer Biol.* **4**: 251–258.

45. **Morris, B. J.** 1992. Molecular biology of renin. I: Gene and protein structure, synthesis and processing. *J. Hypertens.* **10**: 209–214.

46. **Mort, J. S., A. D. Recklies, and A. R. Poole.** 1985. Release of cathepsin B precursors from human and murine tumours. *Prog. Clin. Biol. Res.* **180**: 243–245.

47. **Nishimura, Y.** 1988. [Intracellular processing and activation of lysosomal cathepsins]. *Seikagaku* **60**: 1271–1278.

48. **Nishimura, Y., T. Kawabata, K. Furuno, and K. Kato.** 1989. Intracellular processing and activation of lysosomal cathepsin L. In *Intracellular Proteolysis. Mechanisms and Regulations*, N. Katunuma and E. Kominami (eds), pp. 61–73. Japan Scientific Societies Press, Tokyo.

49. **Pitot, H. C. and E. Gohda.** 1987. Cathepsin T. *Methods Enzymol.* **142**: 279–289.

50. **Rich, D. H.** 1986. Inhibitors of cysteine proteinases. In *Proteinase Inhibitors*, A. J. Barrett and G. Salvesen (eds), pp. 153–217. Elsevier, Amsterdam.

51. **Rochefort, H., F. Capony, M. Garcia, V. Cavailles, G. Freiss, M. Chambon, M. Morisset, and F. Vignon.** 1987. Estrogen-induced lysosomal proteases secreted by breast cancer cells: a role in carcinogenesis? *J. Cell Biochem.* **35**: 17–29.

52. **Schmitt, M., F. Janicke, N. Moniwa, N. Chucholowski, L. Pache, and H. Graeff.** 1992. Tumor-associated urokinase-type plasminogen activator: biological and clinical significance. *Biol. Chem. Hoppe Seyler* **373**: 611–622.

53. **Shaw, E.** 1990. Cysteinyl proteinases and their selective inactivation. *Adv. Enzymol. Relat. Areas Mol. Biol.* **63**: 271–347.

54. **Shaw, E.** 1994. Peptidyl diazomethanes as inhibitors of cysteine and serine proteinases. *Methods Enzymol.* **244**: 649–656.

55. **Sloane, B. F.** 1990. Cathepsin B and cystatins: evidence for a role in cancer progression. *Semin. Cancer Biol.* **1**: 137–152.

56. **Sloane, B. F. and K. V. Honn.** 1984. Cysteine proteinases and metastasis. *Cancer Metastasis Rev.* **3**: 249–263.

57. **Sloane, B. F., K. Moin, E. Krepela, and J. Rozhin.** 1990. Cathepsin B and its endogenous inhibitors: the role in tumor malignancy. *Cancer Metastasis Rev.* **9**: 333–352.

58. **Sloane, B. F., J. Rozhin, E. Krepela, G. Ziegler, and M. Sameni.** 1991. The malignant phenotype and cysteine proteinases. *Biomed. Biochim. Acta* **50**: 549–554.

59. **Sloane, B. F., J. Rozhin, T. T. Lah, N. A. Day, M. Buck, R. E. Ryan, J. D. Crissman, and K. V. Honn.** 1988. Tumor cathepsin B and its endogenous inhibitors in metastasis. *Adv. Exp. Med. Biol.* **233**: 259–268.

60. **Sloane, B. F., J. Rozhin, D. Robinson, and K. V. Honn.** 1990. Role for cathepsin B and cystatins in tumor growth and progression. *Biol. Chem. Hoppe Seyler* **371**(Suppl.): 193–198.

61. **Steer, M. L.** 1992. How and where does acute pancreatitis begin? *Arch. Surg.* **127**: 1350–1353.

62. **Steiner, D. F., K. Docherty, and R. Carroll.** 1984. Golgi/granule processing of peptide hormone and neuropeptide precursors: a minireview. *J. Cell Biochem.* **24**: 121–130.

63. **Takahashi, T.** 1988. [Cathepsin B – recent studies and perspective]. *Seikagaku* **60**: 1279–1284.

64. **Turk, V., J. Brzin, M. Kotnik, B. Lenarcic, T. Popovic, A. Ritonja, M. Trstenjak, L. Begic Odobasic, and W. Machleidt.** 1986. Human cysteine proteinases and their protein inhibitors stefins, cystatins and kininogens. *Biomed. Biochim. Acta* **45**: 1375–1384.

65. **Turk, V., J. Brzin, B. Lenarcic, P. Locnikar, T. Popovic, A. Ritonja, J. Babnik, W. Bode, and W. Machleidt.** 1985. Structure and function of lysosomal cysteine proteinases and their protein inhibitors. *Prog. Clin. Biol. Res.* **180**: 91–103.

66. **Tutel'ian, V. A. and A. V. Vasil'ev.** 1990. [Lysosomes in cellular activity. Physiology and pathology]. *Vestn. Akad. Med. Nauk. SSSR* 14–21.

67. **Vaes, G., J. M. Delaisse, and Y. Eeckhout.** 1992. Relative roles of collagenase and lysosomal cysteine-proteinases in bone resorption. *Matrix Suppl.* **1**: 383–388.

68. **Virca, G. D. and H. P. Schnebli.** 1984. The elastase/α_1-proteinase inhibitor balance in the lung. A review. *Schweiz. Med. Wochenschr.* **114**: 895–898.

69. **Wildenthal, K. and J. S. Crie.** 1980. The role of lysosomes and lysosomal enzymes in cardiac protein turnover. *Fed. Proc.* **39**: 37–41.

70. **Woolley, D. E.** 1984. Collagenolytic mechanisms in tumor cell invasion. *Cancer Metastasis Rev.* **3**: 361–372.

Catalytic properties

71. **Agarwal, S. K. and M. Y. Khan.** 1987. Does cathepsin B play a role in intracellular protein degradation? *Biochem. Int.* **15**: 785–792.

72. **Agarwal, S. K. and M. Y. Khan.** 1988. A probable mechanism of inactivation by urea of goat spleen cathepsin B. Unfolding and refolding studies. *Biochem. J.* **256**: 609–613.

73. **Anderson, A. J.** 1969. Effects of lysosomal collagenolytic enzymes, anti-inflammatory drugs and other substances on some properties of insoluble collagen. *Biochem. J.* **113**: 457–463.

74. **Aronson, N. N. J. and A. J. Barrett.** 1978. The specificity of cathepsin B. Hydrolysis of glucagon at the C-terminus by a peptidyldipeptidase mechanism. *Biochem. J.* **171**: 759–765.

75. **Assfalg Machleidt, I., G. Rothe, S. Klingel, R. Banati, W. F. Mangel, G. Valet, and W. Machleidt.** 1992. Membrane permeable fluorogenic rhodamine substrates for selective determination of cathepsin L. *Biol. Chem. Hoppe Seyler* **373**: 433–440.

76. **Azarian, A. V. and A. A. Galoian.** 1987. [Brain cathepsin as dipeptidylcarboxypeptidase transforming provasopressor, pro-opioid and model peptides]. *Vopr. Med. Khim.* **33**: 78–81.

77. **Bajkowski, A. S. and A. Frankfater.** 1983. The pH dependency of bovine spleen cathepsin B-catalyzed transfer of N α-benzyloxycarbonyl-L-lysine from p-nitrophenol to water and dipeptide nucleophiles. Comparisons with papain. *J. Biol. Chem.* **258**: 1650–1655.

78. **Bajkowski, A. S. and A. Frankfater.** 1983. Steady state kinetic evidence for an acyl-enzyme intermediate in reactions catalyzed by bovine spleen cathepsin B. *J. Biol. Chem.* **258**: 1645–1649.

79. **Bansal, R., N. Ahmad, and J. R. Kidwai.** 1980. *In vitro* conversion of proinsulin to insulin by cathepsin B in isolated islets and its inhibition by cathepsin B antibodies. *Acta Diabetol. Lat.* **17**: 255–266.

80. **Baricos, W. H., S. L. Cortez, Q. C. Le, Y. W. Zhou, R. M. Dicarlo, S. E. O'Connor, and S. V. Shah.** 1990. Glomerular basement membrane degradation by endogenous cysteine proteinases in isolated rat glomeruli. *Kidney Int.* **38**: 395–401.

81. **Baricos, W. H., Y. Zhou, R. W. Mason, and A. J. Barrett.** 1988. Human kidney cathepsins B and L. Characterization and potential role in degradation of glomerular basement membrane. *Biochem. J.* **252**: 301–304.

82. **Baricos, W. H., Y. W. Zhou, R. S. Fuerst, A. J. Barrett, and S. V. Shah.** 1987. The role of aspartic and cysteine proteinases in albumin degradation by rat kidney cortical lysosomes. *Arch. Biochem. Biophys.* **256**: 687–691.

83. **Bechet, D., A. Obled, and C. Deval.** 1986. Species variations amongst proteinases in liver lysosomes. *Biosci. Rep.* **6**: 991–997.

84. **Blair, H. C., S. L. Teitelbaum, L. E. Grosso, D. L. Lacey, H. L. Tan, D. W. McCourt, and J. J. Jeffrey.** 1993. Extracellular-matrix degradation at acid pH. Avian osteoclast acid collagenase isolation and characterization. *Biochem. J.* **290**: 873–884.

85. **Bohley, P., H. Kirschke, J. Langner, B. Wiederanders, and S. Ansorge.** 1976. Intrazellulärer Proteinabbau. VIII. Einsatz doppeltmarkierter Substratproteine. *Acta Biol. Med. Ger.* **35**: 301–307.

86. **Bohley, P. and P. O. Seglen.** 1992. Proteases and proteolysis in the lysosome. *Experientia* **48**: 151–157.

87. **Bohmer, F. D., H. Kirschke, P. Bohley, and R. Schon.** 1979. Membrane sialoglycoprotein from human erythrocytes activates lysosomal proteinases. *Acta Biol. Med. Ger.* **38**: 1521–1526.

88. **Bond, J. S. and A. J. Barrett.** 1980. Degradation of fructose-1,6-bisphosphate aldolase by cathepsin B. *Biochem. J.* **189**: 17–25.

89. **Bowser, R. and R. F. Murphy.** 1990. Kinetics of hydrolysis of endocytosed substrates by mammalian cultured cells: early introduction of lysosomal enzymes into the endocytic pathway. *J. Cell Physiol.* **143**: 110–117.

90. **Brocklehurst, K., D. Kowlessur, M. O'Driscoll, G. Patel, S. Quenby, E. Salih, W. Templeton, E. W. Thomas, and F. Willenbrock.** 1987. Substrate-derived two-protonic-state electrophiles as sensitive kinetic specificity probes for cysteine proteinases. Activation of 2-pyridyl disulphides by hydrogen-bonding. *Biochem. J.* **244**: 173–181.

91. **Bromme, D., K. Bescherer, H. Kirschke, and S. Fittkau.** 1987. Enzyme–substrate interactions in the hydrolysis of peptides by cathepsins B and H from rat liver. *Biochem. J.* **245**: 381–385.

92. **Bromme, D., A. Steinert, S. Fittkau, and H. Kirschke.** 1987. Action of rat liver cathepsin B on bradykinin and on the oxidized insulin A-chain. *FEBS Lett.* **219**: 441–444.

93. **Bromme, D., A. Steinert, S. Friebe, S. Fittkau, B. Wiederanders, and H. Kirschke.** 1989. The specificity of bovine spleen cathepsin S. A comparison with rat liver cathepsins L and B. *Biochem. J.* **264**: 475–481.

94. **Buck, M. R., D. G. Karustis, N. A. Day, K. V. Honn, and B. F. Sloane.** 1992. Degradation of extracellular-matrix proteins by human cathepsin B from normal and tumour tissues. *Biochem. J.* **282**: 273–278.

95. **Bulychev, A. G., O. A. Assinovskaia, and E. G. Semenova.** 1987. [Lysosomotropic agents as regulators of the activity of lysosomal hydrolases]. *Vopr. Med. Khim.* **33**: 20–24.

96. **Burleigh, M. C., A. J. Barrett, and G. S. Lazarus.** 1974. Cathepsin B_1. A lysosomal enzyme that degrades native collagen. *Biochem. J.* **137**: 387–398.

97. **Bushell, G., C. Nelson, H. Chiu, C. Grimley, W. Henzel, J. Burnier, and S. Fong.** 1993. Evidence supporting a role for cathepsin B in the generation of T cell antigenic epitopes of human growth hormone. *Mol. Immunol.* **30**: 587–591.

98. **Buttle, D. J., B. C. Bonner, D. Burnett, and A. J. Barrett.** 1988. A catalytically active high-M_r form of human cathepsin B from sputum. *Biochem. J.* **254**: 693–699.

99. **Buttle, D. J. and J. Saklatvala.** 1992. Lysosomal cysteine endopeptidases mediate interleukin 1-stimulated cartilage proteoglycan degradation. *Biochem. J.* **287**: 657–661.

100. **Buus, S. and O. Werdelin.** 1986. A group-specific inhibitor of lysosomal cysteine proteinases selectively inhibits both proteolytic degradation and presentation of the antigen dinitrophenyl-poly-L-lysine by guinea pig accessory cells to T cells. *J. Immunol.* **136**: 452–458.

101. **Calam, D. H. and H. J. Thomas.** 1972. Water-insoluble enzymes for peptide sequencing: dipeptidyl aminopeptidase I (cathepsin C), an enzyme with subunit structure. *Biochim. Biophys. Acta* **276**: 328–332.

102. **Callahan, P. X., J. K. McDonald, and S. Ellis.** 1972. Dipeptidyl aminopeptidase I: application in sequencing of peptides. *Fed. Proc.* **31**: 1105–1113.

103. **Callahan, P. X., J. K. McDonald, and S. Ellis.** 1972. Sequencing of peptides with dipeptidyl aminopeptidase I. *Methods Enzymol.* **25**: 282–298.

104. **Caprioli, R. M., W. E. J. Seifert, and D. E. Sutherland.** 1973. Polypeptide sequencing: use of dipeptidylaminopeptidase I and gas chromatography-mass spectrometry. *Biochem. Biophys. Res. Commun.* **55**: 67–75.

105. **Cardozo, C., C. Kurtz, and M. Lesser.** 1992. Degradation of rat lung collagens by cathepsin B. *J. Lab. Clin. Med.* **119**: 169–175.

106. **Carey, P. R., R. H. Angus, H. H. Lee, and A. C. Storer.** 1984. Identity of acyl group conformations in the active sites of papain and cathepsin B by resonance Raman spectroscopy. *J. Biol. Chem.* **259**: 14357–14360.

107. **Chapman, C. B. and G. F. Mitchell.** 1982. Proteolytic cleavage of immunoglobulin by enzymes released by Fasciola hepatica. *Vet. Parasitol.* **11**: 165–178.

108. **Chatterjee, R. and G. Kalnitsky.** 1986. The specificity of rabbit lung cathepsin I on biopeptides. *Biomed. Biochim. Acta* **45**: 1447–1455.

109. **Chopra, R. K., C. H. Pearson, G. A. Pringle, D. S. Fackre, and P. G. Scott.** 1985. Dermatan sulphate is located on serine-4 of bovine skin proteodermatan sulphate. Demonstration that most molecules possess only one glycosaminoglycan chain and comparison of amino acid sequences around glycosylation sites in different proteoglycans. *Biochem. J.* **232**: 277–279.

110. **Coffey, J. W. and C. De Duve.** 1968. Digestive activity of lysosomes. I. The digestion of proteins by extracts of rat liver lysosomes. *J. Biol. Chem.* **243**: 3255–3263.

111. **Crivellaro, O., P. S. Lazo, O. Tsolas, S. Pontremoli, and B. L. Horecker.** 1978. Properties of a fructose 1,6-bisphosphatase converting enzyme in rat liver lysosomes. *Arch. Biochem. Biophys.* **189**: 490–498.

112. **Davies, M., K. T. Hughes, and G. J. Thomas.** 1980. Evidence that kidney lysosomal proteinases degrade the collagen of glomerular basement membrane. *Renal. Physiol.* **3**: 116–119.

113. **Delaisse, J. M., Y. Eeckhout, and G. Vaes.** 1980. Inhibition of bone resorption in culture by inhibitors of thiol proteinases. *Biochem. J.* **192**: 365–368.

114. **Dennison, C., R. Pike, T. Coetzer, and K. Kirk.** 1992. Characterisation of the activity and stability of single-chain cathepsin L and of proteolytically active cathepsin L/cystatin complexes. *Biol. Chem. Hoppe Seyler* **373**: 419–425.

115. **Deval, C., D. Bechet, A. Obled, and M. Ferrara.** 1990. Purification and properties of different isoforms of bovine cathepsin B. *Biochem. Cell Biol.* **68**: 822–826.

116. **Docherty, K., R. J. Carroll, and D. F. Steiner.** 1982. Conversion of proinsulin to insulin: Involvement of a 31,500 molecular weight thiol protease. *Proc. Natl Acad. Sci. U. S. A.* **79**: 4613–4617.

117. **Dufour, E. and B. Ribadeau Dumas.** 1988. Proteolytic specificity of chicken cathepsin L on bovine beta-casein. *Biosci. Rep.* **8**: 185–191.

118. **Dunn, A. D., H. E. Crutchfield, and J. T. Dunn.** 1991. Thyroglobulin processing by thyroidal proteases. Major sites of cleavage by cathepsins B, D, and L. *J. Biol. Chem.* **266**: 20198–20204.

119. **Dunn, A. D. and J. T. Dunn.** 1988. Cysteine proteinases from human thyroids and their actions on thyroglobulin. *Endocrinology* **123**: 1089–1097.

120. **Eeckhout, Y. and G. Vaes.** 1977. Further studies on the activation of procollagenase, the latent precursor of bone collagenase. Effects of lysosomal cathepsin B, plasmin and kallikrein, and spontaneous activation. *Biochem. J.* **166**: 21–31.

121. **Ehrlich, P. H., G. R. Matsueda, M. N. Margolies, S. S. Husain, and E. Haber.** 1980. Isolation of an active heavy-chain variable domain from a homogeneous rabbit antibody by cathepsin B digestion of the aminoethylated heavy chain. *Biochemistry* **19**: 4091–4096.

122. **Erickson Viitanen, S., E. Balestreri, M. McDermott, and B. L. Horecker.** 1986. Sites of cleavage of rabbit muscle aldolase by purified cathepsin M from rabbit liver. *Biochem. Med. Metab. Biol.* **35**: 191–198.

123. **Etherington, D. J.** 1976. Bovine spleen cathepsin B$_1$ and collagenolytic cathepsin. A comparative study of the properties of the two enzymes in the degradation of native collagen. *Biochem. J.* **153**: 199–209.

124. **Etherington, D. J. and H. Birkedahl Hansen.** 1987. The influence of dissolved calcium salts on the degradation of hard-tissue collagens by lysosomal cathepsins. *Coll. Relat. Res.* **7**: 185–199.

125. **Etherington, D. J. and P. J. Evans.** 1977. The action of cathepsin B and collagenolytic cathepsin in the degradation of collagen. *Acta Biol. Med. Ger.* **36**: 1555–1563.

126. **Evans, P. and D. J. Etherington.** 1978. Characterisation of cathepsin B and collagenolytic cathepsin from human placenta. *Eur. J. Biochem.* **83**: 87–97.

127. **Evans, P. and D. J. Etherington.** 1979. Action of cathepsin N on the oxidized B-chain of bovine insulin. *FEBS Lett.* **99**: 55–58.

128. **Everts, V., W. Beertsen, and W. Tigchelaar Gutter.** 1985. The digestion of phagocytosed collagen is inhibited by the proteinase inhibitors leupeptin and E-64. *Coll. Relat. Res.* **5**: 315–336.

129. **Fosang, A. J., P. J. Neame, K. Last, T. E. Hardingham, G. Murphy, and J. A. Hamilton.** 1992. The interglobular domain of cartilage aggrecan is cleaved by PUMP, gelatinases, and cathepsin B. *J. Biol. Chem.* **267**: 19470–19474.

130. **Fouchier, F., J. L. Mego, J. Dang, and C. Simon.** 1983. Intralysosomal hydrolysis of thyroglobulin. I. Modulation by lysosomal membrane permeability and exogenous factors. *Acta Endocrinol. Copenh.* **103**: 53–61.

131. **Freeman, S. J. and N. A. Brown.** 1985. Comparative effects of cathepsin inhibitors on rat embryonic development *in vitro*. Evidence that cathepsin D is unimportant in the proteolytic function of yolk sac. *J. Embryol. Exp. Morphol.* **86**: 271–281.

132. **Fujita, T.** 1988. Parathyroid hormone – *quo vadis*? *Nippon Naibunpi. Gakkai Zasshi* **64**: 1250–1257.

133. **Gabrijelcic, D., R. Gollwitzer, T. Popovic, and V. Turk.** 1988. Proteolytic cleavage of human fibrinogen by cathepsin B. *Biol. Chem. Hoppe Seyler* **369**(Suppl.): 287–292.

134. **Gal, S. and M. M. Gottesman.** 1986. The major excreted protein (MEP) of transformed mouse cells and cathepsin L have similar protease specificity. *Biochem. Biophys. Res. Commun.* **139**: 156–162.

135. **Gal, S. and M. M. Gottesman.** 1986. The major excreted protein of transformed fibroblasts is an activable acid-protease. *J. Biol. Chem.* **261**: 1760–1765.

136. **Gauthier, F., T. Moreau, G. Lalmanach, M. Brillard Bourdet, M. Ferrer Di Martino, and L. Juliano.** 1993. A new, sensitive fluorogenic substrate for papain based on the sequence of the cystatin inhibitory site. *Arch. Biochem. Biophys.* **306**: 304–308.

137. **Gohda, E. and H. C. Pitot.** 1980. Purification and characterization of a factor catalyzing the conversion of the multiple forms of tyrosine aminotransferase from rat liver. *J. Biol. Chem.* **255**: 7371–7379.

138. **Gohda, E. and H. C. Pitot.** 1981. A new thiol proteinase from rat liver. *J. Biol. Chem.* **256**: 2567–2572.

139. **Gordon, J. I., H. F. Sims, C. Edelstein, A. M. Scanu, and A. W. Strauss.** 1985. Extracellular processing of proapolipoprotein A-II in Hep G2 cell cultures is mediated by a 54-kDa protease immunologically related to cathepsin B. *J. Biol. Chem.* **260**: 14824–14831.

140. **Guinec, N., V. Dalet Fumeron, and M. Pagano.** 1990. [Digestion *in vitro* of basal membrane, bovine lens capsule, by human liver lysosome cathepsins B, H and L and ascitic fluid tumor cathepsin B from ovarian adenocarcinoma]. *Pathol. Biol. Paris.* **38**: 988–992.

141. **Guinec, N., V. Dalet Fumeron, and M. Pagano.** 1992. Quantitative study of the binding of cysteine proteinases to basement membranes. *FEBS Lett.* **308**: 305–308.

142. **Guinec, N., M. Pagano, V. Dalet Fumeron, and R. Engler.** 1990. "*In vitro*" digestion of intact bovine lens capsules by four human lysosomal cysteine-proteinases. *Biol. Chem. Hoppe Seyler* **371**(Suppl.): 239–254.

143. **Hakansson, H. O., A. Borgstrom, and K. Ohlsson.** 1991. Porcine pancreatic cationic pro-elastase. Studies on the activation, turnover and interaction with plasma proteinase inhibitors. *Biol. Chem. Hoppe Seyler* **372**: 465–472.

144. **Hannappel, E., J. S. MacGregor, S. Davoust, and B. L. Horecker.** 1982. Limited proteolysis of liver and muscle aldolases: effects of subtilisin, cathepsin B, and Staphylococcus aureus protease. *Arch. Biochem. Biophys.* **214**: 293–298.

145. **Hargrove, J. L., E. Gohda, H. C. Pitot, and D. K. Granner.** 1982. Cathepsin T (convertase) generates the multiple forms of tyrosine aminotransferase by limited proteolysis. *Biochemistry* **21**: 283–289.

146. **Harvima, R. J., K. Yabe, J. E. Fräki, K. Fukuyama, and W. L. Epstein.** 1988. Hydrolysis of histones by proteinases. *Biochem. J.* **250**: 859–864.

147. **Hasnain, S., T. Hirama, A. Tam, and J. S. Mort.** 1992. Characterization of recombinant rat cathepsin B and nonglycosylated mutants expressed in yeast. New insights into the pH dependence of cathepsin B-catalyzed hydrolyses. *J. Biol. Chem.* **267**: 4713–4721.

148. **Himmelreich, G., G. Dooijewaard, P. Breinl, W. O. Bechstein, P. Neuhaus, C. Kluft, and H. Riess.** 1993. Evolution of urokinase-type plasminogen activator (u-PA) and tissue-type plasminogen activator (t-PA) in orthotopic liver transplantation (OLT). *Thromb. Haemost.* **69**: 56–59.

149. **Hiwasa, T., S. Sakiyama, S. Yokoyama, J. M. Ha, J. Fujita, S. Noguchi, Y. Bando, E. Kominami, and N. Katunuma.** 1988. Inhibition of cathepsin L-induced degradation of epidermal growth factor receptors by c-Ha-ras gene products. *Biochem. Biophys. Res. Commun.* **151**: 78–85.

150. **Hiwasa, T., S. Sakiyama, S. Yokoyama, J. M. Ha, S. Noguchi, Y. Bando, E. Kominami, and N. Katunuma.** 1988. Degradation of epidermal growth factor receptors by cathepsin L-like protease: inhibition of the degradation by c-Ha-ras gene products. *FEBS Lett.* **233**: 367–370.

151. **Huisman, W., L. Lanting, H. J. Doddema, J. M. Bouma, and M. Gruber.** 1974. Role of individual cathepsins in lysosomal protein digestion as tested by specific inhibitors. *Biochim. Biophys. Acta* **370**: 297–307.

152. **Hutchinson, D. W. and A. Tunnicliffe.** 1987. The preparation and properties of immobilised dipeptidyl-aminopeptidase I (cathepsin C). *Biochim. Biophys. Acta* **916**: 1–4.

153. **Inagami, T., H. Okamoto, K. Ohtsuki, K. Shimamoto, J. Chao, and H. S. Margolius.** 1982. Human plasma inactive renin: purification and activation by proteases. *J. Clin. Endocrinol. Metab.* **55**: 619–627.

154. **Isemura, M., Z. Yosizawa, K. Takahashi, H. Kosaka, N. Kojima, and T. Ono.** 1981. Characterization of porcine plasma fibronectin and its fragmentation by porcine liver cathepsin B. *J. Biochem. Tokyo* **90**: 1–9.

155. **Ishii, Y., Y. Hashizume, E. Kominami, and Y. Uchiyama.** 1991. Changes in immunoreactivity for cathepsin H in rat type II alveolar epithelial cells and its proteolytic activity in bronchoalveolar lavage fluid over 24 hours. *Anat. Rec.* **230**: 519–523.

156. **Iwanaga, M., E. Yamamoto, and M. Fukumoto.** 1985. Cathepsin activity in cholesteatoma. *Ann. Otol. Rhinol. Laryngol.* **94**: 309–312.

157. **Jadot, M., C. Colmant, S. Wattiaux De Coninck, and R. Wattiaux.** 1984. Intralysosomal hydrolysis of glycyl-L-phenylalanine 2-naphthylamide. *Biochem. J.* **219**: 965–970.

158. **Jarvinen, M.** 1976. α-N-Benzoylarginine-2-naphthylamide hydrolase (cathepsin B_1?) from rat skin. III. Substrate specificity, modifier characteristics, and transformation of the enzyme at acidic pH. *Acta Chem. Scand. B.* **30**: 53–60.

159. **Johnson, D. A., A. J. Barrett, and R. W. Mason.** 1986. Cathepsin L inactivates α_1-proteinase inhibitor by cleavage in the reactive site region. *J. Biol. Chem.* **261**: 14748–14751.

160. **Kakegawa, H., T. Nikawa, K. Tagami, H. Kamioka, K. Sumitani, T. Kawata, M. Drobnic Kosorok, B. Lenarcic, V. Turk, and N. Katunuma.** 1993. Participation of cathepsin L on bone resorption. *FEBS Lett.* **321**: 247–250.

161. **Kargel, H. J., R. Dettmer, G. Etzold, H. Kirschke, P. Bohley, and J. Langner.** 1980. Action of cathepsin L on the oxidized B-chain of bovine insulin. *FEBS Lett.* **114**: 257–260.

162. **Kargel, H. J., R. Dettmer, G. Etzold, H. Kirschke, P. Bohley, and J. Langner.** 1981. Action of rat liver cathepsin L on glucagon. *Acta Biol. Med. Ger.* **40**: 1139–1143.

163. **Katunuma, N., E. Kominami, S. Hashida, and N. Wakamatsu.** 1982. Modification of rat liver fructose biphosphate aldolase by lysosomal proteinases. *Adv. Enzyme Regul.* **20**: 337–350.

164. Katunuma, N., T. Towatari, M. Tamai, and K. Hanada. 1983. Use of new synthetic substrates for assays of cathepsin L and cathepsin B. *J. Biochem. Tokyo* **93**: 1129–1135.

165. Keilova, H. and B. Keil. 1969. Isolation and specificity of cathepsin B. *FEBS Lett.* **4**: 295–298.

166. Khan, M. Y. and S. Ahmad. 1987. Anomalous behaviour of cathepsin B: dependence of activity and stability on salt concentration. *Biochem. Int.* **15**: 111–115.

167. Khouri, H. E., C. Plouffe, S. Hasnain, T. Hirama, A. C. Storer, and R. Menard. 1991. A model to explain the pH-dependent specificity of cathepsin B-catalysed hydrolyses. *Biochem. J.* **275**: 751–757.

168. Khouri, H. E., T. Vernet, R. Menard, F. Parlati, P. Laflamme, D. C. Tessier, B. Gour Salin, D. Y. Thomas, and A. C. Storer. 1991. Engineering of papain: selective alteration of substrate specificity by site-directed mutagenesis. *Biochemistry* **30**: 8929–8936.

169. Kirschke, H., J. Langner, B. Wiederanders, S. Ansorge, P. Bohley, and U. Broghammer. 1976. Intrazellulärer Proteinabbau. VII. Kathepsin L und H: Zwei neue Proteinasen aus Rattenleberlysosomen. *Acta Biol. Med. Ger.* **35**: 285–299.

170. Kirschke, H., P. Locnikar, and V. Turk. 1984. Species variations amongst lysosomal cysteine proteinases. *FEBS Lett.* **174**: 123–127.

171. Kirschke, H., M. Pepperle, I. Schmidt, and B. Wiederanders. 1986. Are there species differences amongst the lysosomal cysteine proteinases? *Biomed. Biochim. Acta* **45**: 1441–1446.

172. Kirschke, H., I. Schmidt, and B. Wiederanders. 1986. Cathepsin S. The cysteine proteinase from bovine lymphoid tissue is distinct from cathepsin L (EC 3.4.22.15). *Biochem. J.* **240**: 455–459.

173. Knight, C. G. 1980. Human cathepsin B. Application of the substrate N-benzyloxycarbonyl-L-arginyl-L-arginine 2-naphthylamide to a study of the inhibition by leupeptin. *Biochem. J.* **189**: 447–453.

174. Kobayashi, H., H. Ohi, M. Sugimura, H. Shinohara, T. Fujii, and T. Terao. 1992. Inhibition of *in vitro* ovarian cancer cell invasion by modulation of urokinase-type plasminogen activator and cathepsin B. *Cancer Res.* **52**: 3610–3614.

175. Koga, H., N. Mori, H. Yamada, Y. Nishimura, K. Tokuda, K. Kato, and T. Imoto. 1991. Rat cathepsin H-catalyzed transacylation: comparisons of the mechanism and the specificity with papain-superfamily proteases. *J. Biochem. Tokyo* **110**: 939–944.

176. Koga, H., N. Mori, H. Yamada, Y. Nishimura, K. Tokuda, K. Kato, and T. Imoto. 1992. Endo-and aminopeptidase activities of rat cathepsin H. *Chem. Pharm. Bull. Tokyo* **40**: 965–970.

177. Koga, H., H. Yamada, Y. Nishimura, K. Kato, and T. Imoto. 1990. Comparative study on specificities of rat cathepsin L and papain: amino acid differences at substrate-binding sites are involved in their specificities. *J. Biochem. Tokyo* **108**: 976–982.

178. Koga, H., H. Yamada, Y. Nishimura, K. Kato, and T. Imoto. 1991. Multiple proteolytic action of rat liver cathepsin B: specificities and pH-dependences of the endo- and exopeptidase activities. *J. Biochem. Tokyo* **110**: 179–188.

179. Kolar, Z., E. Krepela, J. Kovarik, M. Kohoutek, M. Rypka, and J. Vicar. 1987. Clinical experience with the testing of serum proteinase activity by cathepsin B-like sensitive amino acid derivatives of 7–amino-4-methylcoumarine. *Neoplasma* **34**: 601–608.

180. Komada, F., K. Okumura, and R. Hori. 1985. Fate of porcine and human insulin at the subcutaneous injection site. II. *In vitro* degradation of insulins in the subcutaneous tissue of the rat. *J. Pharmacobiodyn.* **8**: 33–40.

181. Kominami, E., S. Hashida, and N. Katunuma. 1981. Properties of fructose-1,6-bisphosphate aldolase inactivating enzymes in rat liver lysosomes. *Biochim. Biophys. Acta* **659**: 390–400.

182. Kominami, E., I. Kunio, and N. Katunuma. 1987. Activation of the intramyofibral autophagic-lysosomal system in muscular dystrophy. *Am. J. Pathol.* **127**: 461–466.

183. Kominami, E., T. Tsukahara, Y. Bando, and N. Katunuma. 1987. Autodegradation of lysosomal cysteine proteinases. *Biochem. Biophys. Res. Commun.* **144**: 749–756.

184. Konno, N., T. Yanagishita, E. Geshi, and T. Katagiri. 1987. Degradation of the cardiac sarcoplasmic reticulum in acute myocardial ischemia. *Jpn. Circ. J.* **51**: 411–420.

185. Kregar, I., V. Turk, and D. Lebez. 1967. Some properties of the cathepsin from the mucosa of rat colon. *Z. Naturforsch. B* **22**: 1360.

186. Kropshofer, H., H. Max, T. Halder, M. Kalbus, C. A. Muller, and H. Kalbacher. 1993. Self-peptides from four HLA-DR alleles share hydrophobic anchor residues near the NH_2-terminal including proline as a stop signal for trimming. *J. Immunol.* **151**: 4732–4742.

187. Kucharz, E. J. and M. Kucharz. 1986. [Serum cathepsin activity in chronic liver diseases]. *Z. Gesamte. Inn. Med.* **41**: 636–638.

188. Kucharz, J. 1986. Serum cathepsin activity in pregnancy. *Zentralbl. Gynakol.* **108**: 321–323.

189. Kudo, T., E. Q. Wei, and R. Inoki. 1990. Is cathepsin B an enkephalin processing enzyme in the tooth pulp? *Prog. Clin. Biol. Res.* **328**: 271–274.

190. Kuribayashi, M., H. Yamada, T. Ohmori, M. Yanai, and T. Imoto. 1993. Endopeptidase activity of cathepsin C, dipeptidyl aminopeptidase I, from bovine spleen. *J. Biochem. Tokyo* **113**: 441–449.

191. **Lah, T. T., M. R. Buck, K. V. Honn, J. D. Crissman, N. C. Rao, L. A. Liotta, and B. F. Sloane.** 1989. Degradation of laminin by human tumor cathepsin B. *Clin. Exp. Metastasis* **7**: 461–468.

192. **Laszlo, A., I. Sohar, and K. Gyurkovits.** 1987. Activity of the lysosomal cysteine proteinases (cathepsin B,H,L) and a metalloproteinase (MMP-7–ase) in the serum of cystic fibrosis homozygous children. *Acta Paediatr. Hung.* **28**: 175–178.

193. **Laumas, S., M. Abdel Ghany, K. Leister, R. Resnick, A. Kandrach, and E. Racker.** 1989. Decreased susceptibility of a 70-kDa protein to cathepsin L after phosphorylation by protein kinase C. *Proc. Natl Acad. Sci. U. S. A.* **86**: 3021–3025.

194. **Lauritzen, C., E. Tuchsen, P. E. Hansen, and O. Skovgaard.** 1991. BPTI and *N*-terminal extended analogues generated by factor Xa cleavage and cathepsin C trimming of a fusion protein expressed in Escherichia coli. *Protein. Expr. Purif.* **2**: 372–378.

195. **Leake, D. S., S. M. Rankin, and J. Collard.** 1990. Macrophage proteases can modify low density lipoproteins to increase their uptake by macrophages. *FEBS Lett.* **269**: 209–212.

196. **Lebez, D., V. Turk, and I. Kregar.** 1968. The possibility of differentiation of cathepsins on the basis of their ability to degrade various protein substrates. *Enzymologia* **34**: 344–348.

197. **Lindley, H.** 1972. The specificity of dipeptidyl aminopeptidase I (cathepsin C) and its use in peptide sequence studies. *Biochem. J.* **126**: 683–685.

198. **Lokshina, L. A., T. P. Egorova, and V. N. Orekhovich.** 1983. [Action of two thiol proteinases from the spleen which are active in neutral media on vasoactive peptides]. *Biokhimiia* **48**: 951–958.

199. **Lokshina, L. A., N. V. Golubeva, F. S. Baranova, and V. N. Orekhovich.** 1987. [Inactivation of blood plasma α_1-proteinase inhibitor as affected by 2 splenic thiol proteinases active in a neutral medium]. *Biull. Eksp. Biol. Med.* **103**: 662–664.

200. **Lokshina, L. A., O. N. Lubkova, T. A. Gureeva, and V. N. Orekhovich.** 1985. [Properties and specificity of cathepsin H from the bovine spleen]. *Vopr. Med. Khim.* **31**: 125–130.

201. **Luetscher, J. A., J. W. Bialek, and G. Grislis.** 1982. Human kidney cathepsins B and H activate and lower the molecular weight of human inactive renin. *Clin. Exp. Hypertens. A* **4**: 2149–2158.

202. **Lynch, G. W. and S. L. Pfueller.** 1988. Thrombin-independent activation of platelet factor XIII by endogenous platelet acid protease. *Thromb. Haemost.* **59**: 372–377.

203. **MacGregor, R. R., J. W. Hamilton, G. N. Kent, R. E. Shofstall, and D. V. Cohn.** 1979. The degradation of proparathormone and parathormone by parathyroid and liver cathepsin B. *J. Biol. Chem.* **254**: 4428–4433.

204. **Maciewicz, R. A., R. J. Wardale, S. F. Wotton, V. C. Duance, and D. J. Etherington.** 1990. Mode of activation of the precursor to cathepsin L: implication for matrix degradation in arthritis. *Biol. Chem. Hoppe Seyler* **371**(Suppl.): 223–228.

205. **Maciewicz, R. A. and S. F. Wotton.** 1991. Degradation of cartilage matrix components by the cysteine proteinases, cathepsins B and L. *Biomed. Biochim. Acta* **50**: 561–564.

206. **Maciewicz, R. A., S. F. Wotton, D. J. Etherington, and V. C. Duance.** 1990. Susceptibility of the cartilage collagens types II, IX and XI to degradation by the cysteine proteinases, cathepsins B and L. *FEBS Lett.* **269**: 189–193.

207. **Marks, N. and M. J. Berg.** 1987. Rat brain cathepsin L: characterization and differentiation from cathepsin B utilizing opioid peptides. *Arch. Biochem. Biophys.* **259**: 131–143.

208. **Marks, N., M. J. Berg, and M. Benuck.** 1986. Preferential action of rat brain cathepsin B as a peptidyl dipeptidase converting pro-opioid oligopeptides. *Arch. Biochem. Biophys.* **249**: 489–499.

209. **Marks, N., M. J. Berg, and W. Danho.** 1989. Enkephalin analogs as substrates for the assay of brain cysteine proteinase (cathepsin L) and its endogenous inhibitors. *Peptides* **10**: 391–394.

210. **Marks, N., M. Kopitar, F. Stern, and M. J. Berg.** 1985. Proenkephalin and enkephalin metabolism by rat brain cathepsin B: conversion, inactivation, and suppression by an endogenous inhibitor. *Prog. Clin. Biol. Res.* **180**: 247–249.

211. **Marks, N., A. Suhar, and M. Benuck.** 1980. Breakdown of opiate peptides by brain cathepsins and dipeptidyl carboxypeptidase. *Adv. Biochem. Psychopharmacol.* **22**: 205–217.

212. **Mason, R. W.** 1986. Species variations amongst lysosomal cysteine proteinases. *Biomed. Biochim. Acta* **45**: 1433–1440 [published erratum appears in 1987 *Biomed. Biochim. Acta* **46**: following 650].

213. **Mason, R. W.** 1986. Species variants of cathepsin L and their immunological identification. *Biochem. J.* **240**: 285–288.

214. **Mason, R. W., S. Gal, and M. M. Gottesman.** 1987. The identification of the major excreted protein (MEP) from a transformed mouse fibroblast cell line as a catalytically active precursor form of cathepsin L. *Biochem. J.* **248**: 449–454.

215. **Mason, R. W., D. A. Johnson, A. J. Barrett, and H. A. Chapman.** 1986. Elastinolytic activity of human cathepsin L. *Biochem. J.* **233**: 925–927.

216. Mason, R. W., M. A. Taylor, and D. J. Etherington. 1982. Purification and characterisation of collagenolytic cathepsins from rabbit liver. *FEBS Lett.* **146**: 33–36.

217. Matsuda, Y., F. Ogushi, K. Ogawa, and N. Katunuma. 1986. Structure and properties of albumin Tokushima and its proteolytic processing by cathepsin B *in vitro*. *J. Biochem. Tokyo* **100**: 375–379.

218. Matsuishi, M., T. Matsumoto, A. Okitani, and H. Kato. 1992. Mode of action of rabbit skeletal muscle cathepsin B towards myofibrillar proteins and the myofibrillar structure. *Int. J. Biochem.* **24**: 1967–1978.

219. Matsukura, U., A. Okitani, T. Nishimuro, and H. Kato. 1981. Mode of degradation of myofibrillar proteins by an endogenous protease, cathepsin L. *Biochim. Biophys. Acta* **662**: 41–47.

220. Matsunaga, Y., T. Saibara, H. Kido, and N. Katunuma. 1993. Participation of cathepsin B in processing of antigen presentation to MHC class II. *FEBS Lett.* **324**: 325–330.

221. McCoy, K., S. Gal, R. H. Schwartz, and M. M. Gottesman. 1988. An acid protease secreted by transformed cells interferes with antigen processing. *J. Cell Biol.* **106**: 1879–1884.

222. McCoy, K. L. and R. H. Schwartz. 1988. The role of intracellular acidification in antigen processing. *Immunol. Rev.* **106**: 129–147.

223. McDonald, J. K., P. X. Callahan, B. B. Zeitman, and S. Ellis. 1969. Inactivation and degradation of glucagon by dipeptidyl aminopeptidase I (cathepsin C) of rat liver. *J. Biol. Chem.* **244**: 6199–6208.

224. McDonald, J. K. and S. Ellis. 1975. On the substrate specificity of cathepsins B_1 and B_2 including a new fluorogenic substrate for cathepsin B1. *Life Sci.* **17**: 1269–1276.

225. McDonald, J. K., T. J. Reilly, B. B. Zeitman, and S. Ellis. 1966. Cathepsin C: a chloride-requiring enzyme. *Biochem. Biophys. Res. Commun.* **22**: 771–775.

226. McDonald, J. K., B. B. Zeitman, P. X. Callahan, and S. Ellis. 1974. Angiotensinase activity of dipeptidyl aminopeptidase I (cathepsin C) of rat liver. *J. Biol. Chem.* **249**: 234–240.

227. McDonald, J. K., B. B. Zeitman, and S. Ellis. 1972. Detection of a lysosomal carboxypeptidase and a lysosomal dipeptidase in highly-purified dipeptidyl aminopeptidase I (cathepsin C) and the elimination of their activities from preparations used to sequence peptides. *Biochem. Biophys. Res. Commun.* **46**: 62–70.

228. McDonald, J. K., B. B. Zeitman, T. J. Reilly, and S. Ellis. 1969. New observations on the substrate specificity of cathepsin C (dipeptidyl aminopeptidase I). Including the degradation of beta-corticotropin and other peptide hormones. *J. Biol. Chem.* **244**: 2693–2709.

229. McGuire, M. J., P. E. Lipsky, and D. L. Thiele. 1992. Purification and characterization of dipeptidyl peptidase I from human spleen. *Arch. Biochem. Biophys.* **295**: 280–288.

230. McGuire, M. J., P. E. Lipsky, and D. L. Thiele. 1993. Generation of active myeloid and lymphoid granule serine proteases requires processing by the granule thiol protease dipeptidyl peptidase I. *J. Biol. Chem.* **268**: 2458–2467.

231. McKay, M. J., M. W. Marsh, H. Kirschke, and J. S. Bond. 1984. Inactivation of fructose-1,6-bisphosphate aldolase by cathepsin L stimulation by ATP. *Biochim. Biophys. Acta* **784**: 9–15.

232. McKay, M. J., M. K. Offermann, A. J. Barrett, and J. S. Bond. 1983. Action of human liver cathepsin B on the oxidized insulin B chain. *Biochem. J.* **213**: 467–471.

233. Melloni, E., S. Pontremoli, F. Salamino, B. Sparatore, M. Michetti, and B. L. Horecker. 1981. Characterization of three rabbit liver lysosomal proteinases with fructose 1,6-bisphosphatase converting enzyme activity. *Arch. Biochem. Biophys.* **208**: 175–183.

234. Melloni, E., S. Pontremoli, F. Salamino, B. Sparatore, M. Michetti, and B. L. Horecker. 1981. Changes during fasting in the activity of a specific lysosomal proteinase, fructose-1,6-bisphosphatase converting enzyme. *Proc. Natl Acad. Sci. U. S. A.* **78**: 1499–1502.

235. Menard, R., E. Carmona, C. Plouffe, D. Bromme, Y. Konishi, J. Lefebvre, and A. C. Storer. 1993. The specificity of the S1′ subsite of cysteine proteases. *FEBS Lett.* **328**: 107–110.

236. Mort, J. S. and M. S. Leduc. 1984. The combined action of two enzymes in human serum can mimic the activity of cathepsin B. *Clin. Chim. Acta* **140**: 173–182.

237. Mouritsen, S., M. Meldal, O. Werdelin, A. S. Hansen, and S. Buus. 1992. MHC molecules protect T cell epitopes against proteolytic destruction. *J. Immunol.* **149**: 1987–1993.

238. Murphy, G., R. Ward, J. Gavrilovic, and S. Atkinson. 1992. Physiological mechanisms for metalloproteinase activation. *Matrix Suppl.* **1**: 224–230.

239. Nakai, N., Y. Fujii, K. Kobashi, and J. Hase. 1983. Effect of fructose 1,6-bisphosphate on the activity of liver pyruvate kinase after limited proteolysis with cathepsin B. *Biochem. Biophys. Res. Commun.* **110**: 682–687.

240. Nakai, N., K. Wada, K. Kobashi, and J. Hase. 1978. The limited proteolysis of rabbit muscle aldolase by cathepsin B_1. *Biochem. Biophys. Res. Commun.* **83**: 881–885.

241. Nakashima, K. and K. Ogino. 1974. Regulation of rabbit liver fructose-1,6-diphosphatase. II. Modification by lysosomal cathepsin B1 from the same cell. *J. Biochem. Tokyo* **75**: 355–365.

242. **Noda, T., K. Isogai, H. Hayashi, and N. Katunuma.** 1981. Susceptibilities of various myofibrillar proteins to cathepsin B and morphological alteration of isolated myofibrils by this enzyme. *J. Biochem. Tokyo* **90**: 371–379.

243. **Okazaki, H., C. Tani, M. Ando, K. Ishii, S. Ishibashi, Y. Nishimura, and K. Kato.** 1992. Possible involvement of cathepsin L in processing of rat liver hexokinase to eliminate mitochondria-binding ability. *J. Biochem. Tokyo* **112**: 409–413.

244. **Otto, K. and U. Baur.** 1972. [Influence of cathepsin B_1 on phosphofructokinase and hexosediphosphatase]. *Hoppe Seylers Z. Physiol. Chem.* **353**: 741–742.

245. **Otto, K. and S. Bhakdi.** 1969. [Studies on cathepsin B': specificity and properties]. *Hoppe Seylers Z. Physiol. Chem.* **350**: 1577–1588.

246. **Otto, K. and P. Schepers.** 1967. [On catheptic inactivation of some enzymes in rat liver, particularly glucokinase]. *Hoppe Seylers Z. Physiol. Chem.* **348**: 482–490.

247. **Page, A. E., A. R. Hayman, L. M. Andersson, T. J. Chambers, and M. J. Warburton.** 1993. Degradation of bone matrix proteins by osteoclast cathepsins. *Int. J. Biochem.* **25**: 545–550.

248. **Perez, H. D., O. Ohtani, D. Banda, R. Ong, K. Fukuyama, and I. M. Goldstein.** 1983. Generation of biologically active, complement-(C5) derived peptides by cathepsin H. *J. Immunol.* **131**: 397 402.

249. **Pohl, J., S. Davinic, I. Blaha, P. Strop, and V. Kostka.** 1987. Chromophoric and fluorophoric peptide substrates cleaved through the dipeptidyl carboxypeptidase activity of cathepsin B. *Anal. Biochem.* **165**: 96–101.

250. **Polgar, L. and C. Csoma.** 1987. Dissociation of ionizing groups in the binding cleft inversely controls the endo- and exopeptidase activities of cathepsin B. *J. Biol. Chem.* **262**: 14448–14453.

251. **Pontremoli, S., E. Melloni, M. Michetti, F. Salamino, B. Sparatore, and B. L. Horecker.** 1982. Limited proteolysis of liver aldolase and fructose 1,6-bisphosphatase by lysosomal proteinases: effect on complex formation. *Proc. Natl Acad. Sci. U. S. A.* **79**: 2451–2454.

252. **Pontremoli, S., E. Melloni, F. Salamino, B. Sparatore, M. Michetti, and B. L. Horecker.** 1982. Cathepsin M: a lysosomal proteinase with aldolase-inactivating activity. *Arch. Biochem. Biophys.* **214**: 376–385.

253. **Pote, M. S. and W. Altekar.** 1981. Muscle aldolase: the stress-dependent modification of catalytic and structural properties by rat muscle lysosomal cathepsin B. *Biochim. Biophys. Acta* **661**: 303–314.

254. **Puri, R. B., K. Anjaneyulu, J. R. Kidwai, and V. K. Mohan Rao.** 1978. *In vitro* conversion of proinsulin to insulin by cathepsin B and role of C-peptide. *Acta Diabetol. Lat.* **15**: 243–250.

255. **Quinn, P. S. and J. D. Judah.** 1978. Calcium-dependent Golgi-vesicle fusion and cathepsin B in the conversion of proalbumin into albumin in rat liver. *Biochem. J.* **172**: 301–309.

256. **Roche, P. A. and P. Cresswell.** 1991. Proteolysis of the class II-associated invariant chain generates a peptide binding site in intracellular HLA-DR molecules. *Proc. Natl Acad. Sci. U. S. A.* **88**: 3150–3154.

257. **Sano, K., S. Waguri, N. Sato, E. Kominami, and Y. Uchiyama.** 1993. Coexistence of renin and cathepsin B in secretory granules of granular duct cells in male mouse submandibular gland. *J. Histochem. Cytochem.* **41**: 433–438.

258. **Santoro, L., A. Reboul, A. M. Journet, and M. G. Colomb.** 1993. Major involvement of cathepsin B in the intracellular proteolytic processing of exogenous IgGs in U937 cells. *Mol. Immunol.* **30**: 1033–1039.

259. **Sarobe, P., J. J. Lasarte, E. Larrea, J. J. Golvano, I. Prieto, A. Gullon, J. Prieto, and F. Borras Cuesta.** 1993. Enhancement of peptide immunogenicity by insertion of a cathepsin B cleavage site between determinants recognized by B and T cells. *Res. Immunol.* **144**: 257–262.

260. **Schauer, P., H. Hren Vencelj, and M. Likar.** 1974. On the activity of cathepsin C in human embryonic kidney cell cultures infected with *Herpesvirus hominis* (herpes simplex). *Experientia* **30**: 232–233.

261. **Schmidt, M. T. and H. Kirschke.** 1990. Degradation of proteoglycans by cathepsin L and H. *Ann. Jagiellonian Univ. Mol. Biol. Ser.* **19**: 117–126.

262. **Schwartz, W. and J. W. Bird.** 1977. Degradation of myofibrillar proteins by cathepsins B and D. *Biochem. J.* **167**: 811–820.

263. **Shcherbak, I. G., N. N. Nikandrov, V. P. Faenkova, and L. N. Kirpichenok.** 1987. [Products of the proteolytic effect of cathepsin on various protein substrates]. *Vopr. Med. Khim.* **33**: 115–119.

264. **Singh, H. and G. Kalnitsky.** 1980. α-N-Benzoylarginine-β-naphthylamide hydrolase, an aminoendopeptidase from rabbit lung. *J. Biol. Chem.* **255**: 369–374.

265. **Slater, E. E., A. D. Gounaris, and E. Haber.** 1979. Isolation of a renal thiol protease that activates inactive plasma renin. *Clin. Sci.* **57**(Suppl. 5): 93s–96s.

266. **Smarel, A. M., S. W. Worobec, A. G. Ferguson, R. S. Decker, and M. Lesch.** 1986. Limited proteolysis of rabbit cardiac procathepsin D in a cell-free system. *Am. J. Physiol.* **250**: C589–C596.

267. **Smedsrod, B., S. Johansson, and H. Pertoft.** 1985. Studies *in vivo* and *in vitro* on the uptake and degradation of soluble collagen α1(I) chains in rat liver endothelial and Kupffer cells. *Biochem. J.* **228**: 415–424.

268. **Snellman, O.** 1969. Cathepsin B, the lysosomal thiol proteinase of calf liver. *Biochem. J.* **114**: 673–678.

269. **Storer, A. C. and R. Ménard.** 1994. Catalytic mechanism in papain family of cysteine peptidases. *Methods Enzymol.* **244**: 486–500.

270. **Suhar, A. and N. Marks.** 1979. Purification and properties of brain cathepsin B. Evidence for cleavage of pituitary lipotropins. *Eur. J. Biochem.* **101**: 23–20.

271. **Tagawa, K., T. Kunishita, K. Maruyama, K. Yoshikawa, E. Kominami, T. Tsuchiya, K. Suzuki, T. Tabira, H. Sugita, and S. Ishiura.** 1991. Alzheimer's disease amyloid β-clipping enzyme (APP secretase): identification, purification, and characterization of the enzyme. *Biochem. Biophys. Res. Commun.* **177**: 377–387.

272. **Tagawa, K., K. Maruyama, and S. Ishiura.** 1992. Amyloid β/A4 precursor protein (APP) processing in lysosomes. *Ann. N. Y. Acad. Sci.* **674**: 129–137.

273. **Takahashi, H., K. B. Cease, and J. A. Berzofsky.** 1989. Identification of proteases that process distinct epitopes on the same protein. *J. Immunol.* **142**: 2221–2229.

274. **Takahashi, S., K. Murakami, and Y. Miyake.** 1982. Activation of kidney prorenin by kidney cathepsin B isozymes. *J. Biochem. Tokyo* **91**: 419–422.

275. **Takahashi, T., A. H. Dehdarani, and J. Tang.** 1988. Porcine spleen cathepsin H hydrolyzes oligopeptides solely by aminopeptidase activity. *J. Biol. Chem.* **263**: 10952–10957.

276. **Takahashi, T., A. H. Dehdarani, S. Yonezawa, and J. Tang.** 1986. Porcine spleen cathepsin B is an exopeptidase. *J. Biol. Chem.* **261**: 9375–9381.

277. **Takeda, A., T. Jimi, Y. Wakayama, N. Misugi, S. Miyake, and T. Kumagai.** 1992. Demonstration of cathepsins B, H and L in xenografts of normal and Duchenne-muscular-dystrophy muscles transplanted into nude mice. *Biochem. J.* **288**: 643–648.

278. **Taugner, R., C. P. Buhrle, R. Nobiling, and H. Kirschke.** 1985. Coexistence of renin and cathepsin B in epithelioid cell secretory granules. *Histochemistry* **83**: 103–108.

279. **Tchoupe, J. R., T. Moreau, F. Gauthier, and J. G. Bieth.** 1991. Photometric or fluorometric assay of cathepsin B, L and H and papain using substrates with an aminotrifluoromethylcoumarin leaving group. *Biochim. Biophys. Acta* **1076**: 149–151.

280. **Thiele, D. L. and P. E. Lipsky.** 1990. The action of leucyl-leucine methyl ester on cytotoxic lymphocytes requires uptake by a novel dipeptide-specific facilitated transport system and dipeptidyl peptidase I-mediated conversion to membranolytic products. *J. Exp. Med.* **172**: 183–194.

281. **Thiele, D. L. and P. E. Lipsky.** 1990. Mechanism of L-leucyl-L-leucine methyl ester-mediated killing of cytotoxic lymphocytes: dependence on a lysosomal thiol protease, dipeptidyl peptidase I, that is enriched in these cells. *Proc. Natl Acad. Sci. U. S. A.* **87**: 83–87.

282. **Thomas, G. J. and M. Davies.** 1989. The potential role of human kidney cortex cysteine proteinases in glomerular basement membrane degradation. *Biochim. Biophys. Acta* **990**: 246–253.

283. **Thomas, G. J., K. T. Hughes, and M. Davies.** 1980. The isolation of three thiol proteinases from human kidney that attack glomerular basement membrane. *Biochem. Soc. Trans.* **8**: 596–597.

284. **Towatari, T. and N. Katunuma.** 1983. Selective cleavage of peptide bonds by cathepsins L and B from rat liver. *J. Biochem. Tokyo* **93**: 1119–1128.

285. **Towatari, T., K. Tanaka, D. Yoshikawa, and N. Katunuma.** 1978. Purification and properties of a new cathepsin from rat liver. *J. Biochem. Tokyo* **84**: 659–672.

286. **Uchiyama, Y., M. Nakajima, T. Watanabe, S. Waguri, N. Sato, M. Yamamoto, Y. Hashizume, and E. Kominami.** 1991. Immunocytochemical localization of cathepsin B in rat anterior pituitary endocrine cells, with special reference to its co-localization with renin and prorenin in gonadotrophs. *J. Histochem. Cytochem.* **39**: 1199–1205.

287. **Valiulis, R. A. and V. M. Stepanov.** 1971. [Use of cathepsin C for determination of the amino acid sequence of peptides]. *Biokhimiia* **36**: 886–887.

288. **Wada, K. and T. Tanabe.** 1985. Proteolysis of liver acetyl coenzyme A carboxylase by cathepsin B. *FEBS Lett.* **180**: 74–76.

289. **Wang, P. H., Y. S. Do, L. Macaulay, T. Shinagawa, P. W. Anderson, J. D. Baxter, and W. A. Hsueh.** 1991. Identification of renal cathepsin B as a human prorenin-processing enzyme. *J. Biol. Chem.* **266**: 12633–12638.

290. **Watari, E. and K. Yokomuro.** 1992. T-cell hybridomas recognizing the envelope proteins of Semliki forest virus: their sensitivity to endo/lysosomal protease and the antigenicity. *Jpn. J. Med. Sci. Biol.* **45**: 113–125.

291. **Willenbrock, F. and K. Brocklehurst.** 1984. Natural structural variation in enzymes as a tool in the study of mechanism exemplified by a comparison of the catalytic-site structure and characteristics of cathepsin B and papain. pH-dependent kinetics of the reactions of cathepsin B from bovine spleen and from rat liver with a thiol-specific two-protonic-state probe (2,2′-dipyridyl disulphide) and with a specific synthetic substrate (N-α-benzyloxycarbonyl-L-arginyl-L-arginine 2-naphthylamide). *Biochem. J.* **222**: 805–814.

292. **Willenbrock, F. and K. Brocklehurst.** 1985. A general framework of cysteine-proteinase mechanism deduced from studies on enzymes with structurally different analogous catalytic-site residues Asp-158 and -161 (papain and actinidin), Gly-196 (cathepsin B) and Asn-165 (cathepsin H). Kinetic studies up to pH 8 of the hydrolysis of N-α-benzyloxycarbonyl-L-arginyl-L-arginine 2-naphthylamide catalysed by cathepsin B and of L-arginine 2-naphthylamide catalysed by cathepsin H. *Biochem. J.* **227**: 521–528.

293. **Willenbrock, F. and K. Brocklehurst.** 1985. Preparation of cathepsins B and H by covalent chromatography and characterization of their catalytic sites by reaction with a thiol-specific two-protonic-state reactivity probe. Kinetic study of cathepsins B and H extending into alkaline media and a rapid spectroscopic titration of cathepsin H at pH 3–4. *Biochem. J.* **227**: 511–519.

294. **Willenbrock, F. and K. Brocklehurst.** 1986. Chemical evidence for the pH-dependent control of ion-pair geometry in cathepsin B. Benzofuroxan as a reactivity probe sensitive to differences in the mutual disposition of the thiolate and imidazolium components of cysteine proteinase catalytic sites. *Biochem. J.* **238**: 103–107.

295. **Xin, X. Q., B. Gunesekera, and R. W. Mason.** 1992. The specificity and elastinolytic activities of bovine cathepsins S and H. *Arch. Biochem. Biophys.* **299**: 334–339.

296. **Yamamoto, H., Y. Murawaki, and H. Kawasaki.** 1992. Collagenolytic cathepsin B and L activity in experimental fibrotic liver and human liver. *Res. Commun. Chem. Pathol. Pharmacol.* **76**: 95–112.

297. **Yanagisawa, K., S. Sato, T. Miyatake, E. Kominami, and N. Katsunuma.** 1984. Degradation of myelin proteins by cathepsin B and inhibition by E-64 analogue. *Neurochem. Res.* **9**: 691–694.

298. **Yokota, K., F. Shono, S. Yamamoto, E. Kominami, and N. Katunuma.** 1983. Transformation of leukotriene D$_4$ catalyzed by lysosomal cathepsin H of rat liver. *J. Biochem. Tokyo* **94**: 1173–1178.

Determination and purification

299. **Ahmad, S. and M. Y. Khan.** 1990. Further characterization of buffalo spleen cathepsin B. *Biochem. Int.* **22**: 951–958.

300. **Aranishi, F., K. Hara, and T. Ishihara.** 1992. Purification and characterization of cathepsin H from hepatopancreas of carp *Cyprinus carpio. Comp. Biochem. Physiol. B* **102**: 499–505.

301. **Azaryan, A., N. Barkhudaryan, and A. Galoyan.** 1985. Some properties of human and bovine brain cathepsin B. *Neurochem. Res.* **10**: 1511–1524.

302. **Azaryan, A. and A. Galoyan.** 1987. Human and bovine brain cathepsin L and cathepsin H: purification, physico-chemical properties, and specificity. *Neurochem. Res.* **12**: 207–213.

303. **Bajkowski, A. S. and A. Frankfater.** 1975. Specific spectrophotometric assays for cathepsin B$_1$. *Anal. Biochem.* **68**: 119–127.

304. **Bando, Y., E. Kominami, and N. Katunuma.** 1986. Purification and tissue distribution of rat cathepsin L. *J. Biochem. Tokyo* **100**: 35–42.

305. **Barcelo, F., N. Vives, and J. Bozal.** 1980. [Purification and properties of dipeptidyl amino-peptidase I from chicken liver]. *Rev. Esp. Fisiol.* **36**: 321–330.

306. **Barrett, A. J.** 1973. Human cathepsin B$_1$. Purification and some properties of the enzyme. *Biochem. J.* **131**: 809–822.

307. **Barrett, A. J.** 1976. An improved color reagent for use in Barrett's assay of Cathepsin B. *Anal. Biochem.* **76**: 374–376.

308. **Barrett, A. J.** 1980. Fluorimetric assays for cathepsin B and cathepsin H with methylcoumarylamide substrates. *Biochem. J.* **187**: 909–912.

309. **Benenson, A. M. and M. G. Osipova.** 1969. [A simplified procedure for the isolation of cathepsin C from bovine spleen]. *Biokhimiia* **34**: 372–375.

310. **Bradley, J. D. and J. N. Whitaker.** 1986. Isolation and characterization of cathepsin B from bovine brain. *Neurochem. Res.* **11**: 851–867.

311. **Bromme, D., P. R. Bonneau, P. Lachance, B. Wiederanders, H. Kirschke, C. Peters, D. Y. Thomas, A. C. Storer, and T. Vernet.** 1993. Functional expression of human cathepsin S in *Saccharomyces cerevisiae*. Purification and characterization of the recombinant enzyme. *J. Biol. Chem.* **268**: 4832–4838.

312. **Butterworth, J. and J. J. Duncan.** 1980. The hydrolysis of N-benzoyl-L-phenylalanyl-L-valyl-L-arginine-p-nitroanilide and its use as a substrate for the assay of cathepsin B. *Anal. Biochem.* **106**: 156–162.

313. **Coetzer, T. H., R. N. Pike, and C. Dennison.** 1992. Localization of an immunoinhibitory epitope of the cysteine proteinase, cathepsin L. *Immunol. Invest.* **21**: 495–506.

314. **Cox, S. W. and B. M. Eley.** 1987. Preliminary studies on cysteine and serine proteinase activities in inflamed human gingiva using different 7-amino-4-trifluoromethyl coumarin substrates and protease inhibitors. *Arch. Oral Biol.* **32**: 599–605.

315. **Dalet Fumeron, V., N. Guinec, and M. Pagano.** 1991. High-performance liquid chromatographic method for the simultaneous purification of cathepsins B, H and L from human liver. *J. Chromatogr.* **568**: 55–68.

316. **Delaisse, J. M., P. Ledent, and G. Vaes.** 1991. Collagenolytic cysteine proteinases of bone tissue. Cathepsin B, (pro)cathepsin L and a cathepsin L-like 70 kDa proteinase. *Biochem. J.* **279**: 167–174.

317. **Dennison, C. and R. N. Pike.** 1991. A peptide antibody that specifically inhibits cathepsin L. *Adv. Exp. Med. Biol.* **303**: 285–288.

318. **Dilakian, E. A., L. A. Lokshina, and V. N. Orekhovich.** 1986. [Characteristics of cysteine peptidohydrolases from the human kidney cortex]. *Vopr. Med. Khim.* **32**: 72–76.

319. **Ducastaing, A. and D. J. Etherington.** 1978. Purification of bovine spleen collagenolytic cathepsin, (cathepsin N). *Biochem. Soc. Trans.* **6**: 938–940.

320. **Eley, B. M. and S. W. Cox.** 1992. Correlation of gingival crevicular fluid proteases with clinical and radiological measurements of periodontal attachment loss. *J. Dent.* **20**: 90–99.

321. **Erickson Viitanen, S., E. Balestreri, M. J. McDermott, B. L. Horecker, E. Melloni, and S. Pontremoli.** 1985. Purification and properties of rabbit liver cathepsin M and cathepsin B. *Arch. Biochem. Biophys.* **243**: 46–61.

322. **Etherington, D. J.** 1974. The purification of bovine cathepsin B_1 and its mode of action on bovine collagens. *Biochem. J.* **137**: 547–557.

323. **Evans, B. and E. Shaw.** 1983. Inactivation of cathepsin B by active site-directed disulfide exchange. Application in covalent affinity chromatography. *J. Biol. Chem.* **258**: 10227–10232.

324. **Fazili, K. M. and M. A. Qasim.** 1986. Purification and some properties of buffalo liver cathepsin B. *J. Biochem. Tokyo* **100**: 293–299.

325. **Funabiki, R., K. Yagasaki, C. Shirakawa, H. Sugita, and S. Ishiura.** 1990. Activity measurement of lysosomal cysteine proteinases, cathepsins B, H and L, in crude tissue extracts, and their relation to the fractional rate of protein degradation. *Int. J. Biochem.* **22**: 1303–1306.

326. **Gabrijelcic, D., A. Annan Prah, B. Rodic, B. Rozman, V. Cotic, and V. Turk.** 1990. Determination of cathepsins B and H in sera and synovial fluids of patients with different joint diseases. *J. Clin. Chem. Clin. Biochem.* **28**: 149–153.

327. **Gohda, E. and H. C. Pitot.** 1981. Purification and characterization of a new thiol proteinase from rat kidney. *Biochim. Biophys. Acta* **659**: 114–122.

328. **Gottesman, M. M. and F. Cabral.** 1981. Purification and characterization of a transformation-dependent protein secreted by cultured murine fibroblasts. *Biochemistry* **20**: 1659–1665.

329. **Gounaris, A. D. and E. E. Slater.** 1982. Cathepsin B from human renal cortex. *Biochem. J.* **205**: 295–302.

330. **Graf, F. M. and P. Strauli.** 1983. Use of the avidin−biotin−peroxidase complex (ABC) method for the localization of rabbit cathepsin B in cells and tissues. *J. Histochem. Cytochem.* **31**: 803–810.

331. **Graf, M. F.** 1984. [Use of avidin−biotin−peroxidase complex (ABC) method for localization of rabbit cathepsin B in light and electron microscopy]. *Acta Histochem. Suppl.* **29**: 169–174.

332. **Hardy, M. F. and R. J. Pennington.** 1979. Separation of cathepsin B_1 and related enzymes from rat skeletal muscle. *Biochim. Biophys. Acta* **577**: 253–266.

333. **Higashi, T., M. Hashimoto, M. Watanabe, Y. Yamauchi, M. Fujiwara, H. Nakatsukasa, M. Kobayashi, A. Watanabe, and H. Nagashima.** 1986. Assay procedures for cathepsin B, H and L activities in rat tissue homogenates. *Acta Med. Okayama.* **40**: 27–32.

334. **Hirao, T., K. Hara, and K. Takahashi.** 1984. Purification and characterization of cathepsin B from monkey skeletal muscle. *J. Biochem. Tokyo* **95**: 871–879.

335. **Huang, F. L. and A. L. Tappel.** 1972. Properties of cathepsin C from rat liver. *Biochim. Biophys. Acta* **268**: 527–538.

336. **Ishidoh, K., M. Takeda Ezaki, and E. Kominami.** 1993. Procathepsin L-specific antibodies that recognize procathepsin L but not cathepsin L. *FEBS Lett.* **322**: 79–82.

337. **Joronen, I. A. and V. K. Hopsu Havu.** 1987. Separation and partial characterization of four cysteine proteinases from a human epidermal cell line. *Arch. Dermatol. Res.* **279**: 524–529.

338. **Katunuma, N., T. Towatari, E. Kominami, S. Hashida, K. Takio, and K. Titani.** 1981. Rat liver thiol proteinases: cathepsin B, cathepsin H and cathepsin L. *Acta Biol. Med. Ger.* **40**: 1419–1425.

339. **Kirschke, H.** 1977. Cathepsin H: an endoaminopeptidase. *Acta Biol. Med. Ger.* **36**: 1547–1548.

340. **Kirschke, H., J. Langner, B. Wiederanders, S. Ansorge, and P. Bohley.** 1972. Intrazellulärer Proteinabbau. IV. Isolierung und Charakterisierung von Peptidasen aus Rattenleberlysosomen. *Acta Biol. Med. Ger.* **28**: 305–322.

341. **Kirschke, H., J. Langner, B. Wiederanders, S. Ansorge, and P. Bohley.** 1977. Cathepsin L. A new proteinase from rat-liver lysosomes. *Eur. J. Biochem.* **74**: 293–301.

342. **Kirschke, H., J. Langner, B. Wiederanders, S. Ansorge, P. Bohley, and U. Broghammer.** 1976. [Intracellular protein breakdown. VII. Cathepsin L and H; two new proteinases from rat liver lysosomes]. *Acta Biol. Med. Ger.* **35**: 285–299.

343. **Kirschke, H., J. Langner, B. Wiederanders, S. Ansorge, P. Bohley, and H. Hanson.** 1974. Cathepsin L and proteinases with cathepsin B$_1$-like activity from rat liver lysosomes. In Intracellular Protein Catabolism, H. Hanson and P. Bohley (eds), pp. 210–217. J. A. Barth, Leipzig.

344. **Kirschke, H., J. Langner, B. Wiederanders, S. Ansorge, P. Bohley, and H. Hanson.** 1977. Cathepsin H: an endoaminopeptidase from rat liver lysosomes. *Acta Biol. Med. Ger.* **36**: 185–199.

345. **Kirschke, H. and B. Wiederanders.** 1994. Cathepsin S and related lysosomal endopeptidases. *Methods Enzymol.* **244**: 500–511.

346. **Kirschke, H., B. Wiederanders, D. Bromme, and A. Rinne.** 1989. Cathepsin S from bovine spleen. Purification, distribution, intracellular localization and action on proteins. *Biochem. J.* **264**: 467–473.

347. **Kominami, E. and N. Katunuma.** 1982. Immunological studies on cathepsins B and H from rat liver. *J. Biochem. Tokyo* **91**: 67–71.

348. **Kominami, E., T. Tsukahara, Y. Bando, and N. Katunuma.** 1985. Distribution of cathepsins B and H in rat tissues and peripheral blood cells. *J. Biochem. Tokyo* **98**: 87–93.

349. **Koohmaraie, M. and D. H. Kretchmar.** 1990. Comparisons of four methods for quantification of lysosomal cysteine proteinase activities. *J. Anim. Sci.* **68**: 2362–2370.

350. **Kregar, I., P. Locnikar, T. Popovic, A. Suhar, T. Lah, A. Ritonja, F. Gubensek, and V. Turk.** 1981. Bovine intracellular cysteine proteinases. *Acta Biol. Med. Ger.* **40**: 1433–1438.

351. **Kregar, I., V. Turk, and D. Lebez.** 1967. Purification and some properties of the cathepsin from the small intestinal mucosa of pig. *Enzymologia* **33**: 80–88.

352. **Kucharz, E.** 1984. A modified micromethod for determination of cathepsin activity in blood serum. *Z. Med. Lab. Diagn.* **25**: 282–284.

353. **Liao, J. C. and J. F. Lenney.** 1984. Cathepsins J and K: high molecular weight cysteine proteinases from human tissues. *Biochem. Biophys. Res. Commun.* **124**: 909–916.

354. **Liao, J. C. and J. F. Lenney.** 1985. Cathepsin J: a high molecular weight cysteine proteinase from human tissues. *Prog. Clin. Biol. Res.* **180**: 225–227.

355. **Locnikar, P., T. Popovic, T. Lah, I. Kregar, J. Babnik, M. Kopitar, and V. Turk.** 1981. The bovine cysteine proteinases, cathepsin B, H and S. In *Proteinases and Their Inhibitors. Structure, Function and Applied Aspects*, V. Turk and L. Vitale (eds), pp. 109–116. Mladinska Knjiga and Pergamon Press, Ljubljana and Oxford.

356. **Lokshina, L. A., T. A. Gureeva, O. N. Lubkova, and V. N. Orekhovich.** 1982. [Characteristics of two thiol proteinases from spleen active in neutral media]. *Biokhimiia* **47**: 1299–1307.

357. **Lokshina, L. A., I. A. Tarkhanova, O. N. Lubkova, N. V. Golubeva, and T. A. Gureeva.** 1983. [Immunochemical study of bovine spleen thiol proteinases: cathepsin S B, H and L]. *Biull. Eksp. Biol. Med.* **96**: 50–53.

358. **Lynn, K. R. and R. S. Labow.** 1984. A comparison of four sulfhydryl cathepsins (B, C, H, and L) from porcine spleen. *Can. J. Biochem. Cell Biol.* **62**: 1301–1308.

359. **MacGregor, R. R., J. W. Hamilton, R. E. Shofstall, and D. V. Cohn.** 1979. Isolation and characterization of porcine parathyroid cathepsin B. *J. Biol. Chem.* **254**: 4423–4427.

360. **Maciewicz, R. A. and D. J. Etherington.** 1988. A comparison of four cathepsins (B, L, N and S) with collagenolytic activity from rabbit spleen. *Biochem. J.* **256**: 433–440.

361. **Maciewicz, R. A. and P. J. Knight.** 1988. Transmission densitometry of stained nitrocellulose paper. *Anal. Biochem.* **175**: 85–90.

362. **Mason, R. W., G. D. Green, and A. J. Barrett.** 1985. Human liver cathepsin L. *Biochem. J.* **226**: 233–241.

363. **Mason, R. W., M. A. Taylor, and D. J. Etherington.** 1984. The purification and properties of cathepsin L from rabbit liver. *Biochem. J.* **217**: 209–217.

364. **Metrione, R. M., A. G. Neves, and J. S. Fruton.** 1966. Purification and properties of dipeptidyl transferase (cathepsin C). *Biochemistry* **5**: 1597–1604.

365. **Moin, K., N. A. Day, M. Sameni, S. Hasnain, T. Hirama, and B. F. Sloane.** 1992. Human tumour cathepsin B. Comparison with normal liver cathepsin B. *Biochem. J.* **285**: 427–434.

366. **Mort, J. S. and M. Leduc.** 1982. A simple, economical method for staining gels for cathepsin B-like activity. *Anal. Biochem.* **119**: 148–152.

367. **Nakagawa, H. and S. Ohtaki.** 1984. Partial purification and characterization of two thiol proteases from hog thyroid lysosomes. *Endocrinology* **115**: 33–40.

368. **Nikawa, T., T. Towatari, and N. Katunuma.** 1992. Purification and characterization of cathepsin J from rat liver. *Eur. J. Biochem.* **204**: 381–393.

369. **Obled, A., A. Ouali, and C. Valin.** 1984. Cysteine proteinase content of rat muscle lysosomes. Evidence for an unusual proteinase activity. *Biochimie* **66**: 609–616.

370. **Ogino, K. and K. Nakashima.** 1974. Purification of rabbit liver cathepsin B1. *J. Biochem. Tokyo* **75**: 723–730.

371. **Okitani, A., M. Matsuishi, T. Matsumoto, E. Kamoshida, M. Sato, U. Matsukura, M. Watanabe, H. Kato, and M. Fujimaki.** 1988. Purification and some properties of cathepsin B from rabbit skeletal muscle. *Eur. J. Biochem.* **171**: 377–381.

372. **Okitani, A., U. Matsukura, H. Kato, and M. Fujimaki.** 1980. Purification and some properties of a myofibrillar protein-degrading protease, cathepsin L, from rabbit skeletal muscle. *J. Biochem. Tokyo* **87**: 1133–1143.

373. **Otto, K.** 1967. [On a new cathepsin. Purification from bovine spleen, properties, as well as comparison with cathepsin B]. *Hoppe Seylers Z. Physiol. Chem.* **348**: 1449–1460.

374. **Otto, K. and H. Riesenkonig.** 1975. Improved purification of cathepsin B1 and cathepsin B2. *Biochim. Biophys. Acta* **379**: 462–475.

375. **Page, A., M. J. Warburton, T. J. Chambers, and A. R. Hayman.** 1991. Purification and characterisation of cysteine proteinases from human osteoclastomas. *Biochem. Soc. Trans.* **19**: 286S.

376. **Paszkowski, A., H. Singh, and G. Kalnitsky.** 1983. Properties of dipeptidyl peptidase I from rabbit (Oryctolagus cuniculus) lungs. *Acta Biochim. Pol.* **30**: 363–380.

377. **Pierart Gallois, M., A. Trouet, and P. Tulkens.** 1977. Production of rabbit antibodies against active rat cathepsin B. *Acta Biol. Med. Ger.* **36**: 1887–1891.

378. **Pike, R. and C. Dennison.** 1989. A high yield method for the isolation of sheep's liver cathepsin L. *Prep. Biochem.* **19**: 231–245.

379. **Pontremoli, S., E. Melloni, G. Damiani, M. Michetti, F. Salamino, B. Sparatore, and B. L. Horecker.** 1984. Binding of monoclonal antibody to cathepsin M located on the external surface of rabbit lysosomes. *Arch. Biochem. Biophys.* **233**: 267–271.

380. **Popovic, T., J. Brzin, J. Kos, B. Lenarcic, W. Machleidt, A. Ritonja, K. Hanada, and V. Turk.** 1988. A new purification procedure of human kidney cathepsin H, its properties and kinetic data. *Biol. Chem. Hoppe Seyler* **369**(Suppl.): 175–183.

381. **Popovic, T., J. Brzin, A. Ritonja, B. Svetic, and V. Turk.** 1993. Rapid affinity chromatographic method for the isolation of human cathepsin H. *J. Chromatogr.* **615**: 243–249.

382. **Pupyshev, A. B. and T. A. Korolenko.** 1989. [Effective extraction of lysosomal enzymes with digitonin]. *Ukr. Biokhim. Zh.* **61**: 49–54.

383. **Recklies, A. D. and J. S. Mort.** 1982. A radioimmunoassay for total human cathepsin B. *Clin. Chim. Acta* **123**: 127–138.

384. **Rich, D. H., M. A. Brown, and A. J. Barrett.** 1986. Purification of cathepsin B by a new form of affinity chromatography. *Biochem. J.* **235**: 731–734.

385. **Rothe, G., S. Klingel, I. Assfalg Machleidt, W. Machleidt, C. Zirkelbach, R. B. Banati, W. F. Mangel, and G. Valet.** 1992. Flow cytometric analysis of protease activities in vital cells. *Biol. Chem. Hoppe Seyler* **373**: 547–554.

386. **Rowan, A. D., L. Mach, and J. S. Mort.** 1992. Antibodies to rat procathepsin B recognize the active mature enzyme. *Biol. Chem. Hoppe Seyler* **373**: 427–432.

387. **Sato, E., K. Matsuda, and Y. Kanaoka.** 1990. New fluorogenic substrates for microdetermination of cathepsin C. *Chem. Pharm. Bull. Tokyo* **38**: 2043–2044.

388. **Sawicki, G. and M. Warwas.** 1989. Cathepsin H from human placenta. *Acta Biochim. Pol.* **36**: 343–351.

389. **Schwartz, W. N. and A. J. Barrett.** 1980. Human cathepsin H. *Biochem. J.* **191**: 487–497.

390. **Scott, R. P., V. Ninjoor, and P. N. Srivastava.** 1987. Isolation and characterization of cathepsin B from rabbit testis. *J. Reprod. Fertil.* **79**: 67–74.

391. **Seeler, B. J., M. J. Horton, C. M. Szego, and R. J. DeLange.** 1988. Monoclonal antibody toward lysosomal cathepsin B cross-reacts preferentially with distinct histone classes. *Int. J. Biochem.* **20**: 1089–1106.

392. **Singh, H. and G. Kalnitsky.** 1978. Separation of a new α-N-benzoylarginine-β-naphthylamide hydrolase from cathepsin B1. Purification, characterization, and properties of both enzymes from rabbit lung. *J. Biol. Chem.* **253**: 4319–4326.

393. **Stauber, W., V. Fritz, B. Dahlmann, F. Gauthier, H. Kirschke, and R. Ulrich.** 1985. Fluorescence methods for localizing proteinases and proteinase inhibitors in skeletal muscle. *Histochem. J.* **17**: 787–796.

394. **Swanson, A. A., B. J. Martin, and S. S. Spicer.** 1974. Human placental cathepsin B1. Isolation and some physical properties. *Biochem. J.* **137**: 223–228.

395. **Takahashi, K., M. Isemura, and T. Ikenaka.** 1979. Isolation and characterization of three forms of cathepsin B from porcine liver. *J. Biochem. Tokyo* **85**: 1053–1060.

396. **Takahashi, S., K. Murakami, and Y. Miyake.** 1981. Purification and characterization or porcine kidney cathepsin B. *J. Biochem. Tokyo* **90**: 1677–1684.

397. **Takahashi, T., A. H. Dehdarani, P. G. Schmidt, and J. Tang.** 1984. Cathepsins B and H from porcine spleen. Purification, polypeptide chain arrangements, and carbohydrate content. *J. Biol. Chem.* **259**: 9874–9882.

398. **Takahashi, T., P. G. Schmidt, and J. Tang.** 1984. Novel carbohydrate structures of cathepsin B from porcine spleen. *J. Biol. Chem.* **259**: 6059–6062.

399. **Takahashi, T., S. Yonezawa, A. H. Dehdarani, and J. Tang.** 1986. Comparative studies of two cathepsin B isozymes from porcine spleen. Isolation, polypeptide chain arrangements, and enzyme specificity. *J. Biol. Chem.* **261**: 9368–9374.

400. **Takeda, A., Y. Nakamura, and Y. Aoki.** 1992. Enzyme-linked immunosorbent assay for the detection of cathepsin–kininogen complexes in human plasma. *J. Immunol. Methods* **147**: 217–223.

401.	**Towatari, T., K. Tanaka, D. Yoshikawa, and N. Katunuma.** 1976. Separation of a new protease from cathepsin B1 of rat liver lysosomes. *FEBS Lett.* **67**: 284–288.

402.	**Turnsek, T., I. Kregar, and D. Lebez.** 1975. Acid sulphydryl protease from calf lymph nodes. *Biochim. Biophys. Acta* **403**: 514–520.

403.	**van Noorden, C. J. and I. M. Vogels.** 1986. Enzyme histochemical reactions in unfixed and undecalcified cryostat sections of mouse knee joints with special reference to arthritic lesions. *Histochemistry* **86**: 127–133.

404.	**Wada, K. and T. Tanabe.** 1988. Purification and characterization of chicken liver cathepsin B. *J. Biochem. Tokyo* **104**: 472–476.

405.	**Wardale, R. J., R. A. Maciewicz, and D. J. Etherington.** 1986. Monoclonal antibodies to rabbit liver cathepsin B. *Biosci. Rep.* **6**: 639–646.

406.	**Wheeler, T. L. and M. Koohmaraie.** 1991. A modified procedure for simultaneous extraction and subsequent assay of calcium-dependent and lysosomal protease systems from a skeletal muscle biopsy. *J. Anim. Sci.* **69**: 1559–1565.

407.	**Wiederanders, B., H. Kirschke, and S. Schaper.** 1986. The azocasein–urea–pepstatin assay discriminates between lysosomal proteinases. *Biomed. Biochim. Acta* **45**: 1477–1483.

408.	**Yamamoto, K., O. Kamata, and Y. Kato.** 1984. Separation and properties of three forms of cathepsin H-like cysteine proteinase from rat spleen. *J. Biochem. Tokyo* **95**: 477–484.

409.	**Yamashita, M. and S. Konagaya.** 1990. Purifi-cation and characterization of cathepsin L from the white muscle of chum salmon, *Oncorhynchus keta. Comp. Biochem. Physiol. B* **96**: 247–252.

410.	**Yokozeki, H., T. Hibino, and K. Sato.** 1987. Partial purification and characterization of cysteine proteinases in eccrine sweat. *Am. J. Physiol.* **252**: R1119-R1129.

411.	**Zhloba, A. A.** 1986. [Purification, identification and properties of cysteine cathepsins from animal tissues]. *Ukr. Biokhim. Zh.* **58**: 100–113.

412.	**Zvonar, T., I. Kregar, and V. Turk.** 1979. Isolation of cathepsin B and α-N-benzoylarginine-β-naphthylamide hydrolase by covalent chromatography on activated thiol sepharose. *Croat. Chem. Acta* **52**: 411–416.

Reaction with inhibitors

413.	**Abe, K., H. Kondo, H. Watanabe, Y. Emori, and S. Arai.** 1991. Oryzacystatins as the first well-defined cystatins of plant origin and their target proteinases in rice seeds. *Biomed. Biochim. Acta* **50**: 637–641.

414.	**Abrahamson, M., A. J. Barrett, G. Salvesen, and A. Grubb.** 1986. Isolation of six cysteine proteinase inhibitors from human urine. Their physicochemical and enzyme kinetic properties and concentrations in biological fluids. *J. Biol. Chem.* **261**: 11282–11289.

415.	**Abrahamson, M., R. W. Mason, H. Hansson, D. J. Buttle, A. Grubb, and K. Ohlsson.** 1991. Human cystatin C. role of the *N*-terminal segment in the inhibition of human cysteine proteinases and in its inactivation by leucocyte elastase. *Biochem. J.* **273**: 621–626.

416.	**Ahmed, N. K., L. A. Martin, L. M. Watts, J. Palmer, L. Thornburg, J. Prior, and R. E. Esser.** 1992. Peptidyl fluoromethyl ketones as inhibitors of cathepsin B. Implication for treatment of rheumatoid arthritis. *Biochem. Pharmacol.* **44**: 1201–1207.

417.	**Anastasi, A., M. A. Brown, A. A. Kembhavi, M. J. Nicklin, C. A. Sayers, D. C. Sunter, and A. J. Barrett.** 1983. Cystatin, a protein inhibitor of cysteine proteinases. Improved purification from egg white, characterization, and detection in chicken serum. *Biochem. J.* **211**: 129–138.

418.	**Angliker, H., J. Anagli, and E. Shaw.** 1992. Inactivation of calpain by peptidyl fluoromethyl ketones. *J. Med. Chem.* **35**: 216–220.

419.	**Angliker, H., P. Wikstrom, H. Kirschke, and E. Shaw.** 1989. The inactivation of the cysteinyl exopeptidases cathepsin H and C by affinity-labelling reagents. *Biochem. J.* **262**: 63–68.

420.	**Angliker, H., P. Wikstrom, P. Rauber, and E. Shaw.** 1987. The synthesis of lysylfluoromethanes and their properties as inhibitors of trypsin, plasmin and cathepsin B. *Biochem. J.* **241**: 871–875.

421.	**Angliker, H., A. Zumbrunn, and E. Shaw.** 1991. Synthesis of histidine-containing dipeptide affinity-labelling agents. Relative inactivation rates of cathepsins B and L. *Int. J. Pept. Protein. Res.* **38**: 346–349.

422.	**Assfalg Machleidt, I., A. Billing, D. Frohlich, D. Nast Kolb, T. Joka, M. Jochum, and W. Machleidt.** 1992. The role of the kininogens as cysteine proteinase inhibitors in local and systemic inflammation. *Agents Actions Suppl.* **38**: 312–321.

423.	**Auerswald, E. A., G. Genenger, I. Assfalg Machleidt, J. Kos, and W. Bode.** 1989. Synthesis of a (desSer1 Ile29 Leu89) chicken cystatin gene, expression in *E. coli* as fusion protein and its isolation. *FEBS Lett.* **243**: 186–192.

424.	**Auerswald, E. A., G. Genenger, I. Assfalg Machleidt, W. Machleidt, R. A. Engh, and H. Fritz.** 1992. Recombinant chicken egg white cystatin variants of the QLVSG region. *Eur. J. Biochem.* **209**: 837–845.

425.	**Auerswald, E. A., G. Genenger, R. Mentele, S. Lenzen, I. Assfalg Machleidt, L. Mitschang, H. Oschkinat, and H. Fritz.** 1991. Purification and characterization of a chicken egg white cystatin variant expressed in an *Escherichia coli* pIN-III−ompA system. *Eur. J. Biochem.* **200**: 131–138.

426. **Auerswald, E. A., D. Rossler, R. Mentele, and I. Assfalg Machleidt.** 1993. Cloning, expression and characterization of human kininogen domain 3. *FEBS Lett.* **321**: 93–97.

427. **Baici, A. and M. Gyger Marazzi.** 1982. The slow, tight-binding inhibition of cathepsin B by leupeptin. A hysteretic effect. *Eur. J. Biochem.* **129**: 33–41.

428. **Balbin, M., J. P. Freije, M. Abrahamson, G. Velasco, A. Grubb, and C. Lopez Otin.** 1993. A sequence variation in the human cystatin D gene resulting in an amino acid (Cys/Arg) polymorphism at the protein level. *Hum. Genet.* **90**: 668–669.

429. **Balbin, M., A. Grubb, and M. Abrahamson.** 1993. An Ala/Thr variation in the coding region of the human cystatin C gene (CST3) detected as a SstII polymorphism. *Hum. Genet.* **92**: 206–207.

430. **Barrett, A. J., M. E. Davies, and A. Grubb.** 1984. The place of human γ-trace (cystatin C) amongst the cysteine proteinase inhibitors. *Biochem. Biophys. Res. Commun.* **120**: 631–636.

431. **Barrett, A. J., A. A. Kembhavi, M. A. Brown, H. Kirschke, C. G. Knight, M. Tamai, and K. Hanada.** 1982. L-*trans*-Epoxysuccinyl-leucylamido(4-guanidino)butane (E-64) and its analogues as inhibitors of cysteine proteinases including cathepsins B, H and L. *Biochem. J.* **201**: 189–198.

432. **Barrett, A. J., A. A. Kembhavi, and K. Hanada.** 1981. E-64 [L-*trans*-Epoxysuccinyl-leucyl-amido(4-guanidino)butane] and related epoxides as inhibitors of cysteine proteinases. *Acta Biol. Med. Ger.* **40**: 1513–1517.

433. **Berg, M. J. and N. Marks.** 1991. Brain cysteine proteinase inhibitors II: evidence that a 21 kDa papain-binding component resembles ras p21. *J. Neurosci. Res.* **30**: 391–397.

434. **Bige, L., A. Ouali, and C. Valin.** 1985. Purification and characterization of a low molecular weight cysteine proteinase inhibitor from bovine muscle. *Biochim. Biophys. Acta* **843**: 269–275.

435. **Bjork, I., K. Ylinenjarvi, and P. Lindahl.** 1990. Equilibrium and kinetic studies of the interaction of chicken cystatin with four cysteine proteinases. *Biol. Chem. Hoppe Seyler* **371**(Suppl.): 119–124.

436. **Bromme, D., B. Bartels, H. Kirschke, and S. Fittkau.** 1989. Peptide methyl ketones as reversible inhibitors of cysteine proteinases. *J. Enzym. Inhib.* **3**: 13–21.

437. **Bromme, D. and H. U. Demuth.** 1994. *N,O*-Diacyl hydroxamates as selective and irreversible inhibitors of cysteine proteinases. *Methods Enzymol.* **244**: 671–685.

438. **Bromme, D. and H. Kirschke.** 1993. *N*-Peptidyl-*O*-carbamoyl amino acid hydroxamates: irreversible inhibitors for the study of the S2′ specificity of cysteine proteinases. *FEBS Lett.* **322**: 211–214.

439. **Bromme, D., U. Neumann, H. Kirschke, and H. U. Demuth.** 1993. Novel *N*-peptidyl-*O*-acyl hydroxamates: selective inhibitors of cysteine proteinases. *Biochim. Biophys. Acta* **1202**: 271–276.

440. **Bromme, D., R. Rinne, and H. Kirschke.** 1991. Tight-binding inhibition of cathepsin S by cystatins. *Biomed. Biochim. Acta* **50**: 631–635.

441. **Bromme, D., A. Schierhorn, H. Kirschke, B. Wiederanders, A. Barth, S. Fittkau, and H. U. Demuth.** 1989. Potent and selective inactivation of cysteine proteinases with *N*-peptidyl-*O*-acyl hydroxylamines. *Biochem. J.* **263**: 861–866.

442. **Brzin, J., M. Kopitar, V. Turk, and W. Machleidt.** 1983. Protein inhibitors of cysteine proteinases. I. Isolation and characterization of stefin, a cytosolic protein inhibitor of cysteine proteinases from human polymorphonuclear granulocytes. *Hoppe Seylers Z. Physiol. Chem.* **364**: 1475–1480.

443. **Brzin, J., T. Popovic, M. Drobnic Kosorok, M. Kotnik, and V. Turk.** 1988. Inhibitors of cysteine proteinases from potato. *Biol. Chem. Hoppe Seyler* **369**(Suppl.): 233–238.

444. **Brzin, J., T. Popovic, V. Turk, U. Borchart, and W. Machleidt.** 1984. Human cystatin, a new protein inhibitor of cysteine proteinases. *Biochem. Biophys. Res. Commun.* **118**: 103–109.

445. **Buttle, D. J., M. Murata, C. G. Knight, and A. J. Barrett.** 1992. CA074 methyl ester: a proinhibitor for intracellular cathepsin B. *Arch. Biochem. Biophys.* **299**: 377–380.

446. **Buttle, D. J., J. Saklatvala, and A. J. Barrett.** 1993. The inhibition of interleukin 1-stimulated cartilage proteoglycan degradation by cysteine endopeptidase inactivators. *Agents Actions Suppl.* **39**: 161–165.

447. **Buys, C. H., J. M. Bouma, M. Gruber, and E. Wisse.** 1978. Induction of lysosomal storage by suramin. *Naunyn Schmiedebergs Arch. Pharmacol.* **304**: 183–190.

448. **Crawford, C., R. W. Mason, P. Wikstrom, and E. Shaw.** 1988. The design of peptidyldiazomethane inhibitors to distinguish between the cysteine proteinases calpain II, cathepsin L and cathepsin B. *Biochem. J.* **253**: 751–758.

449. **Cullen, B. M., A. McGinty, B. Walker, J. Nelson, I. Halliday, J. R. Bailie, and G. Kay.** 1990. Synthesis and activity of a novel, irreversible inhibitor of cathepsin B. *Biochem. Soc. Trans.* **18**: 315–316.

450. **Dahlmann, B., L. Kuehn, P. C. Heinrich, H. Kirschke, and B. Wiederanders.** 1989. ATP-activated, high-molecular-mass proteinase-I from rat skeletal muscle is a cysteine proteinase-α_1-macroglobulin complex. *Biochim. Biophys. Acta* **991**: 253–262.

451. **Demuth, H. U., A. Stockel, A. Schierhorn, S. Fittkau, H. Kirschke, and D. Bromme.** 1993. Inhibition of aminopeptidases by *N*-aminoacyl-*O*-4-nitrobenzoyl hydroxamates. *Biochim. Biophys. Acta* **1202**: 265–270.

452. **Dzodzuashvili, V. U.** 1988. [Effect of contrical on the lysosomal proteinase activity of the gastric mucosa of rats in experimental ulcer]. *Vopr. Med. Khim.* **34**: 46–49.

453. **Ermer, A., H. Baumann, G. Steude, K. Peters, S. Fittkau, P. Dolaschka, and N. C. Genov.** 1990. Peptide diazomethyl ketones are inhibitors of subtilisin-type serine proteases. *J. Enzym. Inhib.* **4**: 35–42.

454. **Esser, R. E., L. M. Watts, R. A. Angelo, L. P. Thornburg, J. J. Prior, and J. T. Palmer.** 1993. The effects of fluoromethyl ketone inhibitors of cathepsin B on adjuvant induced arthritis. *J. Rheumatol.* **20**: 1176–1183.

455. **Evans, H. J. and A. J. Barrett.** 1987. A cystatin-like cysteine proteinase inhibitor from venom of the African puff adder (*Bitis arietans*). *Biochem. J.* **246**: 795–797.

456. **Fox, T., E. de Miguel, J. S. Mort, and A. C. Storer.** 1992. Potent slow-binding inhibition of cathepsin B by its propeptide. *Biochemistry* **31**: 12571–12576.

457. **Freije, J. P., M. Balbin, M. Abrahamson, G. Velasco, H. Dalboge, A. Grubb, and C. Lopez Otin.** 1993. Human cystatin D. cDNA cloning, characterization of the *Escherichia coli* expressed inhibitor, and identification of the native protein in saliva. *J. Biol. Chem.* **268**: 15737–15744.

458. **Gauthier, F., M. Pagano, F. Esnard, H. Mouray, and R. Engler.** 1983. A heat stable low molecular weight inhibitor of lysosomal cysteine proteinases in human serum. *Biochem. Biophys. Res. Commun.* **110**: 449–455.

459. **Genenger, G., S. Lenzen, R. Mentele, I. Assfalg Machleidt, and E. A. Auerswald.** 1991. Recombinant Q53E- and Q53N–chicken egg white cystatin variants inhibit papain, actinidin and cathepsin B. *Biomed. Biochim. Acta* **50**: 621–625.

460. **Giordano, C., R. Calabretta, C. Gallina, V. Consalvi, and R. Scandurra.** 1991. 1-Peptidyl-2-haloacetyl hydrazines as active site directed inhibitors of papain and cathepsin B. *Farmaco.* **46**: 1497–1516.

461. **Giraldi, T., G. Sava, L. Perissin, and S. Zorzet.** 1984. Primary tumor growth and formation of spontaneous lung metastases in mice bearing Lewis carcinoma treated with proteinase inhibitors. *Anticancer Res.* **4**: 221–224.

462. **Gounaris, A. D., M. A. Brown, and A. J. Barrett.** 1984. Human plasma α-cysteine proteinase inhibitor. Purification by affinity chromatography, characterization and isolation of an active fragment. *Biochem. J.* **221**: 445–452.

463. **Gour Salin, B. J., P. Lachance, C. Plouffe, A. C. Storer, and R. Menard.** 1993. Epoxysuccinyl dipeptides as selective inhibitors of cathepsin B. *J. Med. Chem.* **36**: 720–725.

464. **Green, G. D., A. A. Kembhavi, M. E. Davies, and A. J. Barrett.** 1984. Cystatin-like cysteine proteinase inhibitors from human liver. *Biochem. J.* **218**: 939–946.

465. **Green, G. D. and E. Shaw.** 1981. Peptidyl diazomethyl ketones are specific inactivators of thiol proteinases. *J. Biol. Chem.* **256**: 1923–1928.

466. **Greenbaum, L. M. and K. Yamafuji.** 1966. The *in vitro* inactivation and formation of plasma kinins by spleen cathepsins. *Br. J. Pharmacol.* **27**: 230–238.

467. **Grinde, B., I. J. Galpin, A. H. Wilby, and R. J. Beynon.** 1983. Inhibition of hepatic protein degradation by synthetic analogues of chymostatin. *J. Biol. Chem.* **258**: 10821–10823.

468. **Grubb, A., M. Abrahamson, I. Olafsson, J. Trojnar, R. Kasprzykowska, F. Kasprzykowski, and Z. Grzonka.** 1990. Synthesis of cysteine proteinase inhibitors structurally based on the proteinase interacting N-terminal region of human cystatin C. *Biol. Chem. Hoppe Seyler* **371**(Suppl.): 137–144.

469. **Hall, A., M. Abrahamson, A. Grubb, J. Trojnar, P. Kania, R. Kasprzykowska, and F. Kasprzykowski.** 1992. Cystatin C based peptidyl diazomethanes as cysteine proteinase inhibitors: influence of the peptidyl chain length. *J. Enzym. Inhib.* **6**: 113–123.

470. **Hall, A., H. Dalboge, A. Grubb, and M. Abrahamson.** 1993. Importance of the evolutionarily conserved glycine residue in the N-terminal region of human cystatin C (Gly-11) for cysteine endopeptidase inhibition. *Biochem. J.* **291**: 123–129.

471. **Hanada, K., M. Tamai, M. Yamagishi, S. Ohmura, J. Sawada, and I. Tanaka.** 1978. Studies on thiol protease inhibitors. Part I. Isolation and characterization of E-64, a new thiol proteinase inhibitor. *Agric. Biol. Chem.* **42**: 523–528.

472. **Harth, G., N. Andrews, A. A. Mills, J. C. Engel, R. Smith, and J. H. McKerrow.** 1993. Peptide-fluoromethyl ketones arrest intracellular replication and intercellular transmission of Trypanosoma cruzi. *Mol. Biochem. Parasitol.* **58**: 17–24.

473. **Hashida, S., E. Kominami, and N. Katunuma.** 1982. Inhibitions of cathepsin B and cathepsin L by E-64 *in vivo*. II. Incorporation of [3H]E-64 into rat liver lysosomes *in vivo*. *J. Biochem. Tokyo* **91**: 1373–1380.

474. **Hashida, S., T. Towatari, E. Kominami, and N. Katunuma.** 1980. Inhibitions by E-64 derivatives of rat liver cathepsin B and cathepsin L *in vitro* and *in vivo*. *J. Biochem. Tokyo* **88**: 1805–1811.

475. **Hibino, T., K. Fukuyama, and W. L. Epstein.** 1980. *In vitro* and *in vivo* inhibition of rat liver cathepsin L by epidermal proteinase inhibitor. *Biochem. Biophys. Res. Commun.* **93**: 440–447.

476. **Hirado, M., D. Iwata, M. Niinobe, and S. Fujii.** 1981. Purification and properties of thiol protease inhibitor from rat liver cytosol. *Biochim. Biophys. Acta* **669**: 21–27.

477. **Hirado, M., M. Niinobe, and S. Fujii.** 1983. Isolation and immunological studies of high and low molecular weight cysteine proteinase inhibitors in bovine serum. *Biochim. Biophys. Acta* **757**: 196–201.

478. **Hirado, M., M. Niinobe, and S. Fujii.** 1984. Purification and characterization of a bovine colostrum low molecular weight cysteine proteinase inhibitor. *J. Biochem. Tokyo* **96**: 51–58.

479. **Hiwasa, T., J. Fujita Yoshigaki, M. Shirouzu, H. Koide, T. Sawada, S. Sakiyama, and S. Yokoyama.** 1993. c-Ha-Ras mutants with point mutations in Gln-Val-Val region have reduced inhibitory activity toward cathepsin B. *Cancer Lett.* **69**: 161–165.

480. **Hiwasa, T., S. Sakiyama, S. Noguchi, J. M. Ha, T. Miyazawa, and S. Yokoyama.** 1987. Degradation of a cAMP-binding protein is inhibited by human c-Ha-ras gene products. *Biochem. Biophys. Res. Commun.* **146**: 731–738.

481. **Hiwasa, T., S. Yokoyama, J. Fujita, E. Kominami, N. Katunuma, and S. Sakiyama.** 1989. Potentiation of cysteine proteinase-inhibitor activity of full-length Ha-ras oncogene products by denaturation-renaturation. *Biochem. Int.* **19**: 593–601.

482. **Hiwasa, T., S. Yokoyama, J. M. Ha, S. Noguchi, and S. Sakiyama.** 1987. c-Ha-ras gene products are potent inhibitors of cathepsins B and L. *FEBS Lett.* **211**: 23–26.

483. **Hu, L. Y. and R. H. Abeles.** 1990. Inhibition of cathepsin B and papain by peptidyl α-keto esters, α-keto amides, α-diketones, and α-keto acids. *Arch. Biochem. Biophys.* **281**: 271–274.

484. **Isemura, S., E. Saitoh, S. Ito, M. Isemura, and K. Sanada.** 1984. Cystatin S: a cysteine proteinase inhibitor of human saliva. *J. Biochem. Tokyo* **96**: 1311–1314.

485. **Isemura, S., E. Saitoh, and K. Sanada.** 1986. Characterization of a new cysteine proteinase inhibitor of human saliva, cystatin SN, which is immunologically related to cystatin S. *FEBS Lett.* **198**: 145–149.

486. **Jarvinen, M.** 1979. Purification and some characteristics of two human serum proteins inhibiting papain and other thiol proteinases. *FEBS Lett.* **108**: 461–464.

487. **Kalsheker, N. A., A. R. Bradwell, and D. Burnett.** 1981. The inhibition of cathepsin B by plasma haptoglobin biochemistry (enzymes, metabolism). *Experientia* **37**: 447–448.

488. **Kirpichenok, L. N. and I. G. Shcherbak.** 1987. [Endogenous inhibitor of cysteine proteases from the human kidney]. *Ukr. Biokhim. Zh.* **59**: 10–15.

489. **Kirschke, H. and E. Shaw.** 1981. Rapid interaction of cathepsin L by Z-Phe-PheCHN12 and Z-Phe-AlaCHN2. *Biochem. Biophys. Res. Commun.* **101**: 454–458.

490. **Kirschke, H., P. Wikstrom, and E. Shaw.** 1988. Active center differences between cathepsins L and B: the S1 binding region. *FEBS Lett.* **228**: 128–130.

491. **Knight, C.G.** 1986. The characterization of enzyme inhibition. In *Proteinase Inhibitors*, A. J. Barrett and G. Salvesen (eds), pp. 23–51. Elsevier, Amsterdam.

492. **Kominami, E., S. Hashida, and N. Katunuma.** 1980. Inhibitions of degradation of rat liver aldolase and lactic dehydrogenase by N-[N-(L-3-*trans*-carboxyoxirane-2-carbonyl)-L-leucyl] agmatine or leupeptin *in vivo*. *Biochem. Biophys. Res. Commun.* **93**: 713–719.

493. **Kondo, H., K. Abe, I. Nishimura, H. Watanabe, Y. Emori, and S. Arai.** 1990. Two distinct cystatin species in rice seeds with different specificities against cysteine proteinases. Molecular cloning, expression, and biochemical studies on oryzacystatin-II. *J. Biol. Chem.* **265**: 15832–15837.

494. **Kopitar, M., J. Brzin, P. Locnikar, and V. Turk.** 1981. Inhibitory mechanism of sericystatin, an intracellular proteinase inhibitor, reacting with cysteine proteinases. *Hoppe Seylers Z. Physiol. Chem.* **362**: 1411–1414.

495. **Kopitar, M., J. Brzin, T. Zvonar, P. Locnikar, I. Kregar, and V. Turk.** 1978. Inhibition studies of an intracellular inhibitor on thiol proteinases. *FEBS Lett.* **91**: 355–359.

496. **Kopitar, M., A. Ritonja, T. Popovic, D. Gabrijelcic, I. Krizaj, and V. Turk.** 1989. A new type of low-molecular mass cysteine proteinase inhibitor from pig leukocytes. *Biol. Chem. Hoppe Seyler* **370**: 1145–1151.

497. **Kopitar, M., F. Stern, and N. Marks.** 1983. Cerebrocystatin suppresses degradation of myelin basic protein by purified brain cysteine proteinase. *Biochem. Biophys. Res. Commun.* **112**: 1000–1006.

498. **Korolenko, T. A., A. B. Pupyshev, and A. V. Mudrakovskaia.** 1987. [Study of intralysosomal protein catabolism using lysosomotropic agents–proteolysis and protease inhibitors]. *Vopr. Med. Khim.* **33**: 93–96.

499. **Kos, J., M. Dolinar, and V. Turk.** 1992. Isolation and characterisation of chicken L- and H-kininogens and their interaction with chicken cysteine proteinases and papain. *Agents Actions Suppl.* **38**: 331–339.

500. **Krantz, A.** 1994. Peptidyl (acyloxy)methanes as quiescent affinity labels for cysteine proteinases. *Methods Enzymol.* **244**: 656–671.

501. **Krantz, A., L. J. Copp, P. J. Coles, R. A. Smith, and S. B. Heard.** 1991. Peptidyl (acyloxy)methyl ketones and the quiescent affinity label concept: the departing group as a variable structural element in the design of inactivators of cysteine proteinases. *Biochemistry* **30**: 4678–4687.

502. **Kruze, K., K. Fehr, and A. Boni.** 1976. Effect of antirheumatic drugs on cathepsin B_1 from bovine spleen. *Z. Rheumatol.* **35**: 95–102.

503. **Kuratomi, Y., S. Akiyama, M. Ono, N. Shiraishi, T. Shimada, S. Ohkuma, and M. Kuwano.** 1986. Thioridazine enhances lysosomal accumulation of epidermal growth factor and toxicity of conjugates of epidermal growth factor with *Pseudomonas* exotoxin. *Exp. Cell Res.* **162**: 436–448.

504. **Leary, R. and E. Shaw.** 1977. Inactivation of cathepsin B$_1$ by diazomethyl ketones. *Biochem. Biophys. Res. Commun.* **79**: 926–931.

505. **Lenarcic, B., A. Ritonja, B. Turk, I. Dolenc, and V. Turk.** 1992. Characterization and structure of pineapple stem inhibitor of cysteine proteinases. *Biol. Chem. Hoppe Seyler* **373**: 459–464.

506. **Lenney, J. F., J. R. Tolan, W. J. Sugai, and A. G. Lee.** 1979. Thermostable endogenous inhibitors of cathepsins B and H. *Eur. J. Biochem.* **101**: 153–161.

507. **Li, Z., G. S. Patil, Z. E. Golubski, H. Hori, K. Tehrani, J. E. Foreman, D. D. Eveleth, R. T. Bartus, and J. C. Powers.** 1993. Peptide α-keto ester, α-keto amide, and α-keto acid inhibitors of calpains and other cysteine proteases. *J. Med. Chem.* **36**: 3472–3480.

508. **Machleidt, W., U. Thiele, I. Assfalg Machleidt, D. Forger, and E. A. Auerswald.** 1991. Molecular mechanism of inhibition of cysteine proteinases by their protein inhibitors: kinetic studies with natural and recombinant variants of cystatins and stefins. *Biomed. Biochim. Acta* **50**: 613–620.

509. **Marks, N., M. J. Berg, R. C. Makofske, and W. Danho.** 1990. Synthetic domains of cystatins linked to enkephalins are novel inhibitors of brain cathepsins L/B. *Peptides* **11**: 679–682.

510. **Marks, N., M. J. Berg, R. C. Makofske, J. Swistok, E. J. Simon, D. Ofri, K. Del Compare, and W. Danho.** 1992. Chimeric enkephalin–cystatins: opioid binding and structure-activity relationships of inhibitory domains. *Pept. Res.* **5**: 194–200.

511. **Marks, N., F. Stern, L. M. Chi, and M. J. Berg.** 1988. Diversity of rat brain cysteine proteinase inhibitors: isolation of low-molecular-weight cystatins and a higher-molecular weight T-kininogenlike glycoprotein. *Arch. Biochem. Biophys.* **267**: 448–458.

512. **Mason, R. W.** 1989. Interaction of lysosomal cysteine proteinases with α$_2$-macroglobulin: conclusive evidence for the endopeptidase activities of cathepsins B and H. *Arch. Biochem. Biophys.* **273**: 367–374.

513. **Mason, R. W., L. T. Bartholomew, and B. S. Hardwick.** 1989. The use of benzyloxycarbonyl [^{125}I]iodotyrosylalanyldiazomethane as a probe for active cysteine proteinases in human tissues. *Biochem. J.* **263**: 945–949.

514. **McConnell, R. M., J. L. York, D. Frizzell, and C. Ezell.** 1993. Inhibition studies of some serine and thiol proteinases by new leupeptin analogues. *J. Med. Chem.* **36**: 1084–1089.

515. **Michaud, D., B. Nguyen Quoc, and S. Yelle.** 1993. Selective inhibition of Colorado potato beetle cathepsin H by oryzacystatins I and II. *FEBS Lett.* **331**: 173–176.

516. **Minakata, K. and M. Asano.** 1985. Acidic cysteine proteinase inhibitor in seminal plasma. *Biol. Chem. Hoppe Seyler* **366**: 15–18.

517. **Mittal, S., N. Raghav, S. Pal, R. C. Kamboj, and H. Singh.** 1993. Effect of some pesticides/weedicides on cathepsin B activity and lysosomal membrane. *Indian J. Biochem. Biophys.* **30**: 187–190.

518. **Moreau, T., F. Esnard, N. Gutman, P. Degand, and F. Gauthier.** 1988. Cysteine-proteinase-inhibiting function of T kininogen and of its proteolytic fragments. *Eur. J. Biochem.* **173**: 185–190.

519. **Moreau, T., N. Gutman, D. Faucher, and F. Gauthier.** 1989. Limited proteolysis of T-kininogen (thiostatin). Release of comparable fragments by different endopeptidases. *J. Biol. Chem.* **264**: 4298–4303.

520. **Moreau, T., J. Hoebeke, G. Lalamanach, M. Hattab, and F. Gauthier.** 1990. Simulation of the inhibitory cystatin surface by a synthetic peptide. *Biochem. Biophys. Res. Commun.* **167**: 117–122.

521. **Muller Esterl, W., H. Fritz, W. Machleidt, A. Ritonja, J. Brzin, M. Kotnik, V. Turk, J. Kellermann, and F. Lottspeich.** 1985. Human plasma kininogens are identical with α-cysteine proteinase inhibitors. Evidence from immunological, enzymological and sequence data. *FEBS Lett.* **182**: 310–314.

522. **Murata, M., S. Miyashita, C. Yokoo, M. Tamai, K. Hanada, K. Hatayama, T. Towatari, T. Nikawa, and N. Katunuma.** 1991. Novel epoxysuccinyl peptides. Selective inhibitors of cathepsin B, in vitro. *FEBS Lett.* **280**: 307–310.

523. **Nakao, H., Y. Kurita, R. Tsuboi, K. Takamori, and H. Ogawa.** 1986. Induction and inhibition of cathepsin B and hemoglobin-hydrolase activity in murine B16 melanoma by thiol protease inhibitors. *Comp. Biochem. Physiol. B* **85**: 435–437.

524. **Nicklin, M. J. and A. J. Barrett.** 1984. Inhibition of cysteine proteinases and dipeptidyl peptidase I by egg-white cystatin. *Biochem. J.* **223**: 245–253.

525. **Nikawa, T., T. Towatari, Y. Ike, and N. Katunuma.** 1989. Studies on the reactive site of the cystatin superfamily using recombinant cystatin A mutants. Evidence that the QVVAG region is not essential for cysteine proteinase inhibitory activities. *FEBS Lett.* **255**: 309–314.

526. **Ohkubo, I., K. Kurachi, T. Takasawa, H. Shiokawa, and M. Sasaki.** 1984. Isolation of a human cDNA for α$_2$-thiol proteinase inhibitor and its identity with low molecular weight kininogen. *Biochemistry* **23**: 5691–5697.

527. **Ohshita, T., E. Kominami, K. Ii, and N. Katunuma.** 1986. Effect of starvation and refeeding on autophagy and heterophagy in rat liver. *J. Biochem. Tokyo* **100**: 623–632.

528. **Ohshita, T., T. Nikawa, T. Towatari, and N. Katunuma.** 1992. Effects of selective inhibition of cathepsin B and general inhibition of cysteine proteinases on lysosomal proteolysis in rat liver *in vivo* and *in vitro. Eur. J. Biochem.* **209**: 223–231.

529. **Ohtani, O., K. Fukuyama, and W. L. Epstein.** 1982. Differences in the behavior of SH-protease inhibitor of rat epidermis on cathepsins B and H. *Comp. Biochem. Physiol. B* **73**: 231–233.

530. **Ohtani, O., K. Fukuyama, and W. L. Epstein.** 1982. Further characterization of cysteine proteinase inhibitors purified from rat and human epidermis. *Biochim. Biophys. Acta* **707**: 21–27.

531. **Ohtani, O., K. Fukuyama, and W. L. Epstein.** 1982. Biochemical properties of thiol proteinase inhibitor purified from psoriatic scales. *J. Invest. Dermatol.* **78**: 280–284.

532. **Pagano, M. and R. Engler.** 1982. Inhibition of human liver cathepsin L by α-thiol proteinase inhibitor. *FEBS Lett.* **138**: 307–310.

533. **Pagano, M. and R. Engler.** 1984. α_2 high molecular mass cysteine proteinase inhibitor: HM_r α_2-CPI. An inhibitor of human liver cathepsin H as probed by kinetic study. *FEBS Lett.* **166**: 62–66.

534. **Pagano, M., R. Engler, M. Gelin, and M. F. Jayle.** 1980. Kinetic study of the interaction between rat haptoglobin and rat liver cathepsin B. *Can. J. Biochem.* **58**: 410–417.

535. **Pagano, M., F. Esnard, R. Engler, and F. Gauthier.** 1984. Inhibition of human liver cathepsin L by α_2 cysteine-proteinase inhibitor and the low-M_r cysteine proteinase inhibitor from human serum. *Biochem. J.* **220**: 147–155.

536. **Pagano, M., M. A. Nicola, and R. Engler.** 1982. Inhibition of cathepsin L and B by haptoglobin, the haptoglobin–hemoglobin complex, and asialohaptoglobin. "*In vitro*" studies in the rat. *Can. J. Biochem.* **60**: 631–637.

537. **Picken, P. P., D. J. Guthrie, and B. Walker.** 1990. Inhibition of bovine cathepsin B by amino acid-derived nitriles. *Biochem. Soc. Trans.* **18**: 316.

538. **Pike, R. N., T. H. Coetzer, and C. Dennison.** 1992. Proteolytically active complexes of cathepsin L and a cysteine proteinase inhibitor; purification and demonstration of their formation *in vitro. Arch. Biochem. Biophys.* **294**: 623–629.

539. **Pliura, D. H., B. J. Bonaventura, R. A. Smith, P. J. Coles, and A. Krantz.** 1992. Comparative behaviour of calpain and cathepsin B toward peptidyl acyloxymethyl ketones, sulphonium methyl ketones and other potential inhibitors of cysteine proteinases. *Biochem. J.* **288**: 759–762.

540. **Pontremoli, S., E. Melloni, F. Salamino, B. Sparatore, M. Michetti, and B. L. Horecker.** 1983. Endogenous inhibitors of lysosomal proteinases. *Proc. Natl Acad. Sci. U. S. A.* **80**: 1261–1264.

541. **Pontremoli, S., E. Melloni, F. Salamino, B. Sparatore, M. Michetti, and B. L. Horecker.** 1984. Interaction of rabbit liver cathepsin M and fructose 1,6-bisphosphatase converting enzyme with their endogenous inhibitors. *Arch. Biochem. Biophys.* **228**: 460–464.

542. **Popovic, T., J. Brzin, A. Ritonja, and V. Turk.** 1990. Different forms of human cystatin C. *Biol. Chem. Hoppe Seyler* **371**: 575–580.

543. **Rasnick, D.** 1985. Synthesis of peptide fluoromethyl ketones and the inhibition of human cathepsin B. *Anal. Biochem.* **149**: 461–465.

544. **Rauber, P., H. Angliker, B. Walker, and E. Shaw.** 1986. The synthesis of peptidylfluoromethanes and their properties as inhibitors of serine proteinases and cysteine proteinases. *Biochem. J.* **239**: 633–640.

545. **Rohrlich, S. T., H. Levy, and D. B. Rifkin.** 1985. Purification and characterization of a low molecular mass cysteine proteinase inhibitor from human amniotic fluid. *Biol. Chem. Hoppe Seyler* **366**: 147–155.

546. **Roughley, P. J., G. Murphy, and A. J. Barrett.** 1978. Proteinase inhibitors of bovine nasal cartilage. *Biochem. J.* **169**: 721–724.

547. **Salvesen, G., C. Parkes, M. Abrahamson, A. Grubb, and A. J. Barrett.** 1986. Human low-M_r kininogen contains three copies of a cystatin sequence that are divergent in structure and in inhibitory activity for cysteine proteinases. *Biochem. J.* **234**: 429–434.

548. **Sasaki, T., M. Kishi, M. Saito, T. Tanaka, N. Higuchi, E. Kominami, N. Katunuma, and T. Murachi.** 1990. Inhibitory effect of di- and tripeptidyl aldehydes on calpains and cathepsins. *J. Enzym. Inhib.* **3**: 195–201.

549. **Sawada, T., S. Sakiyama, and T. Hiwasa.** 1993. v-Ha-Ras insertion/deletion mutants with reduced protease-inhibitory activity have no transforming activity. *FEBS Lett.* **318**: 297–300.

550. **Schultz, R. M., P. Varma Nelson, R. Ortiz, K. A. Kozlowski, A. T. Orawski, P. Pagast, and A. Frankfater.** 1989. Active and inactive forms of the transition-state analog protease inhibitor leupeptin: explanation of the observed slow binding of leupeptin to cathepsin B and papain. *J. Biol. Chem.* **264**: 1497–1507.

551. **Shaw, E.** 1988. Peptidyl sulfonium salts. A new class of protease inhibitors. *J. Biol. Chem.* **263**: 2768–2772.

552. **Shaw, E., H. Angliker, P. Rauber, B. Walker, and P. Wikstrom.** 1986. Peptidyl fluoromethyl ketones as thiol protease inhibitors. *Biomed. Biochim. Acta* **45**: 1397–1403.

553. **Shaw, E. and R. T. Dean.** 1980. The inhibition of macrophage protein turnover by a selective inhibitor of thiol proteinases. *Biochem. J.* **186**: 385–390.

554. **Shaw, E., S. Mohanty, A. Colic, V. Stoka, and V. Turk.** 1993. The affinity-labelling of cathepsin S with peptidyl diazomethyl ketones. Comparison with the inhibition of cathepsin L and calpain. *FEBS Lett.* **334**: 340–342.

555. **Shaw, E., P. Wikstrom, and J. Ruscica.** 1983. An exploration of the primary specificity site of cathepsin B. *Arch. Biochem. Biophys.* **222**: 424–429.

556. **Shikimi, T. and M. Handa.** 1986. Inhibitory effect of human urinary trypsin inhibitor (urinastatin) on lysosomal thiol proteinases. *Jpn. J. Pharmacol.* **42**: 571–574.

557. **Smith, R. A., P. J. Coles, R. W. Spencer, L. J. Copp, C. S. Jones, and A. Krantz.** 1988. Peptidyl *O*-acyl hydroxamates: potent new inactivators of cathepsin B. *Biochem. Biophys. Res. Commun.* **155**: 1201–1206.

558. **Smith, R. A., L. J. Copp, S. L. Donnelly, R. W. Spencer, and A. Krantz.** 1988. Inhibition of cathepsin B by peptidyl aldehydes and ketones: slow-binding behavior of a trifluoromethyl ketone. *Biochemistry* **27**: 6568–6573.

559. **Smith, R. E., D. Rasnick, C. O. Burdick, K. J. Cho, J. C. Rose, and A. Vahratian.** 1988. Visualization of time-dependent inactivation of human tumor cathepsin B isozymes by a peptidyl fluoromethyl ketone using a fluorescent print technique. *Anticancer. Res.* **8**: 525–529.

560. **Spanier, A. M. and J. W. Bird.** 1982. Endogenous cathepsin B inhibitor activity in normal and myopathic red and white skeletal muscle. *Muscle Nerve* **5**: 313–320.

561. **Stone, S. R., D. Rennex, P. Wikstrom, E. Shaw, and J. Hofsteenge.** 1992. Peptidyldiazomethanes. A novel mechanism of interaction with prolyl endopeptidase. *Biochem. J.* **283**: 871–876.

562. **Sueyoshi, T., K. Enjyoji, T. Shimada, H. Kato, S. Iwanaga, Y. Bando, E. Kominami, and N. Katunuma.** 1985. A new function of kininogens as thiol-proteinase inhibitors: inhibition of papain and cathepsins B, H and L by bovine, rat and human plasma kininogens. *FEBS Lett.* **182**: 193–195.

563. **Takahashi, M., T. Tezuka, T. Towatari, and N. Katunuma.** 1990. Properties and nature of a cysteine proteinase inhibitor located in keratohyalin granules of rat epidermis. *FEBS Lett.* **267**: 261–264.

564. **Tamai, M., K. Matsumoto, S. Omura, I. Koyama, Y. Ozawa, and K. Hanada.** 1986. *In vitro* and *in vivo* inhibition of cysteine proteinases by EST, a new analog of E-64. *J. Pharmacobiodyn.* **9**: 672–677.

565. **Teno, N., S. Tsuboi, N. Itoh, H. Okamoto, and Y. Okada.** 1987. Significant effects of Z-Gln-Val-Val-OME, common sequences of thiol proteinase inhibitors on thiol proteinases. *Biochem. Biophys. Res. Commun.* **143**: 749–752.

566. **Towatari, T., T. Nikawa, M. Murata, C. Yokoo, M. Tamai, K. Hanada, and N. Katunuma.** 1991. Novel epoxysuccinyl peptides. A selective inhibitor of cathepsin B, *in vivo*. *FEBS Lett.* **280**: 311–315.

567. **Tsuchida, K., H. Aihara, K. Isogai, K. Hanada, and N. Shibata.** 1986. Degradation of myocardial structural proteins in myocardial infarcted dogs is reduced by Ep459, a cysteine proteinase inhibitor. *Biol. Chem. Hoppe Seyler* **367**: 39–45.

568. **Turk, B., I. Dolenc, V. Turk, and J. G. Bieth.** 1993. Kinetics of the pH-induced inactivation of human cathepsin L. *Biochemistry* **32**: 375–380.

569. **Turk, B., I. Krizaj, B. Kralj, I. Dolenc, T. Popovic, J. G. Bieth, and V. Turk.** 1993. Bovine stefin C, a new member of the stefin family. *J. Biol. Chem.* **268**: 7323–7329.

570. **Umezawa, H.** 1982. Low-molecular-weight enzyme inhibitors of microbial origin. *Annu. Rev. Microbiol.* **36**: 75–99.

571. **van Noorden, C. J. and V. Everts.** 1991. Selective inhibition of cysteine proteinases by Z-PheAlaCH$_2$F suppresses digestion of collagen by fibroblasts and osteoclasts. *Biochem. Biophys. Res. Commun.* **178**: 178–184.

572. **Verbanac, D., M. Zanetti, and D. Romeo.** 1993. Chemotactic and protease-inhibiting activities of antibiotic peptide precursors. *FEBS Lett.* **317**: 255–258.

573. **Wakamatsu, N., E. Kominami, and N. Katunuma.** 1982. Comparison of properties of thiol proteinase inhibitors from rat serum and liver. *J. Biol. Chem.* **257**: 14653–14656.

574. **Wakamatsu, N., E. Kominami, K. Takio, and N. Katunuma.** 1984. Three forms of thiol proteinase inhibitor from rat liver formed depending on the oxidation-reduction state of a sulfhydryl group. *J. Biol. Chem.* **259**: 13832–13838.

575. **Walker, B., B. M. Cullen, G. Kay, I. M. Halliday, A. McGinty, and J. Nelson.** 1992. The synthesis, kinetic characterization and application of a novel biotinylated affinity label for cathepsin B. *Biochem. J.* **283**: 449–453.

576. **Walker, B., N. McCarthy, A. Healy, T. Ye, and M. A. McKervey.** 1993. Peptide glyoxals: a novel class of inhibitor for serine and cysteine proteinases. *Biochem. J.* **293**: 321–323.

577. **Warwas, M. and G. Sawicki.** 1985. Cysteine proteinase inhibitors in human placenta. *Placenta.* **6**: 455–463.

578. **Watanabe, H., G. D. Green, and E. Shaw.** 1979. A comparison of the behavior of chymotrypsin and cathepsin B towards peptidyl diazomethyl ketones. *Biochem. Biophys. Res. Commun.* **89**: 1354–1360.

579. **Wikstrom, P., H. Kirschke, S. Stone, and E. Shaw.** 1989. The properties of peptidyl diazoethanes and chloroethanes as protease inactivators. *Arch. Biochem. Biophys.* **270**: 286–293.

580. **Wilcox, D. and R. W. Mason.** 1992. Inhibition of cysteine proteinases in lysosomes and whole cells. *Biochem. J.* **285**: 495–502.

581. **Yamamoto, K., O. Kamata, and Y. Kato.** 1984. Differential effects of anti-inflammatory agents on lysosomal cysteine proteinases cathepsins B and H from rat spleen. *Jpn. J. Pharmacol.* **35**: 253–258.

582. **Yamamoto, K., M. Takeda, and Y. Kato.** 1985. Characteristics of activation of cathepsin B by sodium salicylate and comparison of catalytic site properties of cathepsins B and H. *Jpn. J. Pharmacol.* **39**: 207–215.

583. **Yamamoto, K., M. Takeda, and Y. Kato.** 1985. Sodium salicylate activates cathepsin B but not cathepsin H from rat spleen. *Jpn. J. Pharmacol.* **38**: 215–218.

584. **Zumbrunn, A., S. Stone, and E. Shaw.** 1988. The synthesis and properties of peptidylmethylsulphonium salts with two cationic residues as potential inhibitors of prohormone processing. *Biochem. J.* **256**: 989–994.

Expression of the enzymes in cancer

585. **Achkar, C., Q. M. Gong, A. Frankfater, and A. S. Bajkowski.** 1990. Differences in targeting and secretion of cathepsins B and L by BALB/3T3 fibroblasts and Moloney murine sarcoma virus-transformed BALB/3T3 fibroblasts. *J. Biol. Chem.* **265**: 13650–13654.

586. **Alberti, C., A. Frattini, and S. Ferretti.** 1993. [Urothelial tumors and the extracellular matrix]. *Minerva Urol. Nefrol.* **45**: 11–18.

587. **Asokan, R., G. K. Reddy, and S. C. Dhar.** 1992. Studies on the intracellular degradation of newly synthesized collagen in 3-methylcholanthrene induced fibrosarcoma cells. *Life. Sci.* **51**: 10651071.

588. **Assfalg Machleidt, I., M. Jochum, W. Klaubert, D. Inthorn, and W. Machleidt.** 1988. Enzymatically active cathepsin B dissociating from its inhibitor complexes is elevated in blood plasma of patients with septic shock and some malignant tumors. *Biol. Chem. Hoppe Seyler* **369**(Suppl.): 263–269.

589. **Baici, A., M. Gyger Marazzi, and P. Strauli.** 1984. Extracellular cysteine proteinase and collagenase activities as a consequence of tumor-host interaction in the rabbit V2 carcinoma. *Invasion. Metastasis* **4**: 13–27.

590. **Baici, A., M. Knopfel, and R. Keist.** 1988. Tumor–host interactions in the rabbit V2 carcinoma: stimulation of cathepsin B in host fibroblasts by a tumor-derived cytokine. *Invasion Metastasis* **8**: 143–158.

591. **Baici, A. and P. Sträuli.** 1985. Release of proteinases by cultures of human cell lines derived from squamous carcinomas of the tongue and larynx. *Exp. Cell Biol.* **53**: 213–219.

592. **Bassalyk, L. S., P. E. Tsanev, S. M. Parshikova, and L. V. Demidov.** 1992. [Lysosomal proteolytic enzymes in the processes of melanoma invasion and metastasis]. *Vopr. Onkol.* **38**: 418–425.

593. **Benitez Bribiesca, L., R. Freyre Horta, and G. Gallegos Vargas.** 1980. Protease and antiprotease concentrations in serum and vaginal fluid of patients with carcinoma of the cervix. *Arch. Invest. Med. Mex.* **11**: 523–545.

594. **Boike, G., T. Lah, B. F. Sloane, J. Rozhin, K. Honn, R. Guirguis, M. L. Stracke, L. A. Liotta, and E. Schiffmann.** 1992. A possible role for cysteine proteinase and its inhibitors in motility of malignant melanoma and other tumour cells. *Melanoma Res.* **1**: 333–340.

595. **Braulke, T., L. Mach, B. Hoflack, and J. Glossl.** 1992. Biosynthesis and endocytosis of lysosomal enzymes in human colon carcinoma SW 1116 cells: impaired internalization of plasma membrane-associated cation-independent mannose 6-phosphate receptor. *Arch. Biochem. Biophys.* **298**: 176–181.

596. **Brodt, P., R. Reich, L. A. Moroz, and A. F. Chambers.** 1992. Differences in the repertoires of basement membrane degrading enzymes in two carcinoma sublines with distinct patterns of site-selective metastasis. *Biochim. Biophys. Acta* **1139**: 77–83.

597. **Cavanaugh, P. G., B. F. Sloane, A. S. Bajkowski, G. J. Gasic, T. B. Gasic, and K. V. Honn.** 1983. Involvement of a cathepsin B-like cysteine proteinase in platelet aggregation induced by tumor cells and their shed membrane vesicles. *Clin. Exp. Metastasis* **1**: 297–307.

598. **Chambers, A. F., R. Colella, D. T. Denhardt, and S. M. Wilson.** 1992. Increased expression of cathepsins L and B and decreased activity of their inhibitors in metastatic, ras-transformed NIH 3T3 cells. *Mol. Carcinog.* **5**: 238–245.

599. **Chauhan, S. S., L. J. Goldstein, and M. M. Gottesman.** 1991. Expression of cathepsin L in human tumors. *Cancer Res.* **51**: 1478–1481.

600. **Chernaia, V. I. and A. D. Reva.** 1989. [Cathepsin H activity in the human brain and human brain neoplasms]. *Ukr. Biokhim. Zh.* **61**: 47–50.

601. **Chung, S. M.** 1990. Variant cathepsin L activity from gastric cancer tissue. *Jpn. J. Cancer Res.* **81**: 813–819.

602. **Chung, S. M. and K. Kawai.** 1989. Variant cathepsin B activity secreted from human pancreatic cancer cell lines into protein-free chemically defined medium. *Gastroenterol. Jpn.* **24**: 699–706.

603. **Chung, S. M. and K. Kawai.** 1990. Protease activities in gastric cancer tissues. *Clin. Chim. Acta* **189**: 205–210.

604. **Chung, S. M., K. Kawai, H. M. Chung, K. Kawamoto, and M. Tada.** 1990. [Protease activities in gastric and colon cancer tissues]. *Nippon Shokakibyo Gakkai Zasshi* **87**: 1678–1685.

605. **Corticchiato, O., J. F. Cajot, M. Abrahamson, S. J. Chan, D. Keppler, and B. Sordat.** 1992. Cystatin C and cathepsin B in human colon carcinoma: expression by cell lines and matrix degradation. *Int. J. Cancer* **52**: 645–652.

606. **Crocker, J. and R. Jenkins.** 1984. A quantitative study of histiocytic reticulum cells in diffuse and follicular non-Hodgkin's lymphomas. *J. Clin. Pathol.* **37**: 1222–1226.

607. **Cullen, B. M., I. M. Halliday, G. Kay, J. Nelson, and B. Walker.** 1992. The application of a novel biotinylated affinity label for the detection of a cathepsin B-like precursor produced by breasttumour cells in culture. *Biochem. J.* **283**: 461–465.

608. **Cullen, B. M., J. Nelson, B. Walker, M. McGivern, and G. Kay.** 1990. Facile solubilization of tumour-associated cathepsin B by acid treatment. *Biochem. Soc. Trans.* **18**: 317.

609. **Dengler, R., T. Lah, D. Gabrijelcic, V. Turk, H. Fritz, and B. Emmerich.** 1991. Detection of cathepsin B in tumor cytosol and urine of breast cancer patients. *Biomed. Biochim. Acta* **50**: 555–560.

610. **Denhardt, D. T., A. H. Greenberg, S. E. Egan, R. T. Hamilton, and J. A. Wright.** 1987. Cysteine proteinase cathepsin L expression correlates closely with the metastatic potential of H-ras-transformed murine fibroblasts. *Oncogene.* **2**: 55–59.

611. **Dilakian, E. A., N. I. Solov'eva, and L. Z. Topol'.** 1991. [Cysteine proteinases at various stages of neoplastic transformation of rat fibroblasts]. *Vopr. Med. Khim.* **37**: 36–39.

612. **Doherty, P. J., L. Hua, G. Liau, S. Gal, D. E. Graham, M. Sobel, and M. M. Gottesman.** 1985. Malignant transformation and tumor promoter treatment increase levels of a transcript for a secreted glycoprotein. *Mol. Cell Biol.* **5**: 466–473.

613. **Dong, J. M., E. M. Prence, and G. G. Sahagian.** 1989. Mechanism for selective secretion of a lysosomal protease by transformed mouse fibroblasts. *J. Biol. Chem.* **264**: 7377–7383.

614. **Dufek, V., V. Jirasek, V. Kral, B. Matous, and E. Drazna.** 1985. Changes in serum cathepsin B-like activity in patients with colorectal cancer. *Neoplasma* **32**: 51–54.

615. **Dufek, V., B. Matous, and V. Kral.** 1984. Serum alkaline-stable acid thiol proteinase–a possible marker for primary liver carcinoma. *Neoplasma* **31**: 99–107.

616. **Dufek, V., B. Matous, V. Kral, and L. Bures.** 1984. Characterization of cathepsin B-like proteinase from ascitic fluid of patients with primary liver cancer. *Neoplasma* **31**: 581–590.

617. **Dufek, V., B. Matous, V. Kral, and L. Bures.** 1985. [Cathepsin B-like and possibilities of its use in the diagnosis of ascites of neoplastic origin]. *Cas. Lek. Cesk.* **124**: 46–49.

618. **Duncan, R., P. Kopeckova Rejmanova, J. Strohalm, I. Hume, H. C. Cable, J. Pohl, J. B. Lloyd, and J. Kopecek.** 1987. Anticancer agents coupled to *N*-(2-hydroxypropyl)methacrylamide copolymers. I. Evaluation of daunomycin and puromycin conjugates *in vitro*. *Br. J. Cancer* **55**: 165–174.

619. **Durdey, P., J. C. Cooper, S. Switala, R. F. King, and N. S. Williams.** 1985. The role of peptidases in cancer of the rectum and sigmoid colon. *Br. J. Surg.* **72**: 378–381.

620. **Erdel, M., G. Trefz, E. Spiess, S. Habermaas, H. Spring, T. Lah, and W. Ebert.** 1990. Localization of cathepsin B in two human lung cancer cell lines. *J. Histochem. Cytochem.* **38**: 1313–1321.

621. **Etherington, D. J. and W. H. Taylor.** 1969. Investigation of the cathepsins of human gastric carcinomata. *Biochem. J.* **115**: 43P–44P.

622. **Ferguson, W. W., M. R. Fidler, C. K. Folkmann, and J. R. Starling.** 1979. Correlation of lysosomal enzymes and cachexia in the tumor-bearing rat. *J. Surg. Res.* **26**: 150–156.

623. **Fraki, J. E., S. Nieminen, and V. K. Hopsu Havu.** 1979. Proteolytic enzymes and plasminogen activator in melanoma. *J. Cutan. Pathol.* **6**: 195–200.

624. **Gabrijelcic, D., B. Svetic, D. Spaic, J. Skrk, J. Budihna, and V. Turk.** 1992. Determination of cathepsins B, H, L and kininogen in breast cancer patients. *Agents Actions Suppl.* **38**: 350–357.

625. **Gabrijelcic, D., B. Svetic, D. Spaic, J. Skrk, M. Budihna, I. Dolenc, T. Popovic, V. Cotic, and V. Turk.** 1992. Cathepsins B, H and L in human breast carcinoma. *Eur. J. Clin. Chem. Clin. Biochem.* **30**: 69–74.

626. **Gal, S., M. C. Willingham, and M. M. Gottesman.** 1985. Processing and lysosomal localization of a glycoprotein whose secretion is transformation stimulated. *J. Cell Biol.* **100**: 535–544.

627. **Gingras, M. C., L. Jarolim, J. Finch, G. T. Bowden, J. A. Wright, and A. H. Greenberg.** 1990. Transient alterations in the expression of protease and extracellular matrix genes during metastatic lung colonization by H-ras-transformed 10T1/2 fibroblasts. *Cancer Res.* **50**: 4061–4066.

628. **Giraldi, T., G. Sava, S. Zorzet, L. Perissin, and P. Piccini.** 1987. Activity and inhibition by cytotoxic and antimetastatic drugs of cathepsin B-like cysteine proteinase in transplantable leukemias in mice. *Anticancer Res.* **7**: 343–346.

629. **Goretzki, L., M. Schmitt, K. Mann, J. Calvete, N. Chucholowski, M. Kramer, W. A. Gunzler, F. Janicke, and H. Graeff.** 1992. Effective activation of the proenzyme form of the urokinase-type plasminogen activator (pro-uPA) by the cysteine protease cathepsin L. *FEBS Lett.* **297**: 112–118.

630. **Gottesman, M. M.** 1978. Transformation-dependent secretion of a low molecular weight protein by murine fibroblasts. *Proc. Natl Acad. Sci. U. S. A.* **75**: 2767–2771.

631. **Grabske, R., A. Azevedo, and R. E. Smith.** 1979. Elevated proteinase activities in mouse lung tumors quantitated by synthetic fluorogenic substrates. *J. Histochem. Cytochem.* **27**: 1505–1508.

632. **Graf, F. M., G. Haemmerli, and P. Strauli.** 1984. Cathepsin B containing cells in the rabbit mesentery during invasion of V2 carcinoma cells. *Histochemistry* **80**: 509–515.

633. **Graf, F. M. and P. Strauli.** 1985. [Immunohistochemical localization of cathepsin B in the rabbit's mesenterium following intraperitoneal implantation of V2 carcinoma cells]. *Acta Histochem. Suppl.* **31**: 263–267.

634. **Graf, M., A. Baici, and P. Strauli.** 1981. Histochemical localization of cathepsin B at the invasion front of the rabbit V2 carcinoma. *Lab. Invest.* **45**: 587–596.

635. **Higashiyama, M., O. Doi, K. Kodama, H. Yokouchi, and R. Tateishi.** 1993. Cathepsin B expression in tumour cells and laminin distribution in pulmonary adenocarcinoma. *J. Clin. Pathol.* **46**: 18–22.

636. **Hirano, T., T. Manabe, and S. Takeuchi.** 1993. Serum cathepsin B levels and urinary excretion of cathepsin B in the cancer patients with remote metastasis. *Cancer Lett.* **70**: 41–44.

637. **Kampschmidt, R. F. and D. Wells.** 1968. Acid hydrolase activity during the growth, necrosis, and regression of the Jensen sarcoma. *Cancer Res.* **28**: 1938–1943.

638. **Kampschmidt, R. F. and D. Wells.** 1969. Acid hydrolase activity during the induction and transplantation of hepatomas in the rat. *Cancer Res.* **29**: 1028–1035.

639. **Kazakova, O. V. and V. N. Orekhovich.** 1967. [A comparative study of the liver cathepsins of normal rats, rats with sarcoma and sarcoma cathepsins in rats]. *Biull. Eksp. Biol. Med.* **63**: 76–80.

640. **Keppler, D., M. C. Fondaneche, V. Dalet Fumeron, M. Pagano, and P. Burtin.** 1988. Immunohistochemical and biochemical study of a cathepsin B-like proteinase in human colonic cancers. *Cancer Res.* **48**: 6855–6862.

641. **Keppler, D., M. Pagano, V. Dalet Fumeron, and R. Engler.** 1985. [Regulation of neoplasm-specific cathepsin B by cysteine-protease inhibitors present in cancerous exudates]. *C. R. Acad. Sci. III.* **300**: 471–474.

642. **Keppler, D., M. Pagano, V. Dalet Fumeron, and R. Engler.** 1988. Purification and characterization of two different precursor forms of the cathepsin B-like proteinase from human malignant ascitic fluid. *Biol. Chem. Hoppe Seyler* **369**(Suppl.): 185–190.

643. **Keren, Z. and S. J. LeGrue.** 1988. Identification of cell surface cathepsin B-like activity on murine melanomas and fibrosarcomas: modulation by butanol extraction. *Cancer Res.* **48**: 1416–1421.

644. **Khokha, R., P. Waterhouse, P. Lala, M. Zimmer, D. T. Denhardt, and R. K. Khokka.** 1991. Increased proteinase expression during tumor progression of cell lines down-modulated for TIMP levels: a new transformation paradigm? *J. Cancer Res. Clin. Oncol.* **117**: 333–338 [published erratum appears in 1991 *J Cancer Res. Clin. Oncol.* **117**: 620].

645. **Kobayashi, H., N. Moniwa, M. Sugimura, H. Shinohara, H. Ohi, and T. Terao.** 1993. Effects of membrane-associated cathepsin B on the activation of receptor-bound prourokinase and subsequent invasion of reconstituted basement membranes. *Biochim. Biophys. Acta* **1178**: 55–62.

646. **Kobayashi, H., N. Moniwa, M. Sugimura, H. Shinohara, H. Ohi, and T. Terao.** 1993. Increased cell-surface urokinase in advanced ovarian cancer. *Jpn. J. Cancer Res.* **84**: 633–640.

647. **Kobayashi, H., M. Schmitt, L. Goretzki, N. Chucholowski, J. Calvete, M. Kramer, W. A. Günzler, F. Jänicke, and H. Graeff.** 1991. Cathepsin B efficiently activates the soluble and the tumor cell receptor-bound form of the proenzyme urokinase-type plasminogen activator (Pro-uPA). *J. Biol. Chem.* **266**: 5147–5152.

648. **Kolar, Z., V. Dufek, E. Krepela, J. Vicar, and V. Kral.** 1985. Changes in tissue and serum activity of cathepsin B-like cysteine proteinase during colorectal carcinogenesis by 1,2-dimethylhydrazine in mice. *Neoplasma* **32**: 571–579.

649. **Koppel, P., A. Baici, R. Keist, S. Matzku, and R. Keller.** 1984. Cathepsin B-like proteinase as a marker for metastatic tumor cell variants. *Exp. Cell Biol.* **52**: 293–299.

650. **Korbelik, M., J. Skrk, P. Schauer, A. Suhar, V. Turk, and M. Likar.** 1986. The effect of intracellular proteinases on transformation of human lymphocytes. *Adv. Exp. Med. Biol.* **198**(Pt B): 113–120.

651. **Krecicki, T. and M. Siewinski.** 1992. Serum cathepsin B-like activity as a potential marker of laryngeal carcinoma. *Eur. Arch. Otorhinolaryngol.* **249**: 293–295.

652. **Krepela, E., J. Bartek, D. Skalkova, J. Vicar, D. Rasnick, J. Taylor Papadimitriou, and R. C. Hallowes.** 1987. Cytochemical and biochemical evidence of cathepsin B in malignant, transformed and normal breast epithelial cells. *J. Cell Sci.* **87**: 145–154.

653. **Krepela, E., E. Kasafirek, K. Novak, and J. Viklicky.** 1990. Increased cathepsin B activity in human lung tumors. *Neoplasma* **37**: 61–70.

654. **Krepela, E., P. Vesely, A. Chaloupkova, D. Zicha, P. Urbanec, D. Rasnick, and J. Vicar.** 1989. Cathepsin B in cells of two rat sarcomas with different rates of spontaneous metastasis. *Neoplasma* **36**: 529–540.

655. **Krepela, E., J. Vicar, and M. Cernoch.** 1989. Cathepsin B in human breast tumor tissue and cancer cells. *Neoplasma* **36**: 41–52.

656. **Krieger, N. S., V. P. Sukhatme, and D. A. Bushinsky.** 1990. Conditioned medium from ras oncogene-transformed NIH 3T3 cells induces bone resorption in vitro. *J. Bone Miner. Res.* **5**: 159–164.

657. **Lage, A., J. W. Diaz, and F. Hernandez.** 1979. Acid hydrolases in the extracellular medium of Ehrlich ascites tumor cells. *Neoplasma* **26**: 57–62.

658. **Lah, T. T., J. L. Clifford, K. M. Helmer, N. A. Day, K. Moin, K. V. Honn, J. D. Crissman, and B. F. Sloane.** 1989. Inhibitory properties of low molecular mass cysteine proteinase inhibitors from human sarcoma. *Biochim. Biophys. Acta* **993**: 63–73.

659. **Lah, T. T., M. Kokalj Kunovar, M. Drobnic Kosorok, J. Babnik, R. Golouh, I. Vrhovec, and V. Turk.** 1992. Cystatins and cathepsins in breast carcinoma. *Biol. Chem. Hoppe Seyler* **373**: 595–604.

660. **Lah, T. T., M. Kokalj Kunovar, B. Strukelj, J. Pungercar, D. Barlic Maganja, M. Drobnic Kosorok, L. Kastelic, J. Babnik, R. Golouh, and V. Turk.** 1992. Stefins and lysosomal cathepsins B, L and D in human breast carcinoma. *Int. J. Cancer* **50**: 36–44.

661. **Luthgens, K., W. Ebert, G. Trefz, D. Gabrijelcic, V. Turk, and T. Lah.** 1993. Cathepsin B and cysteine proteinase inhibitors in bronchoalveolar lavage fluid of lung cancer patients. *Cancer Detect. Prev.* **17**: 387–397.

662. **Maciewicz, R. A., R. J. Wardale, D. J. Etherington, and C. Paraskeva.** 1989. Immunodetection of cathepsins B and L present in and secreted from human pre-malignant and malignant colorectal tumour cell lines. *Int. J. Cancer* **43**: 478–486.

663. **Mangan, C. E., G. L. Flickinger, D. Reed, W. Bergantz, S. I. Rubin, and J. J. Mikuta.** 1984. Levels of urinary cathepsin B-like substance in patients with gynecologic malignancy. *Am. J. Clin. Oncol.* **7**: 481–485.

664. **Mason, R. W., D. Wilcox, P. Wikstrom, and E. N. Shaw.** 1989. The identification of active forms of cysteine proteinases in Kirsten-virus-transformed mouse fibroblasts by use of a specific radiolabelled inhibitor. *Biochem. J.* **257**: 125–129.

665. **Matsuoka, Y., H. Tsushima, Y. Koga, H. Mihara, and V. K. Hopsu Havu.** 1992. An inactive cathepsin B-like enzyme and cysteine proteinase inhibitors in colon cancer ascites. *Neoplasma* **39**: 107–114.

666. **Mikulewicz, W., I. Berdowska, J. Jarmulowicz, and M. Siewinski.** 1993. Decrease *in vivo* of cysteine endopeptidases in blood of patients with tumor of the larynx. *Anticancer Drugs* **4**: 341–344.

667. **Moin, K., J. Rozhin, T. B. McKernan, V. J. Sanders, D. Fong, K. V. Honn, and B. F. Sloane.** 1989. Enhanced levels of cathepsin B mRNA in murine tumors. *FEBS Lett.* **244**: 61–64.

668. **Monsky, W. L. and W. T. Chen.** 1993. Proteases of cell adhesion proteins in cancer. *Semin. Cancer Biol.* **4**: 251–258.

669. **Morland, B. and G. Kaplan.** 1978. Properties of a murine monocytic tumor cell line J-774 *in vitro*. II. Enzyme activities. *Exp. Cell Res.* **115**: 63–72.

670. **Morris, V. L., A. B. Tuck, S. M. Wilson, D. Percy, and A. F. Chambers.** 1993. Tumor progression and metastasis in murine D2 hyperplastic alveolar nodule mammary tumor cell lines. *Clin. Exp. Metastasis* **11**: 103–112.

671. **Mort, J. S., M. Leduc, and A. D. Recklies.** 1981. A latent thiol proteinase from ascitic fluid of patients with neoplasia. *Biochim. Biophys. Acta* **662**: 173–180.

672. **Mort, J. S., M. S. Leduc, and A. D. Recklies.** 1983. Characterization of a latent cysteine proteinase from ascitic fluid as a high molecular weight form of cathepsin B. *Biochim. Biophys. Acta* **755**: 369–375.

673. **Mort, J. S., A. D. Recklies, and A. R. Poole.** 1980. Characterization of a thiol proteinase secreted by malignant human breast tumours. *Biochim. Biophys. Acta* **614**: 134–143.

674. **Murnane, M. J., K. Sheahan, M. Ozdemirli, and S. Shuja.** 1991. Stage-specific increases in cathepsin B messenger RNA content in human colorectal carcinoma. *Cancer Res.* **51**: 1137–1142.

675. **Nakao, H., K. Takamori, and H. Ogawa.** 1989. Interaction of tumor and surrounding tissue of mice inoculated B16 melanoma variants in terms of enzyme activity. *Int. J. Biochem.* **21**: 739–743.

676. **Ohsawa, T., T. Higashi, and T. Tsuji.** 1989. The secretion of high molecular weight cathepsin B from cultured human liver cancers. *Acta Med. Okayama.* **43**: 9–15.

677. **Olstein, A. D. and I. E. Liener.** 1983. Comparative studies of mouse liver cathepsin B and an analogous tumor thiol proteinase. *J. Biol. Chem.* **258**: 11049–11056.

678. **Ozeki, Y., K. Takishima, K. Takagi, S. Aida, S. Tamai, G. Mamiya, and T. Ogata.** 1993. Immunohistochemical analysis of cathepsin B expression in human lung adenocarcinoma: the role in cancer progression. *Jpn. J. Cancer Res.* **84**: 972–975.

679. **Ozen, H.** 1993. Advances in bladder cancer. *Curr. Opin. Oncol.* **5**: 574–580.

680. **Pagano, M., V. Dalet Fumeron, and R. Engler.** 1989. The glycosylation state of the precursors of the cathepsin B-like proteinase from human malignant ascitic fluid: possible implication in the secretory pathway of these proenzymes. *Cancer Lett.* **45**: 13–19.

681. **Pagano, M., D. Keppler, V. Fumeron Dalet, and R. Engler.** 1986. Inhibition of the cathepsin B like proteinase by a low molecular weight cysteine-proteinase inhibitor from ascitic fluid and plasma α_2 macroglobulin. *Biochem. Cell Biol.* **64**: 1218–1225.

682. **Persky, B., L. E. Ostrowski, P. Pagast, A. Ahsan, and R. M. Schultz.** 1986. Inhibition of proteolytic enzymes in the *in vitro* amnion model for basement membrane invasion. *Cancer Res.* **46**: 4129–4134.

683. **Petrova Skalkova, D., E. Krepela, D. Rasnick, and J. Vicar.** 1987. A latent form of cathepsin B in pleural effusions. I. Characterization of the enzyme in breast cancer patients. *Biochem. Med. Metab. Biol.* **38**: 219–227.

684. **Pietras, R. J. and J. A. Roberts.** 1981. Cathepsin B-like enzymes. Subcellular distribution and properties in neoplastic and control cells from human ectocervix. *J. Biol. Chem.* **256**: 8536–8544.

685. **Pietras, R. J., C. M. Szego, C. E. Mangan, B. J. Seeler, and M. M. Burtnett.** 1979. Elevated serum cathepsin B_1-like activity in women with neoplastic disease. *Gynecol. Oncol.* **7**: 1–17.

686. **Pietras, R. J., C. M. Szego, C. E. Mangan, B. J. Seeler, M. M. Burtnett, and M. Orevi.** 1978. Elevated serum cathepsin B_1 and vaginal pathology after prenatal DES exposure. *Obstet. Gynecol.* **52**: 321–327.

687. **Pietras, R. J., C. M. Szego, J. A. Roberts, and B. J. Seeler.** 1981. Lysosomal cathepsin B-like activity: mobilization in prereplicative and neoplastic epithelial cells. *J. Histochem. Cytochem.* **29**: 440–450.

688. **Poole, A. R., K. J. Tiltman, A. D. Recklies, and T. A. Stoker.** 1978. Differences in secretion of the proteinase cathepsin B at the edges of human breast carcinomas and fibroadenomas. *Nature* **273**: 545–547.

689. **Qian, F., A. S. Bajkowski, D. F. Steiner, S. J. Chan, and A. Frankfater.** 1989. Expression of five cathepsins in murine melanomas of varying metastatic potential and normal tissues. *Cancer Res.* **49**: 4870–4875.

690. **Qian, F., S. J. Chan, Q. M. Gong, A. S. Bajkowski, D. F. Steiner, and A. Frankfater.** 1991. The expression of cathepsin B and other lysosomal proteinases in normal tissues and in tumors. *Biomed. Biochim. Acta* **50**: 531–540.

691. **Qian, F., A. Frankfater, D. F. Steiner, A. S. Bajkowski, and S. J. Chan.** 1991. Characterization of multiple cathepsin B mRNAs in murine B16a melanoma. *Anticancer Res.* **11**: 1445–1451.

692. **Rabin, M. S., P. J. Doherty, and M. M. Gottesman.** 1986. The tumor promoter phorbol 12-myristate 13-acetate induces a program of altered gene expression similar to that induced by platelet-derived growth factor and transforming oncogenes. *Proc. Natl Acad. Sci. U. S. A.* **83**: 357360.

693. **Recklies, A. D., A. R. Poole, and J. S. Mort.** 1982. A cysteine proteinase secreted from human breast tumours is immunologically related to cathepsin B. *Biochem. J.* **207**: 633–636.

694. **Redwood, S. M., B. C. Liu, R. E. Weiss, D. E. Hodge, and M. J. Droller.** 1992. Abrogation of the invasion of human bladder tumor cells by using protease inhibitor(s). *Cancer* **69**: 1212–1219.

695. **Rinderknecht, H. and I. G. Renner.** 1980. Increased cathepsin B activity in pancreatic juice from a patient with pancreatic cancer. *N. Engl. J. Med.* **303**: 462–463.

696. **Rozhin, J., D. Robinson, M. A. Stevens, T. T. Lah, K. V. Honn, R. E. Ryan, and B. F. Sloane.** 1987. Properties of a plasma membrane-associated cathepsin B-like cysteine proteinase in metastatic B16 melanoma variants. *Cancer Res.* **47**: 6620–6628.

697. **Rozhin, J., R. L. Wade, K. V. Honn, and B. F. Sloane.** 1989. Membrane-associated cathepsin L: a role in metastasis of melanomas. *Biochem. Biophys. Res. Commun.* **164**: 556–561 [published erratum appears in 1989 *Biochem. Biophys. Res. Commun.* **165**:1444].

698. **Ryan, R. E., J. D. Crissman, K. V. Honn, and B. F. Sloane.** 1985. Cathepsin B-like activity in viable tumor cells isolated from rodent tumors. *Cancer Res.* **45**: 3636–3641.

699. **Sahagian, G. G. and M. M. Gottesman.** 1982. The predominant secreted protein of transformed murine fibroblasts carries the lysosomal mannose 6-phosphate recognition marker. *J. Biol. Chem.* **257**: 11145–11150.

700. **Samuel, S. K., R. A. Hurta, P. Kondaiah, N. Khalil, E. A. Turley, J. A. Wright, and A. H. Greenberg.** 1992. Autocrine induction of tumor protease production and invasion by a metallothionein-regulated TGF-β_1 (Ser223, 225). *EMBO J.* **11**: 1599–1605.

701. **Sano, N., M. Shibata, K. Izumi, and H. Otsuka.** 1988. Histopathological and immunohistochemical studies on nickel sulfide-induced tumors in F344 rats. *Jpn. J. Cancer Res.* **79**: 212–221.

702. **Schlagenhauff, B., C. Klessen, S. Teichmann Dorr, H. Breuninger, and G. Rassner.** 1992. Demonstration of proteases in basal cell carcinomas. A histochemical study using amino acid-4-methoxy-2-naphthylamides as chromogenic substrates. *Cancer* **70**: 1133–1140.

703. **Schmitt, M., L. Goretzki, F. Jänicke, F. Calvete, M. Eulitz, H. Kobayashi, N. Chucholowski, and H. Graeff.** 1991. Biological and clinical relevance of the urokinase-type plasminogen activator (uPA) in breast cancer. *Biomed. Biochim. Acta* **50**: 731–741.

704. **Sedo, A., E. Krepela, and E. Kasafirek.** 1991. Dipeptidyl peptidase IV, prolyl endopeptidase and cathepsin B activities in primary human lung tumors and lung parenchyma. *J. Cancer Res. Clin. Oncol.* **117**: 249–253.

705. **Sekiya, S., T. Oosaki, N. Suzuki, and H. Takamizawa.** 1985. Invasion potential of human choriocarcinoma cell lines and the role of lytic enzymes. *Gynecol. Oncol.* **22**: 324–333.

706. **Shamberger, R. J.** 1969. Lysosomal enzyme changes in growing and regressing mammary tumours. *Biochem. J.* **111**: 375–383.

707. **Shamberger, R. J. and G. Rudolph.** 1967. Increase of lysosomal enzymes in skin cancers. *Nature* **213**: 617–618.

708. **Sheahan, K., S. Shuja, and M. J. Murnane.** 1989. Cysteine protease activities and tumor development in human colorectal carcinoma. *Cancer Res.* **49**: 3809–3814.

709. **Shuja, S., K. Sheahan, and M. J. Murnane.** 1991. Cysteine endopeptidase activity levels in normal human tissues, colorectal adenomas and carcinomas. *Int. J. Cancer* **49**: 341–346.

710. **Silvis, N. G., P. E. Swanson, J. C. Manivel, V. N. Kaye, and M. R. Wick.** 1988. Spindle-cell and pleomorphic neoplasms of the skin. A clinicopathologic and immunohistochemical study of 30 cases, with emphasis on "atypical fibroxanthomas". *Am. J. Dermatopathol.* **10**: 9–19.

711. **Sloane, B. F., J. R. Dunn, and K. V. Honn.** 1981. Lysosomal cathepsin B: correlation with metastatic potential. *Science* **212**: 1151–1153.

712. **Sloane, B. F., J. Rozhin, J. S. Hatfield, J. D. Crissman, and K. V. Honn.** 1987. Plasma membrane-associated cysteine proteinases in human and animal tumors. *Exp. Cell Biol.* **55**: 209–224.

713. **Sloane, B. F., J. Rozhin, K. Johnson, H. Taylor, J. D. Crissman, and K. V. Honn.** 1986. Cathepsin B: association with plasma membrane in metastatic tumors. *Proc. Natl Acad. Sci. U. S. A.* **83**: 2483–2487.

714. **Sloane, B. F., J. Rozhin, K. Moin, G. Ziegler, D. Fong, and R. J. Muschel.** 1992. Cysteine endopeptidases and their inhibitors in malignant progression of rat embryo fibroblasts. *Biol. Chem. Hoppe Seyler* **373**: 589–594.

715. **Stearns, N. A., J. M. Dong, J. X. Pan, D. A. Brenner, and G. G. Sahagian.** 1990. Comparison of cathepsin L synthesized by normal and transformed cells at the gene, message, protein, and oligosaccharide levels. *Arch. Biochem. Biophys.* **283**: 447–457.

716. **Strauli, P. and G. Haemmerli.** 1984. The V2 carcinoma of the rabbit as an integrated model of tumor invasion. *Bull. Cancer Paris* **71**: 447–452.

717. **Swanson, P. E., M. W. Stanley, B. W. Scheithauer, and M. R. Wick.** 1988. Primary cutaneous leiomyosarcoma. A histological and immunohistochemical study of 9 cases, with ultrastructural correlation. *J. Cutan. Pathol.* **15**: 129–141.

718. **Sylven, B.** 1968. Lysosomal enzyme activity in the interstitial fluid of solid mouse tumour transplants. *Eur. J. Cancer* **4**: 463–474.

719. **Sylven, B., O. Snellman, and P. Strauli.** 1974. Immunofluorescent studies on the occurrence of cathepsin B_1 at tumor cell surfaces. *Virchows Arch. B. Cell Pathol.* **17**: 97–112.

720. **Taniguchi, S., Y. Nishimura, T. Takahashi, T. Baba, and K. Kato.** 1990. Augmented excretion of procathepsin L of a fos-transferred highly metastatic rat cell line. *Biochem. Biophys. Res. Commun.* **168**: 520–526.

721. **Tanimoto, K., Y. Murawaki, and C. Hirayama.** 1984. Granulocyte collagenase and cathepsin B in patients with cancer of digestive organs. *Gastroenterol. Jpn.* **19**: 537–542.

722. **Trefz, G., M. Erdel, E. Spiess, and W. Ebert.** 1990. Detection of cathepsin B, plasminogen activators and plasminogen activator inhibitor in human non-small lung cancer cell lines. *Biol. Chem. Hoppe Seyler* **371**: 617–624.

723. **Tsushima, H., F. Hyodoh, E. Yoshida, A. Ueki, and V. K. Hopsu Havu.** 1992. Inactive cathepsin B-like enzyme in human melanoma culture medium. *Melanoma Res.* **1**: 341–347.

724. **Tsushima, H., H. Sumi, K. Hamanaka, N. Toki, H. Sato, and H. Mihara.** 1985. Cysteine protease inhibitors isolated from human malignant melanoma tissue. *J. Lab. Clin. Med.* **106**: 712–717.

725. **Tsushima, H., A. Ueki, Y. Matsuoka, H. Mihara, and V. K. Hopsu-Havu.** 1991. Characterization of a cathepsin-H-like enzyme from a human melanoma cell line. *Int. J. Cancer* **48**: 726–732.

726. **Ueda, M.** 1993. [A study on cathepsin B-like substance in patients with urological cancer]. *Nippon Hinyokika Gakkai Zasshi* **84**: 355–363.

727. **van der Stappen, J. W., C. Paraskeva, A. C. Williams, A. Hague, and R. A. Maciewicz.** 1991. Relationship between the secretion of cysteine proteinases and their inhibitors and malignant potential. *Biochem. Soc. Trans.* **19**: 362S.

728. **Vasishta, A., P. R. Baker, P. E. Preece, R. A. Wood, and A. Cuschieri.** 1984. Serum and tissue proteinase-like peptidase activities in women undergoing total mastectomy for breast cancer. *Eur. J. Cancer Clin. Oncol.* **20**: 203–208.

729. **Vasishta, A., P. R. Baker, P. E. Preece, R. A. Wood, and A. Cuschieri.** 1989. Proteinase-like peptidase activities and oestrogen receptor levels in breast cancer tissue. *J. Cancer Res. Clin. Oncol.* **115**: 89–92.

730. **Watanabe, M., T. Higashi, M. Hashimoto, I. Tomoda, S. Tominaga, N. Hashimoto, S. Morimoto, Y. Yamauchi, H. Nakatsukasa, M. Kobayashi, and et al.** 1987. Elevation of tissue cathepsin B and L activities in gastric cancer. *Hepatogastroenterology.* **34**: 120–122.

731. **Watanabe, M., T. Higashi, A. Watanabe, T. Osawa, Y. Sato, Y. Kimura, S. Tominaga, N. Hashimoto, Y. Yoshida, S. Morimoto, and** *et al.* 1989. Cathepsin B and L activities in gastric cancer tissue: correlation with histological findings. *Biochem. Med. Metab. Biol.* **42**: 21–29.

732. **Weiss, R. E., B. C. Liu, T. Ahlering, L. Dubeau, and M. J. Droller.** 1990. Mechanisms of human bladder tumor invasion: role of protease cathepsin B. *J. Urol.* **144**: 798–804.

733. **Yagel, S., A. H. Warner, H. N. Nellans, P. K. Lala, C. Waghorne, and D. T. Denhardt.** 1989. Suppression by cathepsin L inhibitors of the invasion of amnion membranes by murine cancer cells. *Cancer Res.* **49**: 3553–3557.

734. **Yamaguchi, N., S. M. Chung, O. Shiroeda, K. Koyama, and J. Imanishi.** 1990. Characterization of a cathepsin L-like enzyme secreted from human pancreatic cancer cell line HPC-YP. *Cancer Res.* **50**: 658–663.

735. **Young, P. R. and S. M. Spevacek.** 1993. Substratum acidification and proteinase activation by murine B16F10 melanoma cultures. *Biochim. Biophys. Acta* **1182**: 69–74.

736. **Zhang, J. Y. and R. M. Schultz.** 1992. Fibroblasts transformed by different ras oncogenes show dissimilar patterns of protease gene expression and regulation. *Cancer Res.* **52**: 6682–6689.

Expression of the enzymes in pancreatitis

737. **Adler, G., C. Hahn, H. F. Kern, and K. N. Rao.** 1985. Cerulein-induced pancreatitis in rats: increased lysosomal enzyme activity and autophagocytosis. *Digestion* **32**: 10–18.

738. **Figarella, C., B. Miszczuk Jamska, and A. J. Barrett.** 1988. Possible lysosomal activation of pancreatic zymogens. Activation of both human trypsinogens by cathepsin B and spontaneous acid. Activation of human trypsinogen 1. *Biol. Chem. Hoppe Seyler* **369**(Suppl.): 293–298.

739. **Haber, P. S., J. S. Wilson, M. V. Apte, and R. C. Pirola.** 1993. Fatty acid ethyl esters increase rat pancreatic lysosomal fragility. *J. Lab. Clin. Med.* **121**: 759–764.

740. **Hirano, T. and T. Manabe.** 1992. Effect of short-termed pancreatic duct obstruction on the pancreatic subcellular organellar fragility and pancreatic lysosomal enzyme secretion in rabbits. *Scand. J. Clin. Lab. Invest.* **52**: 523–535.

741. **Hirano, T. and T. Manabe.** 1992. Fragility of subcellular organelles induced by pancreatic duct obstruction in rabbits. *Nippon Geka Hokan* **61**: 334–349.

742. **Hirano, T. and T. Manabe.** 1993. A new experimental model for gallstone pancreatitis: short-termed pancreatico-biliary duct obstruction and exocrine stimulation with systemic hypotension in rats. *Nippon Geka Hokan* **62**: 3–15.

743. **Hirano, T. and T. Manabe.** 1993. A possible mechanism for gallstone pancreatitis: repeated short-term pancreaticobiliary duct obstruction with exocrine stimulation in rats. *Proc. Soc. Exp. Biol. Med.* **202**: 246–252.

744. **Hirano, T. and T. Manabe.** 1993. Effect of ethanol on pancreatic lysosomes in rats: a possible mechanism for alcoholic pancreatitis. *Nippon Geka Hokan* **62**: 16–23.

745. **Hirano, T., T. Manabe, K. Imanishi, and T. Tobe.** 1991. [Changes of lysosomal and digestive enzymes in rat caerulein pancreatitis]. *Nippon Geka Hokan* **60**: 122–134.

746. **Hirano, T., T. Manabe, K. Imanishi, and T. Tobe.** 1993. Protective effect of a cephalosporin, Shiomarin, plus a new potent protease inhibitor, E3123, on rat taurocholate-induced pancreatitis. *J. Gastroenterol. Hepatol.* **8**: 52–59.

747. **Hirano, T., T. Manabe, T. Kyogoku, K. Ando, and T. Tobe.** 1991. Pancreatic lysosomal enzyme secretion via gut-hormone-regulated pathway in rats. *Nippon Geka Hokan* **60**: 415–423.

748. **Hirano, T., T. Manabe, G. Ohshio, and Y. Nio.** 1992. Protective effects of combined therapy with a protease inhibitor, ONO 3307, and a xanthine oxidase inhibitor, allopurinol on temporary ischaemic model of pancreatitis in rats. *Nippon Geka Hokan* **61**: 224–233.

749. **Hirano, T., T. Manabe, H. Printz, A. Saluja, and M. Steer.** 1992. Secretion of lysosomal and digestive enzymes into pancreatic juice under physiological and pathological conditions in rabbits. *Nippon Geka Hokan* **61**: 103–124.

750. **Hirano, T., T. Manabe, M. Steer, H. Printz, R. Calne, and T. Tobe.** 1993. Protective effects of therapy with a protease and xanthine oxidase inhibitor in short form pancreatic biliary obstruction and ischemia in rats. *Surg. Gynecol. Obstet.* **176**: 371–381.

751. **Hirano, T., T. Manabe, and T. Tobe.** 1990. Pancreatic lysosomal enzyme secretion is changed by hepatectomy in rats. *Scand. J. Gastroenterol.* **25**: 1274–1280.

752. **Hirano, T., T. Manabe, and T. Tobe.** 1990. Lysosomal redistribution and increased fragility of pancreatic acinar cells in early stage after hepatectomy in rats. *Nippon Geka Hokan* **59**: 377–382.

753. **Hirano, T., T. Manabe, and T. Tobe.** 1991. Cellular alterations of parotid gland of rats with acute pancreatitis induced by cerulein. *Int. J. Pancreatol.* **10**: 217–227.

754. **Hirano, T., T. Manabe, F. Yotsumoto, K. Ando, K. Imanishi, and T. Tobe.** 1993. Effect of prostaglandin E on the redistribution of lysosomal enzymes in caerulein-induced pancreatitis. *Hepatogastroenterology*. **40**: 155–158.

755. **Hirano, T., A. Saluja, P. Ramarao, M. M. Lerch, M. Saluja, and M. L. Steer.** 1991. Apical secretion of lysosomal enzymes in rabbit pancreas occurs via a secretagogue regulated pathway and is increased after pancreatic duct obstruction. *J. Clin. Invest.* **87**: 865–869.

756. **Lerch, M. M., A. K. Saluja, R. Dawra, M. Saluja, and M. L. Steer.** 1993. The effect of chloroquine administration on two experimental models of acute pancreatitis. *Gastroenterology* **104**: 1768–1779.

757. **Niederau, C. and J. H. Grendell.** 1988. Intracellular vacuoles in experimental acute pancreatitis in rats and mice are an acidified compartment. *J. Clin. Invest.* **81**: 229–236.

758. **Ohshio, G., A. Saluja, and M. L. Steer.** 1991. Effects of short-term pancreatic duct obstruction in rats. *Gastroenterology* **100**: 196–202.

759. **Ohshio, G., A. K. Saluja, U. Leli, A. Sengupta, and M. L. Steer.** 1989. Esterase inhibitors prevent lysosomal enzyme redistribution in two noninvasive models of experimental pancreatitis. *Gastroenterology* **96**: 853–859.

760. **Printz, H., A. Saluja, U. Leli, A. Sengupta, and M. Steer.** 1990. Effects of hemorrhagic shock, aspirin, and ethanol on secretagogue-induced experimental pancreatitis. *Int. J. Pancreatol.* **6**: 207–217.

761. **Rao, K. N., M. F. Zuretti, F. M. Baccino, and B. Lombardi.** 1980. Acute hemorrhagic pancreatic necrosis in mice: the activity of lysosomal enzymes in the pancreas and the liver. *Am. J. Pathol.* **98**: 45–59.

762. **Saluja, A., M. Saluja, A. Villa, U. Leli, P. Rutledge, J. Meldolesi, and M. Steer.** 1989. Pancreatic duct obstruction in rabbits causes digestive zymogen and lysosomal enzyme colocalization. *J. Clin. Invest.* **84**: 1260–1266.

763. **Saluja, M., A. Saluja, M. M. Lerch, and M. L. Steer.** 1991. A plasma protease which is expressed during supramaximal stimulation causes *in vitro* subcellular redistribution of lysosomal enzymes in rat exocrine pancreas. *J. Clin. Invest.* **87**: 1280–1285.

764. **Shikimi, T., D. Yamamoto, and M. Handa.** 1987. Pancreatic lysosomal thiol proteinases and inhibitors in acute pancreatitis induced in rats. *J. Pharmacobiodyn.* **10**: 750–757.

765. **Singh, M.** 1987. Alcoholic pancreatitis in rats fed ethanol in a nutritionally adequate liquid diet. *Int. J. Pancreatol.* **2**: 311–324.

766. **Singh, M.** 1992. Effect of chronic ethanol feeding on factors leading to inappropriate intrapancreatic activation of zymogens in the rat pancreas. *Digestion*. **53**: 114–120.

767. **Steer, M. L. and J. Meldolesi.** 1988. Pathogenesis of acute pancreatitis. *Annu. Rev. Med.* **39**: 95–105.

768. **Willemer, S., R. Bialek, and G. Adler.** 1990. Localization of lysosomal and digestive enzymes in cytoplasmic vacuoles in caerulein-pancreatitis. *Histochemistry* **94**: 161–170.

769. **Wilson, J. S., M. V. Apte, M. C. Thomas, P. S. Haber, and R. C. Pirola.** 1992. Effects of ethanol, acetaldehyde and cholesteryl esters on pancreatic lysosomes. *Gut* **33**: 1099–1104.

770. **Wilson, J. S., M. A. Korsten, M. V. Apte, M. C. Thomas, P. S. Haber, and R. C. Pirola.** 1990. Both ethanol consumption and protein deficiency increase the fragility of pancreatic lysosomes. *J. Lab. Clin. Med.* **115**: 749–755.

771. **Yamaguchi, H., T. Kimura, K. Mimura, and H. Nawata.** 1989. Activation of proteases in cerulein-induced pancreatitis. *Pancreas* **4**: 565–571.

Expression of the enzymes in normal tissues and disorders

772. **Alfieri, S. C., E. M. Pral, E. Shaw, C. Ramazeilles, and M. Rabinovitch.** 1991. *Leishmania amazonensis*: specific labeling of amastigote cysteine proteinases by radioiodinated *N*-benzyloxycarbonyl-tyrosyl-alanyl diazomethane. *Exp. Parasitol.* **73**: 424–432.

773. **Ali, S. Y.** 1967. The presence of cathepsin B in cartilage. *Biochem. J.* **102**: 10C–11C.

774. **Andreasson, S., L. Smith, O. K. Andersen, G. Volden, and B. Risberg.** 1987. Lysosomal enzyme pattern in lung lymph and blood during *E. coli* sepsis in sheep. *Scand. J. Clin. Lab. Invest.* **47**: 355–362.

775. **Aoyagi, T., T. Wada, F. Kojima, M. Nagai, S. Miyoshino, and H. Umezawa.** 1983. Two different modes of enzymatic changes in serum with progression of Duchenne muscular dystrophy. *Clin. Chim. Acta* **129**: 165–173.

776. **Ashmore, C. R., P. J. Summers, and Y. B. Lee.** 1986. Proteolytic enzyme activities and onset of muscular dystrophy in the chick. *Exp. Neurol.* **94**: 585–597.

777. **Assfalg Machleidt, I., M. Jochum, D. Nast Kolb, M. Siebeck, A. Billing, T. Joka, G. Rothe, G. Valet, R. Zauner, H. P. Scheuber, and *et al*.** 1990. Cathepsin B-indicator for the release of lysosomal cysteine proteinases in severe trauma and inflammation. *Biol. Chem. Hoppe Seyler* **371**(Suppl.): 211–222.

778. Baccino, F. M., G. Barrera, G. Bonelli, M. Messina, M. Musi, and L. Tessitore. 1986. Cellular distribution of lysosomal hydrolase activities in the regenerating rat liver. *Cell Biochem. Funct.* **4**: 213–225.

779. Baici, A. and A. Lang. 1990. Cathepsin B secretion by rabbit articular chondrocytes: modulation by cycloheximide and glycosaminoglycans. *Cell Tissue Res.* **259**: 567–573.

780. Baici, A. and A. Lang. 1990. Effect of interleukin-1β on the production of cathepsin B by rabbit articular chondrocytes. *FEBS Lett.* **277**: 93–96.

781. Baici, A., A. Lang, D. Horler, and M. Knopfel. 1988. Cathepsin B as a marker of the dedifferentiated chondrocyte phenotype. *Ann. Rheum. Dis.* **47**: 684–691.

782. Banati, R. B., G. Rothe, G. Valet, and G. W. Kreutzberg. 1993. Detection of lysosomal cysteine proteinases in microglia: flow cytometric measurement and histochemical localization of cathepsin B and L. *Glia* **7**: 183–191.

783. Baricos, W. H., S. L. Cortez, Q. C. Le, L. T. Wu, E. Shaw, K. Hanada, and S. V. Shah. 1991. Evidence suggesting a role for cathepsin L in an experimental model of glomerulonephritis. *Arch. Biochem. Biophys.* **288**: 468–472.

784. Baricos, W. H., S. E. O'Connor, S. L. Cortez, L. T. Wu, and S. V. Shah. 1988. The cysteine proteinase inhibitor, E-64, reduces proteinuria in an experimental model of glomerulonephritis. *Biochem. Biophys. Res. Commun.* **155**: 1318–1323.

785. Baricos, W. H. and S. V. Shah. 1989. Role of cathepsin B and L in anti-glomerular basement membrane nephritis in rats. *Renal. Physiol. Biochem.* **12**: 400–405.

786. Bechet, D. M., M. J. Ferrara, S. B. Mordier, M. P. Roux, C. D. Deval, and A. Obled. 1991. Expression of lysosomal cathepsin B during calf myoblast-myotube differentiation. Characterization of a cDNA encoding bovine cathepsin B. *J. Biol. Chem.* **266**: 14104–14112.

787. Bennett, M. J., L. Chern, K. H. Carpenter, and J. T. Sladky. 1992. Abnormal lysosomal cathepsin activities in leukocytes and cultured skin fibroblasts in late infantile, but not in juvenile neuronal ceroid-lipofuscinosis (Batten disease). *Clin. Chim. Acta* **208**: 111–117.

788. Bernstein, H. G., H. Kirschke, P. Kloss, B. Wiederanders, A. Rinne, and J. Frohlich. 1990. Cathepsin B during early human brain development. *Acta Histochem. Suppl.* **39**: 473–475.

789. Bernstein, H. G., H. Kirschke, B. Wiederanders, D. Schmidt, and A. Rinne. 1990. Antigenic expression of cathepsin B in aged human brain. *Brain Res. Bull.* **24**: 543–549.

790. Bernstein, H. G., R. Sormunen, M. Jarvinen, P. Kloss, H. Kirschke, and A. Rinne. 1989. Cathepsin B immunoreactive neurons in rat brain. A combined light and electron microscopic study. *J. Hirnforsch.* **30**: 313–317.

791. Billing, A., D. Frohlich, I. Assfalg Machleidt, W. Machleidt, and M. Jochum. 1991. Proteolysis of defensive proteins in peritonitis exudate: pathobiochemical aspects and therapeutical approach. *Biomed. Biochim. Acta* **50**: 399–402.

792. Biondi, R. and M. P. Viola Magni. 1983. Behaviour of tyrosine amino transferase and convertase during the first hours after hepatectomy in rats. *Cell Biochem. Funct.* **1**: 97–102.

793. Bird, J. W., T. Berg, A. Milanesi, and W. T. Stauber. 1969. Lysosomal enzymes in aquatic species. I. Distribution and particle properties of muscle lysosomes of the goldfish. *Comp. Biochem. Physiol.* **30**: 457–468.

794. Bird, J. W., F. J. Roisen, G. Yorke, J. A. Lee, M. A. McElligott, D. F. Triemer, and A. St.John. 1981. Lysosomes and proteolytic enzyme activities in cultured striated muscle cells. *J. Histochem. Cytochem.* **29**: 431–439.

795. Bird, J. W., L. Wood, I. Sohar, E. Fekete, R. Colella, G. Yorke, B. Cosentino, and F. J. Roisen. 1985. Localization of cysteine proteinases and an endogenous cysteine proteinase inhibitor in cultured muscle cells. *Biochem. Soc. Trans.* **13**: 1018–1021.

796. Birkenbach, M., K. Josefsen, R. Yalamanchili, G. Lenoir, and E. Kieff. 1993. Epstein-Barr virus-induced genes: first lymphocyte-specific G protein-coupled peptide receptors. *J. Virol.* **67**: 2209–2220.

797. Bouma, J. M. and M. Gruber. 1966. Intracellular distribution of cathepsin B and cathepsin C in rat liver. *Biochim. Biophys. Acta* **113**: 350–358.

798. Bowers, R. R., M. L. Birch, and D. W. Thomas. 1985. A biochemical study of the carrageenan-induced granuloma in the rat lung. *Connect. Tissue. Res.* **13**: 191–206.

799. Braulke, T., R. Bresciani, D. M. Buergisser, and K. von Figura. 1991. Insulin-like growth factor II overexpression does not affect sorting of lysosomal enzymes in NIH-3T3 cells. *Biochem. Biophys. Res. Commun.* **179**: 108–115.

800. Brown, H. H. and H. W. Moon. 1979. Localization and activities of lysosomal enzymes in jejunal and ileal epithelial cells of the young pig. *Am. J. Vet. Res.* **40**: 1573–1577.

801. Burnett, D., J. Crocker, S. C. Afford, C. M. Bunce, G. Brown, and R. A. Stockley. 1986. Cathepsin B synthesis by the HL60 promyelocytic cell line: effects of stimulating agents and anti-inflammatory compounds. *Biochim. Biophys. Acta* **887**: 283–290.

802. **Burnett, D., J. Crocker, and R. A. Stockley.** 1983. Cathepsin B-like cysteine proteinase activity in sputum and immunohistologic identification of cathepsin B in alveolar macrophages. *Am. Rev. Respir. Dis.* **128**: 915–919.

803. **Burnett, D., J. Crocker, and A. T. Vaughan.** 1983. Synthesis of cathepsin B by cells derived from the HL60 promyelocytic leukaemia cell line. *J. Cell Physiol.* **115**: 249–254.

804. **Burnett, D. and R. A. Stockley.** 1985. Cathepsin B-like cysteine proteinase activity in sputum and bronchoalveolar lavage samples: relationship to inflammatory cells and effects of corticosteroids and antibiotic treatment. *Clin. Sci.* **68**: 469–474.

805. **Buttle, D. J., M. Abrahamson, D. Burnett, J. S. Mort, A. J. Barrett, P. M. Dando, and S. L. Hill.** 1991. Human sputum cathepsin B degrades proteoglycan, is inhibited by α_2-macroglobulin and is modulated by neutrophil elastase cleavage of cathepsin B precursor and cystatin C. *Biochem. J.* **276**: 325–331.

806. **Buttle, D. J., D. Burnett, and M. Abrahamson.** 1990. Levels of neutrophil elastase and cathepsin B activities, and cystatins in human sputum: relationship to inflammation. *Scand. J. Clin. Lab. Invest.* **50**: 509–516.

807. **Bylinkina, V. S., N. V. Golubeva, T. A. Gureeva, L. A. Lokshina, A. M. Polianskaia, and R. S. Samoilova.** 1991. [Protein kinase activity in lymphoid cells in various forms of lymphoproliferative disorders]. *Vopr. Med. Khim.* **37**: 30–33.

808. **Cataldo, A. M., P. A. Paskevich, E. Kominami, and R. A. Nixon.** 1991. Lysosomal hydrolases of different classes are abnormally distributed in brains of patients with Alzheimer disease. *Proc. Natl Acad. Sci. U. S. A.* **88**: 10998–11002 [published erratum appears in 1992 *Proc. Natl Acad. Sci. U. S. A.* **89**: 2509].

809. **Chan, M. M. and D. Fong.** 1988. Expression of human cathepsin B protein in *Escherichia coli*. *FEBS Lett.* **239**: 219–222.

810. **Chang, J. C., M. Lesser, O. H. Yoo, and M. Orlowski.** 1986. Increased cathepsin B-like activity in alveolar macrophages and bronchoalveolar lavage fluid from smokers. *Am. Rev. Respir. Dis.* **134**: 538–541.

811. **Chapman, H. A. J., J. J. J. Reilly, R. Yee, and A. Grubb.** 1990. Identification of cystatin C, a cysteine proteinase inhibitor, as a major secretory product of human alveolar macrophages in vitro. *Am. Rev. Respir. Dis.* **141**: 698–705.

812. **Chernaia, V. I. and A. D. Reva.** 1978. [Cathepsin B_1 activity in cat brain tissue]. *Ukr. Biokhim. Zh.* **50**: 411–414.

813. **Chue, C. H., N. Yukioka, E. Yamada, and F. Hazama.** 1993. The possible role of lysosomal enzymes in the pathogenesis of hypertensive cerebral lesions in spontaneously hypertensive rats. *Acta Neuropathol. Berl.* **85**: 383–389.

814. **Codorean, E. and E. Gabrielescu.** 1985. Cytochemical investigation of cathepsin B in rheumatoid synovial membrane and fluid. *Morphol. Embryol. Bucur.* **31**: 269–274.

815. **Colella, R., F. J. Roisen, and J. W. Bird.** 1986. mRNA levels of cathepsins B and D during myogenesis. *Biomed. Biochim. Acta* **45**: 1413–1419.

816. **Cox, S. W. and B. M. Eley.** 1989. Detection of cathepsin B- and L-, elastase-, tryptase-, trypsin-, and dipeptidyl peptidase IV-like activities in crevicular fluid from gingivitis and periodontitis patients with peptidyl derivatives of 7-amino-4-trifluoromethyl coumarin. *J. Periodontal. Res.* **24**: 353–361.

817. **Cox, S. W. and B. M. Eley.** 1992. Cathepsin B/L-, elastase-, tryptase-, trypsin- and dipeptidyl peptidase IV-like activities in gingival crevicular fluid. A comparison of levels before and after basic periodontal treatment of chronic periodontitis patients. *J. Clin. Periodontol.* **19**: 333–339.

818. **Crocker, J., D. Burnett, and E. L. Jones.** 1984. Immunohistochemical demonstration of cathepsin B in the macrophages of benign and malignant lymphoid tissues. *J. Pathol.* **142**: 87–94.

819. **Curreri, P. W., H. V. Kothari, M. J. Bonner, and B. F. Miller.** 1969. Increased activity of lysosomal enzymes in experimental atherosclerosis, and the effect of cortisone. *Proc. Soc. Exp. Biol. Med.* **130**: 1253–1256.

820. **Daston, G. P., D. Baines, J. E. Yonker, and L. D. Lehman McKeeman.** 1991. Effects of lysosomal proteinase inhibition on the development of the rat embryo *in vitro*. *Teratology* **43**: 253–261.

821. **Davis, M. H. and J. Pieringer.** 1987. Regulation of dipeptidyl aminopeptidase I and angiotensin converting enzyme activities in cultured murine brain cells by cortisol and thyroid hormone. *J. Neurochem.* **48**: 447–454.

822. **Dawson, G., S. A. Dawson, and A. N. Siakotos.** 1989. Phospholipases and the molecular basis for the formation of ceroid in Batten disease. *Adv. Exp. Med. Biol.* **266**: 259–270.

823. **Dawson, G. and P. Glaser.** 1987. Apparent cathepsin B deficiency in neuronal ceroid lipofuscinosis can be explained by peroxide inhibition. *Biochem. Biophys. Res. Commun.* **147**: 267–274.

824. **Dawson, G. and P. T. Glaser.** 1988. Abnormal cathepsin B activity in Batten disease. *Am. J. Med. Genet. Suppl.* **5**: 209–220.

825. **Delaisse, J. M., Y. Eeckhout, and G. Vaes.** 1984. *In vivo* and *in vitro* evidence for the involvement of cysteine proteinases in bone resorption. *Biochem. Biophys. Res. Commun.* **125**: 441–447.

826. **Di Cola, D. and G. Federici.** 1983. Absence of tyrosine aminotransferase multiple forms in several mammalian animals. *Comp. Biochem. Physiol. B* **76**: 87–91.

827. **DiPaolo, B. R., R. J. Pignolo, and V. J. Cristofalo.** 1992. Overexpression of the two-chain form of cathepsin B in senescent WI-38 cells. *Exp. Cell Res.* **201**: 500–505.

828. **Docherty, K., R. Carroll, and D. F. Steiner.** 1983. Identification of a 31,500 molecular weight islet cell protease as cathepsin B. *Proc. Natl Acad. Sci. U. S. A.* **80**: 3245–3249.

829. **Docherty, K., J. C. Hutton, and D. F. Steiner.** 1984. Cathepsin B-related proteases in the insulin secretory granule. *J. Biol. Chem.* **259**: 6041–6044.

830. **Docherty, K. and I. D. Phillips.** 1988. Molecular forms of cathepsin B in rat thyroid cells (FRTL5): comparison with molecular forms in liver (Hep G2) and insulin-secreting cells (HIT T15). *Biochim. Biophys. Acta* **964**: 168–174.

831. **Doughty, M. J. and E. I. Gruenstein.** 1987. Cell growth and substrate effects on characteristics of a lysosomal enzyme (cathepsin C) in Duchenne muscular dystrophy fibroblasts. *Biochem. Cell Biol.* **65**: 617–625.

832. **Eisenhauer, D. A., R. Hutchinson, T. Javed, and J. K. McDonald.** 1983. Identification of a cathepsin B-like protease in the crevicular fluid of gingivitis patients. *J. Dent. Res.* **62**: 917–921.

833. **Eley, B. M. and S. W. Cox.** 1991. Cathepsin B- and L-like activities at local gingival sites of chronic periodontitis patients. *J. Clin. Periodontol.* **18**: 499–504.

834. **Eley, B. M. and S. W. Cox.** 1992. Cathepsin B/L-, elastase-, tryptase-, trypsin- and dipeptidyl peptidase IV-like activities in gingival crevicular fluid: correlation with clinical parameters in untreated chronic periodontitis patients. *J. Periodontal. Res.* **27**: 62–69.

835. **Eley, B. M. and S. W. Cox.** 1992. Cathepsin B/L-, elastase-, tryptase-, trypsin- and dipeptidyl peptidase IV-like activities in gingival crevicular fluid: a comparison of levels before and after periodontal surgery in chronic periodontitis patients. *J. Periodontol.* **63**: 412–417.

836. **Eley, B. M., S. W. Cox, and R. M. Watson.** 1991. Protease activities in peri-implant sulcus fluid from patients with permucosal osseointegrated dental implants. Correlation with clinical parameters. *Clin. Oral Implants Res.* **2**: 62–70.

837. **Erickson Lawrence, M., S. D. Zabludoff, and W. W. Wright.** 1991. Cyclic protein-2, a secretory product of rat Sertoli cells, is the proenzyme form of cathepsin L. *Mol. Endocrinol.* **5**: 1789–1798.

838. **Etherington, D. J., R. A. Maciewicz, R. W. Mason, M. A. Taylor, and R. J. Wardale.** 1985. A comparative study of cathepsins N and L and their distribution in different tissues. *Prog. Clin. Biol. Res.* **180**: 223–224.

839. **Etherington, D. J., R. W. Mason, M. A. Taylor, and R. J. Wardale.** 1984. Production of a monospecific antiserum to cathepsin L: the histochemical location of enzyme in rabbit fibroblasts. *Biosci. Rep.* **4**: 121–127.

840. **Etherington, D. J., M. A. Taylor, and B. Henderson.** 1988. Elevation of cathepsin L levels in the synovial lining of rabbits with antigen-induced arthritis. *Br. J. Exp. Pathol.* **69**: 281–289.

841. **Ezaki, J., L. S. Wolfe, K. Ishidoh, D. Muno, T. Ueno, and E. Kominami.** 1994. Degradative pathway of mitochondrial ATP synthase subunit C: analyses in neuronal ceroid lipofuscinosis. *The 10th International Conference on Intracellular Protein Catabolism, Tokyo* L17 (Abstr.).

842. **Fagotto, F.** 1990. Yolk degradation in tick eggs: II. Evidence that cathepsin L-like proteinase is stored as a latent, acid-activable proenzyme. *Arch. Insect. Biochem. Physiol.* **14**: 237–252.

843. **Fagotto, F.** 1990. Yolk degradation in tick eggs: I. Occurrence of a cathepsin L-like acid proteinase in yolk spheres. *Arch. Insect. Biochem. Physiol.* **14**: 217–235.

844. **Felleisen, R. and M. Q. Klinkert.** 1990. *In vitro* translation and processing of cathepsin B of *Schistosoma mansoni. EMBO J.* **9**: 371–377.

845. **Flipo, R. M., F. Heron, G. Huet, M. Balduyck, P. Degand, B. Duquesnoy, and B. Delcambre.** 1991. [Thiol-proteolytic activity in rheumatoid polyarthritis. Assay by spectrofluorimetry]. *Rev. Rhum. Mal. Osteoartic.* **58**: 131–133.

846. **Fraenkel Conrat, J., W. B. Chew, F. Pitlick, and S. Barber.** 1966. Certain properties of leukocytic cathepsins in health and disease. *Cancer* **19**: 1393–1398.

847. **Freimert, C., E. I. Closs, M. Silbermann, V. Erfle, and P. G. Strauss.** 1991. Isolation of a cathepsin B-encoding cDNA from murine osteogenic cells. *Gene* **103**: 259–261.

848. **Frohlich, E., G. Schaumburg Lever, and C. Klessen.** 1993. Immunelectron microscopic localization of cathepsin B in human exocrine glands. *J. Cutan. Pathol.* **20**: 54–60.

849. **Fukuyo, T., S. Hayasaka, S. Hara, M. Nakazawa, and K. Mizuno.** 1984. Acidic lens-protein degrading activity: I. Distribution in bovine ocular tissues. *Jpn. J. Ophthalmol.* **28**: 136–139.

850. **Furuhashi, M., A. Nakahara, H. Fukutomi, E. Kominami, D. Grube, and Y. Uchiyama.** 1991. Immunocytochemical localization of cathepsins B, H, and L in the rat gastro-duodenal mucosa. *Histochemistry* **95**: 231–239.

851. **Furuhashi, M., A. Nakahara, H. Fukutomi, E. Kominami, and Y. Uchiyama.** 1992. Changes in subcellular structures of parietal cells in the rat gastric gland after omeprazole. *Arch. Histol. Cytol.* **55**: 191–201.

852. **Furuno, K., M. N. Goodman, and A. L. Goldberg.** 1990. Role of different proteolytic systems in the degradation of muscle proteins during denervation atrophy. *J. Biol. Chem.* **265**: 8550–8557.

853. **Gairola, C. G., N. I. Galicki, C. Cardozo, Y. L. Lai, and M. Lesser.** 1989. Cigarette smoke stimulates cathepsin B activity in alveolar macrophages of rats. *J. Lab. Clin. Med.* **114**: 419–425.

854. **Gelman, B. B., L. Papa, M. H. Davis, and E. Gruenstein.** 1980. Decreased lysosomal dipeptidyl aminopeptidase I activity in cultured human skin fibroblasts in Duchenne's muscular dystrophy. *J. Clin. Invest.* **65**: 1398–1406.

855. **Gerard, K. W., A. R. Hipkiss, and D. L. Schneider.** 1988. Degradation of intracellular protein in muscle. Lysosomal response to modified proteins and chloroquine. *J. Biol. Chem.* **263**: 18886–18890.

856. **Ghiso, J., E. Saball, J. Leoni, A. Rostagno, and B. Frangione.** 1990. Binding of cystatin C to C4: the importance of sense–antisense peptides in their interaction. *Proc. Natl Acad. Sci. U. S. A.* **87**: 1288–1291.

857. **Goldspink, D. F. and S. E. Lewis.** 1985. Age- and activity-related changes in three proteinase enzymes of rat skeletal muscle. *Biochem. J.* **230**: 833–836.

858. **Goldspink, D. F., S. E. Lewis, and F. J. Kelly.** 1985. Protein turnover and cathepsin B activity in several individual tissues of foetal and senescent rats. *Comp. Biochem. Physiol. B* **82**: 849–853.

859. **Gorthy, W. C. and P. Azari.** 1987. Biochemical and histochemical evidence for lysosomal proteases in rodent lenses. *Exp. Eye Res.* **44**: 747–753.

860. **Gossrau, R.** 1991. Histochemical and biochemical studies of dipeptidyl peptidase I (DPP I) in laboratory rodents. *Acta Histochem.* **91**: 85–100.

861. **Gotz, B., R. Felleisen, E. Shaw, and M. Q. Klinkert.** 1992. Expression of an active cathepsin B-like protein Sm31 from *Schistosoma mansoni* in insect cells. *Trop. Med. Parasitol.* **43**: 282–284.

862. **Gotz, B. and M. Q. Klinkert.** 1993. Expression and partial characterization of a cathepsin B-like enzyme (Sm31) and a proposed 'haemoglobinase' (Sm32) from *Schistosoma mansoni. Biochem. J.* **290**: 801–806.

863. **Govindarajan, K. R. and J. Clausen.** 1974. Regional distribution of cathepsins B_1 and D in human brain. *Brain Res.* **67**: 141–146.

864. **Graf, M., U. Leeman, F. Ruch, and P. Strauli.** 1979. The fluorescence and bright field microscopic demonstration of cathepsin B in human fibroblasts. *Histochemistry* **64**: 319–322.

865. **Griffiths, G. M. and S. Isaaz.** 1993. Granzymes A and B are targeted to the lytic granules of lymphocytes by the mannose-6-phosphate receptor. *J. Cell Biol.* **120**: 885–896.

866. **Grubb, J. D., T. R. Koszalk, J. J. Drabick, and R. M. Metrione.** 1991. The activities of thiol proteases in the rat visceral yolk sac increase during late gestation. *Placenta* **12**: 143–151.

867. **Hall Angeras, M., P. O. Hasselgren, R. V. Dimlich, and J. E. Fischer.** 1991. Myofibrillar proteinase, cathepsin B, and protein breakdown rates in skeletal muscle from septic rats. *Metabolism* **40**: 302–306.

868. **Hamaguchi, Y., M. Ohi, Y. Sakakura, and Y. Miyoshi.** 1986. Significance of lysosomal proteases; cathepsins B and H in maxillary mucosa and nasal polyp with non-atopic chronic inflammation. *Rhinology* **24**: 187–194.

869. **Hamaguchi, Y., K. Sakakura, Y. Majima, and Y. Sakakura.** 1986. Lysosomal thiol proteases in middle ear effusions. *Ann. Otol. Rhinol. Laryngol. Suppl.* **124**: 9–12.

870. **Hamaguchi, Y., K. Sakakura, Y. Majima, and Y. Sakakura.** 1987. Cathepsin B-like thiol proteases and collagenolytic proteases in middle ear effusion from acute and chronic otitis media with effusion. *Acta Otolaryngol. Stockh.* **104**: 119–124.

871. **Hamaguchi, Y., K. Sakakura, Y. Majima, Y. Sakakura, and Y. Miyoshi.** 1986. Lysosomal thiol proteases (cathepsin B-like proteases) in serous middle ear effusions from adult patients. *Acta Otolaryngol. Stockh.* **101**: 257–262.

872. **Hamaguchi, Y., Y. Sakakura, Y. Majima, and S. K. Juhn.** 1987. Kinetics of lysosomal protease activity in human otitis media with effusion. *Am. J. Otolaryngol.* **8**: 194–198.

873. **Hamaguchi, Y., K. Takeuchi, C. S. Jin, Y. Majima, H. Suzumura, Y. Sakakura, and S. K. Juhn.** 1991. The relationship between proteases activity and glycoprotein levels in middle ear effusions from experimental otitis media in cats. *Acta Otolaryngol. Suppl. Stockh.* **483**: 23–29.

874. **Hamaguchi, Y., M. Taya, H. Suzumura, and Y. Sakakura.** 1990. Lysosomal proteases and protease inhibitors in nasal allergy and non-atopic sinusitis. *Am. J. Otolaryngol.* **11**: 37–43.

875. **Hamilton, R. T., K. A. Bruns, M. A. Delgado, J. K. Shim, Y. Fang, D. T. Denhardt, and M. Nilsen Hamilton.** 1991. Developmental expression of cathepsin L and c-rasHa in the mouse placenta. *Mol. Reprod. Dev.* **30**: 285–292.

876. **Hanewinkel, H., J. Glossl, and H. Kresse.** 1987. Biosynthesis of cathepsin B in cultured normal and I-cell fibroblasts. *J. Biol. Chem.* **262**: 12351–12355.

877. **Harris, C. I. and A. G. Baillie.** 1990. The localized elevation of cathepsins B and L in rat gastrocnemius muscle following tenotomy. *Biochem. Soc. Trans.* **18**: 1254–1255.

878. **Hashizume, Y., S. Waguri, T. Watanabe, E. Kominami, and Y. Uchiyama.** 1993. Cysteine proteinases in rat parathyroid cells with special reference to their correlation with parathyroid hormone (PTH) in storage granules. *J. Histochem. Cytochem.* **41**: 273–282.

879. **Hayasaka, S., S. Hara, Y. Takaku, and K. Mizuno.** 1978. Distribution and some properties of cathepsin B in the bovine eyes. *Exp. Eye Res.* **26**: 57–63.

880. **Hayasaka, S., T. Shiono, S. Hara, and T. Fukuyo.** 1983. Partial purification of cathepsin B in the bovine ciliary body and iris. *Invest. Ophthalmol. Vis. Sci.* **24**: 682–686.

881. **Hayes, D. J., C. R. Stubberfield, J. D. McBride, and D. L. Wilson.** 1991. Alterations in cysteine proteinase content of rat lung associated with development of *Pneumocystis carinii* infection. *Infect. Immun.* **59**: 3581–3588.

882. **Heffernan, M., A. Smith, D. Curtain, S. McDonnell, J. Ryan, and J. P. Dalton.** 1991. Characterisation of a cathepsin-B proteinase released by *Fasciola hepatica* (liver fluke). *Biochem. Soc. Trans.* **19**: 27S.

883. **Henze, K. and G. Wolfram.** 1988. [Lysosomal enzyme activity of monocytes/macrophages following incubation with postprandial hyperlipemic serum and its significance for the development of atherosclerosis]. *Klin. Wochenschr.* **66**: 144–148.

884. **Hirano, T., T. Manabe, and T. Tobe.** 1990. [Changes of lysosomal enzymes in pancreatic juice following hepatectomy in rats]. *Nippon Shokakibyo Gakkai Zasshi* **87**: 2503–2508.

885. **Hiwasa, T., T. Sawada, K. Tanaka, T. Chiba, T. Tanaka, E. Kominami, N. Katunuma, and S. Sakiyama.** 1991. Co-localization of ras gene products and cathepsin L in cytoplasmic vesicles in v-Ha-ras-transformed NIH3T3 mouse fibroblasts. *Biomed. Biochim. Acta* **50**: 579–585.

886. **Howie, A. J., D. Burnett, and J. Crocker.** 1985. The distribution of cathepsin B in human tissues. *J. Pathol.* **145**: 307–314.

887. **Huang, S., S. Reisch, L. Schaefer, M. Teschner, A. Heidland, and R. M. Schaefer.** 1992. Effect of dietary protein on glomerular proteinase activities. *Miner. Electrolyte Metab.* **18**: 84–88.

888. **Huet, G., R. M. Flipo, C. Colin, A. Janin, B. Hemon, H. Collyn, R. Lafyatis, B. Duquesnoy, and P. Degand.** 1993. Stimulation of the secretion of latent cysteine proteinase activity by tumor necrosis factor alpha and interleukin-1. *Arthritis Rheum.* **36**: 772–780.

889. **Hummel, R. P., J. H. James, B. W. Warner, P. O. Hasselgren, and J. E. Fischer.** 1988. Evidence that cathepsin B contributes to skeletal muscle protein breakdown during sepsis. *Arch. Surg.* **123**: 221–224.

890. **Hummel, R. P., B. W. Warner, J. H. James, P. O. Hasselgren, and J. E. Fischer.** 1988. Effects of indomethacin and leupeptin on muscle cathepsin B activity and protein degradation during sepsis. *J. Surg. Res.* **45**: 140–144.

891. **Ii, K., K. Hizawa, E. Kominami, Y. Bando, and N. Katunuma.** 1985. Different immunolocalizations of cathepsins B, H, and L in the liver. *J. Histochem. Cytochem.* **33**: 1173–1175.

892. **Ii, K., K. Hizawa, I. Nonaka, H. Sugita, E. Kominami, and N. Katunuma.** 1986. Abnormal increases of lysosomal cysteinine proteinases in rimmed vacuoles in the skeletal muscle. *Am. J. Pathol.* **122**: 193–198.

893. **Ii, K., H. Ito, E. Kominami, and A. Hirano.** 1993. Abnormal distribution of cathepsin proteinases and endogenous inhibitors (cystatins) in the hippocampus of patients with Alzheimer's disease, parkinsonism–dementia complex on Guam, and senile dementia and in the aged. *Virchows Arch. A. Pathol. Anat. Histopathol.* **423**: 185–194.

894. **Im, B., E. Kominami, D. Grube, and Y. Uchiyama.** 1989. Immunocytochemical localization of cathepsins B and H in human pancreatic endocrine cells and insulinoma cells. *Histochemistry* **93**: 111–118.

895. **Ishii, Y., Y. Hashizume, T. Watanabe, S. Waguri, N. Sato, M. Yamamoto, S. Hasegawa, E. Kominami, and Y. Uchiyama.** 1991. Cysteine proteinases in bronchoalveolar epithelial cells and lavage fluid of rat lung. *J. Histochem. Cytochem.* **39**: 461–468.

896. **Ishiura, S., I. Nonaka, H. Nakase, K. Tsuchiya, S. Okada, and H. Sugita.** 1983. Immunocytochemical localization of cathepsin B in degenerating rat skeletal muscle induced by a local anesthetic, bupivacaine. *J. Biochem. Tokyo* **94**: 311–314.

897. **Ishiura, S., I. Nonaka, and H. Sugita.** 1986. Biochemical aspects of bupivacaine-induced acute muscle degradation. *J. Cell Sci.* **83**: 197–212.

898. **Jimi, T., Y. Satoh, A. Takeda, S. Shibuya, Y. Wakayama, and K. Sugita.** 1992. Strong immunoreactivity of cathepsin L at the site of rimmed vacuoles in diseased muscles. *Brain* **115**(Pt 1): 249–260.

899. **Johnson, M. H., C. R. Calkins, R. D. Huffman, D. D. Johnson, and D. D. Hargrove.** 1990. Differences in cathepsin B + L and calcium-dependent protease activities among breed type and their relationship to beef tenderness. *J. Anim. Sci.* **68**: 2371–2379.

900. **Kane, S. E., B. R. Troen, S. Gal, K. Ueda, I. Pastan, and M. M. Gottesman.** 1988. Use of a cloned multidrug resistance gene for coamplification and overproduction of major excreted protein, a transformation-regulated secreted acid protease. *Mol. Cell Biol.* **8**: 3316–3321.

901. **Kar, N. C. and C. M. Pearson.** 1978. Muscular dystrophy and activation of proteinases. *Muscle Nerve* **1**: 308–313.

902. **Karapetian, R. G., S. S. Oganesian, T. N. Akopian, and E. G. Dzhanpoladian.** 1993. [Activity of proteolytic enzymes and their inhibitors in experimental myocardial ischemia]. *Vopr. Med. Khim.* **39**: 29–31.

903. **Karhukorpi, E. K., P. Vihko, and K. Vaananen.** 1992. A difference in the enzyme contents of resorption lacunae and secondary lysosomes of osteoclasts. *Acta Histochem.* **92**: 1–11.

904. **Kasai, M., T. Shirasawa, M. Kitamura, K. Ishido, E. Kominami, and K. Hirokawa.** 1993. Proenzyme from of cathepsin L produced by thymic epithelial cells promotes proliferation of immature thymocytes in the presence of IL-1, IL-7, and anti-CD3 antibody. *Cell Immunol.* **150**: 124–136.

905. **Katona, G., V. Szekessy Hermann, F. Guba, and I. Sohar.** 1991. Effect of vitamin E-deficiency on the activity of some lysosomal and non-lysosomal proteases in rabbit muscles. *Gen. Physiol. Biophys.* **10**: 505–514.

906. **Katunuma, N. and E. Kominami.** 1987. Abnormal expression of lysosomal cysteine proteinases in muscle wasting diseases. *Rev. Physiol. Biochem. Pharmacol.* **108**: 1–20.

907. **Kirschke, H., L. Wood, F. J. Roisen, and J. W. Bird.** 1983. Activity of lysosomal cysteine proteinase during differentiation of rat skeletal muscle. *Biochem. J.* **214**: 871–877.

908. **Kloss, P., H. G. Bernstein, H. Kirschke, and B. Wiederanders.** 1991. Ultrastructural study of cathepsin B immunoreactivity in rat brain neurons: lysosomal and extralysosomal localizations of the antigen. *Acta Anat. Basel* **142**: 138–140.

909. **Komatsu, K., K. Tsukuda, J. Hosoya, and S. Satoh.** 1986. Elevations of cathepsin B and cathepsin L activities in forelimb and hind limb muscles of genetically dystrophic mice. *Exp. Neurol.* **93**: 642–646.

910. **Kominami, E., Y. Bando, K. Ii, K. Hizawa, and N. Katunuma.** 1984. Increases in cathepsins B and L and thiol proteinase inhibitor in muscle of dystrophic hamsters. Their localization in invading phagocytes. *J. Biochem. Tokyo* **96**: 1841–1848.

911. **Kopitz, J., A. Arnold, T. Meissner, and M. Cantz.** 1993. Protein catabolism in fibroblasts cultured from patients with mucolipidosis II and other lysosomal disorders. *Biochem. J.* **295**: 577–580.

912. **Kregar, I., V. Turk, and D. Lebez.** 1967. Acid proteinase from the small intestinal mucosa of rats. *Z. Naturforsch. B* **22**: 992.

913. **Kretchmar, D. H., M. R. Hathaway, R. J. Epley, and W. R. Dayton.** 1989. *In vivo* effect of a β–adrenergic agonist on activity of calcium-dependent proteinases, their specific inhibitor, and cathepsins B and H in skeletal muscle. *Arch. Biochem. Biophys.* **275**: 228–235.

914. **Kugler, P.** 1982. Fluorescent histochemical demonstration of cathepsin B in the rat yolk sac. *Histochemistry* **75**: 215–218.

915. **Kugler, P.** 1985. Demonstration of cathepsin B in the rat kidney using fluorescence histochemistry. *Histochemistry* **82**: 299–300.

916. **Kugler, P. and G. Vornberger.** 1986. Renal cathepsin-B activities in rats after castration and treatment with sex hormones. *Histochemistry* **85**: 157–161.

917. **Kugler, P., G. Wolf, and J. Scherberich.** 1985. Histochemical demonstration of peptidases in the human kidney. *Histochemistry* **83**: 337–341.

918. **Kumamoto, T., S. Araki, S. Watanabe, N. Ikebe, and N. Fukuhara.** 1989. Experimental chloroquine myopathy: morphological and biochemical studies. *Eur. Neurol.* **29**: 202–207.

919. **Kunimatsu, K., K. Yamamoto, E. Ichimaru, Y. Kato, and I. Kato.** 1990. Cathepsins B, H and L activities in gingival crevicular fluid from chronic adult periodontitis patients and experimental gingivitis subjects. *J. Periodontal. Res.* **25**: 69–73.

920. **Lah, T. T., J. Babnik, E. Schiffmann, V. Turk, and U. Skaleric.** 1993. Cysteine proteinases and inhibitors in inflammation: their role in periodontal disease. *J. Periodontol.* **64**: 485–491.

921. **Laszlo, A., I. Sohar, G. Falkay, A. Kovacs, V. Halmos, and J. Szabo.** 1990. Physiological values of cysteine and metalloproteinase activities in chorionic villi. *Acta Obstet. Gynecol. Scand.* **69**: 397–398.

922. **Laszlo, A., I. Sohar, I. Sagi, J. Kovacs, and A. Kovacs.** 1992. Activity of cathepsin H, B and metalloproteinase in the serum of patients with acute myocardial infarction. *Clin. Chim. Acta* **210**: 233–235.

923. **Leites, F. L., I. P. Kudinkina, and B. B. Fuks.** 1969. [Histochemical study of cathepsin C in the spleen during the early stages of immunogenesis]. *Dokl. Akad. Nauk. SSSR* **187**: 193–196.

924. **Lenarcic, B., D. Gabrijelcic, B. Rozman, M. Drobnic Kosorok, and V. Turk.** 1988. Human cathepsin B and cysteine proteinase inhibitors (CPIs) in inflammatory and metabolic joint diseases. *Biol. Chem. Hoppe Seyler* **369**(Suppl.): 257–261.

925. **Lesser, M., J. C. Chang, J. Orlowski, K. H. Kilburn, and M. Orlowski.** 1983. Cathepsin B and prolyl endopeptidase activity in rat peritoneal and alveolar macrophages. Stimulation of peritoneal macrophages by saline lavage. *J. Lab. Clin. Med.* **101**: 327–334.

926. **Lesser, M., N. Galicki, C. Cardozo, and C. G. Gariola.** 1989. Cathepsin L activity in alveolar macrophages of rats: response to cigarette smoke. *Am. J. Respir. Cell Mol. Biol.* **1**: 371–376.

927. **Li, W., R. C. Jaffe, and H. G. Verhage.** 1992. Immunocytochemical localization and messenger ribonucleic acid levels of a progesterone-dependent endometrial secretory protein (cathepsin L) in the pregnant cat uterus. *Biol. Reprod.* **47**: 21–28.

928. **Li, W. G., R. C. Jaffe, A. T. Fazleabas, and H. G. Verhage.** 1991. Progesterone-dependent cathepsin L proteolytic activity in cat uterine flushings. *Biol. Reprod.* **44**: 625–631.

929. **Lokshina, L. A., V. S. Bylinkina, R. S. Samoilova, T. A. Gureeva, N. V. Golubeva, and A. M. Polianskaia.** 1993. [Proteolytic enzymes in human lymphocytic leukemia cells. I. Activity of dipeptidylaminopeptidase IV, plasminogen activator and cathepsins B and L in cells with different immunologic phenotype]. *Biokhimiia* **58**: 1104–1115.

930. **Lonsdale Eccles, J. D. and G. W. Mpimbaza.** 1986. Thiol-dependent proteases of African trypanosomes. Analysis by electrophoresis in sodium dodecyl sulphate/polyacrylamide gels co-polymerized with fibrinogen. *Eur. J. Biochem.* **155**: 469–473.

931. **Lougheed, M., H. F. Zhang, and U. P. Steinbrecher.** 1991. Oxidized low density lipoprotein is resistant to cathepsins and accumulates within macrophages. *J. Biol. Chem.* **266**: 14519–14525.

932. **Madsen, K. M. and C. H. Park.** 1987. Lysosome distribution and cathepsin B and L activity along the rabbit proximal tubule. *Am. J. Physiol.* **253**: F1290-F1301.

933. **Mainferme, F., R. Wattiaux, and K. von Figura.** 1985. Synthesis, transport and processing of cathepsin C in Morris hepatoma 7777 cells and rat hepatocytes. *Eur. J. Biochem.* **153**: 211–216.

934. **Marjomaki, V. S., A. P. Huovila, M. A. Surkka, I. Jokinen, and A. Salminen.** 1990. Lysosomal trafficking in rat cardiac myocytes. *J. Histochem. Cytochem.* **38**: 1155–1164.

935. **Martel Pelletier, J., J. M. Cloutier, and J. P. Pelletier.** 1990. Cathepsin B and cysteine protease inhibitors in human osteoarthritis. *J. Orthop. Res.* **8**: 336–344.

936. **Marzabadi, M. R., R. S. Sohal, and U. T. Brunk.** 1991. Mechanisms of lipofuscinogenesis: effect of the inhibition of lysosomal proteinases and lipases under varying concentrations of ambient oxygen in cultured rat neonatal myocardial cells. *APMIS* **99**: 416–426.

937. **Matsuba, H., T. Watanabe, M. Watanabe, Y. Ishii, S. Waguri, E. Kominami, and Y. Uchiyama.** 1989. Immunocytochemical localization of prorenin, renin, and cathepsins B, H, and L in juxtaglomerular cells of rat kidney. *J. Histochem. Cytochem.* **37**: 1689–1697.

938. **Matsui, K., N. Shirasawa, and Y. Eto.** 1990. Cytoplasmic accumulations in rat primary brain cell cultures following treatment with E-64, a thiol protease inhibitor. *Dev. Neurosci.* **12**: 133–139.

939. **Mazur, M. T., J. J. Shultz, and J. L. Myers.** 1990. Granular cell tumor. Immunohistochemical analysis of 21 benign tumors and one malignant tumor. *Arch. Pathol. Lab. Med.* **114**: 692–696.

940. **Mbawa, Z. R., P. Webster, and J. D. Lonsdale Eccles.** 1991. Immunolocalization of a cysteine protease within the lysosomal system of *Trypanosoma congolense*. *Eur. J. Cell Biol.* **56**: 243–250.

941. **McCormick, D.** 1993. Secretion of cathepsin B by human gliomas *in vitro*. *Neuropathol. Appl. Neurobiol.* **19**: 146–151.

942. **McDonald, J. K., J. T. Culbertson, and N. O. Owers.** 1993. Identification and localization of a novel cathepsin S-like proteinase in guinea pig spermatozoa. *Arch. Biochem. Biophys.* **305**: 1–8.

943. **McGinty, A., M. Moore, D. W. Halton, and B. Walker.** 1993. Characterization of the cysteine proteinases of the common liver fluke *Fasciola hepatica* using novel, active-site directed affinity labels. *Parasitology* **106**: 487–493.

944. **McIntyre, G. F. and A. H. Erickson.** 1991. Procathepsins L and D are membrane-bound in acidic microsomal vesicles. *J. Biol. Chem.* **266**: 15438–15445.

945. **Meheus, L. A., L. M. Fransen, J. G. Raymackers, H. A. Blockx, J. J. Van Beeumen, S. M. Van Bun, and A. Van de Voorde.** 1993. Identification by microsequencing of lipopolysaccharide-induced proteins secreted by mouse macrophages. *J. Immunol.* **151**: 1535–1547.

946. **Miller, B. F. and H. V. Kothari.** 1969. Increased activity of lysosomal enzymes in human atherosclerotic aortas. *Exp. Mol. Pathol.* **10**: 288–294.

947. **Mimura, Y., M. Johnosono, S. Izumo, M. Kuriyama, and M. Osame.** 1992. [Adult-onset neuronal ceroid lipofuscinosis—a case report with biological study]. *Rinsho Shinkeigaku* **32**: 771–773.

948. **Miyata, M.** 1988. [Evaluation of serum cathepsin B-like activity in liver diseases]. *Fukuoka Igaku Zasshi* **79**: 832–838.

949. **Morland, B.** 1979. Studies on selective induction of lysosomal enzyme activities in mouse peritoneal macrophages. *J. Reticuloendothel. Soc.* **26**: 749–762.

950. **Morland, B.** 1985. Cathepsin B activity in human blood monocytes during differentiation in vitro. *Scand. J. Immunol.* **22**: 9–16.

951. **Morland, B. and J. Morland.** 1978. Selective induction of lysosomal enzyme activities in mouse peritoneal macrophages. *J. Reticuloendothel. Soc.* **23**: 469–477.

952. **Morsches, B., H. Holzmann, G. W. Korting, and M. Braun.** 1966. [Contribution on the method of action of chloroquine on the connective tissue, especially on its influence on the serum cathepsin activity]. *Z. Rheumaforsch.* **25**: 398–403.

953. **Mort, J. S. and A. D. Recklies.** 1986. Interrelationship of active and latent secreted human cathepsin B precursors. *Biochem. J.* **233**: 57–63.

954. **Mort, J. S., A. Tam, D. F. Steiner, and S. J. Chan.** 1988. Expression of rat and mouse cathepsin B precursors in *Escherichia coli*. *Biol. Chem. Hoppe Seyler* **369**(Suppl.): 163–167.

955. **Muno, D., N. Sutoh, T. Watanabe, Y. Uchiyama, and E. Kominami.** 1990. Effect of metabolic alterations on the density and the contents of cathepsins B, H and L of lysosomes in rat macrophages. *Eur. J. Biochem.* **191**: 91–98.

956. **Nakao, H., Y. Kurita, R. Tsuboi, K. Takamori, and H. Ogawa.** 1989. Differences in induction of lysosomal protease activity by protease inhibitors in B16 melanoma cell lines. *Int. J. Biochem.* **21**: 139–142.

957. **Nast Kolb, D., C. Waydhas, M. Jochum, K. H. Duswald, W. Machleidt, M. Spannagl, W. Schramm, H. Fritz, and L. Schweiberer.** 1992. [Biochemical factors as objective parameters for assessing the prognosis in polytrauma]. *Unfallchirurg.* **95**: 59–66.

958. **Nemere, I. and A. W. Norman.** 1991. Redistribution of cathepsin B activity from the endosomal–lysosomal pathway in chick intestine within 3 min of calcium absorption. *Mol. Cell Endocrinol.* **78**: 7–16.

959. **Nguyen, Q., J. S. Mort, and P. J. Roughley.** 1990. Cartilage proteoglycan aggregate is degraded more extensively by cathepsin L than by cathepsin B. *Biochem. J.* **266**: 569–573.

960. **Nilsen Hamilton, M., Y. J. Jang, M. Delgado, J. K. Shim, K. Bruns, C. P. Chiang, Y. Fang, C. L. Parfett, D. T. Denhardt, and R. T. Hamilton.** 1991. Regulation of the expression of mitogen-regulated protein (MRP; proliferin) and cathepsin L in cultured cells and in the murine placenta. *Mol. Cell Endocrinol.* **77**: 115–122.

961. **Nishimura, Y. and K. Kato.** 1987. *In vitro* biosynthesis of the lysosomal cathepsin H. *Biochem. Biophys. Res. Commun.* **146**: 159–164.

962. **Nishimura, Y. and K. Kato.** 1987. Intracellular transport and processing of lysosomal cathepsin H. *Biochem. Biophys. Res. Commun.* **148**: 329–334.

963. **Nishimura, Y. and K. Kato.** 1987. Intracellular transport and processing of lysosomal cathepsin B. *Biochem. Biophys. Res. Commun.* **148**: 254–259.

964. **Nishimura, Y. and K. Kato.** 1992. Expression of mouse cathepsin L cDNA in *Saccharomyces cerevisiae*: evidence that cathepsin L is sorted for targeting to yeast vacuole. *Arch. Biochem. Biophys.* **298**: 318–324.

965. **Nonaka, I., S. Ishiura, K. Arahata, K. Ishibashi Ueda, T. Maruyama, and K. Ii.** 1989. Progression in nemaline myopathy. *Acta Neuropathol. Berl.* **78**: 484–491.

966. **Oda, K., Y. Nishimura, Y. Ikehara, and K. Kato.** 1991. Bafilomycin A_1 inhibits the targeting of lysosomal acid hydrolases in cultured hepatocytes. *Biochem. Biophys. Res. Commun.* **178**: 369–377.

967. **Ohsawa, Y., T. Nitatori, S. Higuchi, E. Kominami, and Y. Uchiyama.** 1993. Lysosomal cysteine and aspartic proteinases, acid phosphatase, and an endogenous cysteine proteinase inhibitor, cystatin-β, in rat osteoclasts. *J. Histochem. Cytochem.* **41**: 1075–1083.

968. **Olbricht, C. J.** 1988. Effect of glomerular proteinuria on the activities of lysosomal proteases in isolated segments of rat proximal tubule. *Adv. Exp. Med. Biol.* **240**: 283–291.

969. **Olbricht, C. J.** 1992. Distribution of cathepsins B and L in the kidney and their role in tubular protein absorption. *Eur. J. Clin. Chem. Clin. Biochem.* **30**: 675–681.

970. **Olbricht, C. J., J. K. Cannon, L. C. Garg, and C. C. Tisher.** 1986. Activities of cathepsins B and L in isolated nephron segments from proteinuric and nonproteinuric rats. *Am. J. Physiol.* **250**: F1055–F1062.

971. **Olbricht, C. J., J. K. Cannon, and C. C. Tisher.** 1987. Cathepsin B and L in nephron segments of rats with puromycin aminonucleoside nephrosis. *Kidney Int.* **32**: 354–361.

972. **Olbricht, C. J., M. Fink, and E. Gutjahr.** 1991. Alterations in lysosomal enzymes of the proximal tubule in gentamicin nephrotoxicity. *Kidney Int.* **39**: 639–646.

973. **Olbricht, C. J. and B. Geissinger.** 1992. Renal hypertrophy in streptozotocin diabetic rats: role of proteolytic lysosomal enzymes. *Kidney Int.* **41**: 966–972.

974. **Olbricht, C. J., H. J. Grone, C. Bossaller, E. Gutjahr, and R. F. Hertel.** 1990. Alterations in proteolytic enzymes of the proximal tubule in a rat model of cyclosporine nephrotoxicity. *Transplantation* **50**: 378–381.

975. **Olbricht, C. J., H. J. Grone, E. Gutjahr, and C. Bossaller.** 1990. Potential contribution of lysosomal proteases to cyclosporine-A-induced nephrotoxicity. *Toxicol. Lett.* **53**: 251–252.

976. **Olbricht, C. J., E. Gutjahr, M. Fink, and K. M. Koch.** 1988. Potential role of lysosomal proteases in gentamicin nephrotoxicity. *Adv. Exp. Med. Biol.* **240**: 305–308.

977. **Olbricht, C. J., H. Irmler, E. Gutjahr, and K. M. Koch.** 1993. Effect of low-molecular-weight dextran on proteolytic and nonproteolytic lysosomal enzymes in isolated segments of rat proximal tubule. *Nephron.* **64**: 262–267.

978. **Orlowski, M., J. Orlowski, M. Lesser, and K. H. Kilburn.** 1981. Proteolytic enzymes in bronchopulmonary lavage fluids: cathepsin B-like activity and prolyl endopeptidase. *J. Lab. Clin. Med.* **97**: 467–476.

979. **Osmak, M., M. Korbelik, A. Suhar, J. Skrk, and V. Turk.** 1989. The influence of cathepsin B and leupeptin on potentially lethal damage repair in mammalian cells. *Int. J. Radiat. Oncol. Biol. Phys.* **16**: 707–714.

980. **Ostensen, M., G. Husby, and B. Morland.** 1986. The effect of normal and rheumatic pregnancy sera on intracellular cathepsin B activity in human monocytes. *Clin. Exp. Immunol.* **63**: 434–440.

981. **Ostensen, M., B. Morland, G. Husby, and O. P. Rekvig.** 1983. A serum antibody in patients with rheumatoid arthritis stimulates cathepsin B activity in peritoneal mouse macrophages. *Clin. Exp. Immunol.* **54**: 397–404.

982. **Paczek, L., M. Teschner, R. M. Schaefer, J. Kovar, W. Romen, and A. Heidland.** 1991. Proteinase activity in isolated glomeruli of Goldblatt hypertensive rats. *Clin. Exp. Hypertens. A* **13**: 339–356.

983. **Padilla, M. L., N. I. Galicki, J. Kleinerman, M. Orlowski, and M. Lesser.** 1988. High cathepsin B activity in alveolar macrophages occurs with elastase-induced emphysema but not with bleomycin-induced pulmonary fibrosis in hamsters. *Am. J. Pathol.* **131**: 92–101.

984. **Page, A. E., M. J. Warburton, T. J. Chambers, J. A. Pringle, and A. R. Hayman.** 1992. Human osteoclastomas contain multiple forms of cathepsin B. *Biochim. Biophys. Acta* **1116**: 57–66.

985. **Palmer, D. N.** 1994. The ceroid lipofuscinoses (Batten's disease) are lysosomal proteinoses in which specific hydrophobic proteins are stored. *The 10th International Conference on Intracellular Protein Catabolism, Tokyo* P168 (Abstr.).

986. **Parent, J. B., H. C. Bauer, and K. Olden.** 1982. Tunicamycin treated fibroblasts secrete a cathepsin B-like protease. *Biochem. Biophys. Res. Commun.* **108**: 552–558.

987. **Phillips, I. D., E. G. Black, M. C. Sheppard, and K. Docherty.** 1989. Thyrotrophin, forskolin and ionomycin increase cathepsin B mRNA concentrations in rat thyroid cells in culture. *J. Mol. Endocrinol.* **2**: 207–212.

988. **Pontremoli, S., E. Melloni, M. Michetti, F. Salamino, B. Sparatore, and B. L. Horecker.** 1982. Localization of two lysosomal proteinases on the external surface of the lysosomal membrane. *Biochem. Biophys. Res. Commun.* **106**: 903–909.

989. **Poso, A. R., K. E. Penttila, and K. O. Lindros.** 1991. Heterogenous zonal distribution of lysosomal enzymes in rat liver. *Enzyme* **45**: 174–179.

990. **Prina, E. and J. C. Antoine.** 1990. [Localization and activity of different lysosomal proteases in rat macrophages infected by *Leishmania amazonensis*]. *Pathol. Biol. Paris* **38**: 1020–1022.

991. **Prina, E., J. C. Antoine, B. Wiederanders, and H. Kirschke.** 1990. Localization and activity of various lysosomal proteases in *Leishmania amazonensis*-infected macrophages. *Infect. Immun.* **58**: 1730–1737.

992. **Pringle, T. D., C. R. Calkins, M. Koohmaraie, and S. J. Jones.** 1993. Effects over time of feeding a β-adrenergic agonist to wether lambs on animal performance, muscle growth, endogenous muscle proteinase activities, and meat tenderness. *J. Anim. Sci.* **71**: 636–644.

993. **Punnonen, E. L., S. Autio, H. Kaija, and H. Reunanen.** 1993. Autophagic vacuoles fuse with the prelysosomal compartment in cultured rat fibroblasts. *Eur. J. Cell Biol.* **61**: 54–66.

994. **Punnonen, E. L., S. Autio, V. S. Marjomaki, and H. Reunanen.** 1992. Autophagy, cathepsin L transport, and acidification in cultured rat fibroblasts. *J. Histochem. Cytochem.* **40**: 1579–1587.

995. **Qian, F., A. Frankfater, R. V. Miller, S. J. Chan, and D. F. Steiner.** 1990. Molecular cloning of rat precursor cathepsin H and the expression of five lysosomal cathepsins in normal tissues and in a rat carcinosarcoma. *Int. J. Biochem.* **22**: 1457–1464.

996. **Randell, S. H. and P. L. Sannes.** 1988. Biochemical quantitation and histochemical localization of cathepsin B, dipeptidyl peptidases I and II, and acid phosphatase in pulmonary granulomatosis and fibrosis in rats. *Inflammation* **12**: 67–86.

997. **Reckelhoff, J. F., V. L. Tygart, M. M. Mitias, and J. L. Walcott.** 1993. STZ-induced diabetes results in decreased activity of glomerular cathepsin and metalloprotease in rats. *Diabetes.* **42**: 1425–1432.

998. **Recklies, A. D. and J. S. Mort.** 1985. Characterization of a cysteine proteinase secreted by mouse mammary gland. *Cancer Res.* **45**: 2302–2307.

999. **Recklies, A. D. and J. S. Mort.** 1985. Rat mammary gland in culture secretes a stable high molecular weight form of cathepsin L. *Biochem. Biophys. Res. Commun.* **131**: 402–407.

1000. **Recklies, A. D., C. White, J. Mitchell, and A. R. Poole.** 1985. Secretion of a cysteine proteinase from a hormone-independent cell population of cultured explants of murine mammary gland. *Cancer Res.* **45**: 2294–2301.

1001. **Reilly, J. J., J., P. Chen, L. Z. Sailor, R. W. Mason, and H. A. J. Chapman.** 1990. Uptake of extracellular enzyme by a novel pathway is a major determinant of cathepsin L levels in human macrophages. *J. Clin. Invest.* **86**: 176–183.

1002. **Reilly, J. J., J., P. Chen, L. Z. Sailor, D. Wilcox, R. W. Mason, and H. A. J. Chapman.** 1991. Cigarette smoking induces an elastolytic cysteine proteinase in macrophages distinct from cathepsin L. *Am. J. Physiol.* **261**: L41–L48.

1003. **Reilly, J. J., J., R. W. Mason, P. Chen, L. J. Joseph, V. P. Sukhatme, R. Yee, and H. A. J. Chapman.** 1989. Synthesis and processing of cathepsin L, an elastase, by human alveolar macrophages. *Biochem. J.* **257**: 493–498.

1004. **Ribari, O., I. Sziklai, J. G. Kiss, and I. Sohar.** 1983. Cathepsin-B activity in otosclerosis. *Arch. Otorhinolaryngol.* **238**: 123–125.

1005. **Riess, H., M. Jochum, W. Machleidt, G. Himmelreich, G. Blumhardt, R. Roissaint, and P. Neuhaus.** 1991. Possible role of extracellularly released phagocyte proteinases in coagulation disorder during liver transplantation. *Transplantation* **52**: 482–484.

1006. **Riess, H., M. Jochum, W. Machleidt, G. Himmelreich, R. Slama, and R. Steffen.** 1991. Possible role of the phagocytic proteinases, cathepsin B and elastase, in orthotopic liver transplantation. *Transplant Proc.* **23**: 1947–1948.

1007. **Rifkin, B. R., A. T. Vernillo, A. P. Kleckner, J. M. Auszmann, L. R. Rosenberg, and M. Zimmerman.** 1991. Cathepsin B and L activities in isolated osteoclasts. *Biochem. Biophys. Res. Commun.* **179**: 63–69.

1008. **Rinne, A., M. Jarvinen, H. Kirschke, B. Wiederanders, and V. K. Hopsu Havu.** 1986. Demonstration of cathepsins H and L in rat tissues. *Biomed. Biochim. Acta* **45**: 1465–1476.

1009. **Rinne, A., H. Kirschke, M. Jarvinen, V. K. Hopsu Havu, B. Wiederanders, and P. Bohley.** 1985. Localization of cathepsin H and its inhibitor in the skin and other stratified epithelia. *Arch. Dermatol. Res.* **277**: 190–194.

1010. **Rollag, H. and B. Morland.** 1988. The effect of recombinant interferons on cathepsin B activity in human monocytes. *Scand. J. Rheumatol. Suppl.* **76**: 79–83.

1011. **Rozhin, J., A. P. Gomez, G. H. Ziegler, K. K. Nelson, Y. S. Chang, D. Fong, J. M. Onoda, K. V. Honn, and B. F. Sloane.** 1990. Cathepsin B to cysteine proteinase inhibitor balance in metastatic cell subpopulations isolated from murine tumors. *Cancer Res.* **50**: 6278–6284.

1012. **Ryvniak, V. V., V. S. Gudumak, and E. S. Onia.** 1992. [The electron-histochemical detection of the activity of thiol proteinases in normal and cirrhotically altered liver]. *Biull. Eksp. Biol. Med.* **114**: 212–214.

1013. **Safina, A., T. Korolenko, G. Mynkina, M. Dushkin, and G. Krasnoselskaya.** 1992. Immunomodulators, inflammation and lysosomal proteinases of macrophages. *Agents Actions Suppl.* **38**: 191–197.

1014. **Saint Andre, J. P., V. Rohmer, F. Pinet, M. C. Rousselet, J. C. Bigorgne, and P. Corvol.** 1989. Renin and cathepsin B in human pituitary lactotroph cells. An ultrastructural study. *Histochemistry* **91**: 291–297.

1015. **Sakai, K., H. Moriya, A. Ueyama, and Y. Kishino.** 1991. Morphological heterogeneity among fractionated alveolar macrophages in their release of lysosomal enzymes. *Cell. Mol. Biol.* **37**: 85–94.

1016. **Sakai, K., Y. Nii, A. Ueyama, and Y. Kishino.** 1991. Fluorescence demonstration of cathepsin B activity in fractionated alveolar macrophages. *Cell. Mol. Biol.* **37**: 353–358.

1017. **Sakakura, K., Y. Hamaguchi, T. Harada, M. Yamagiwa, and Y. Sakakura.** 1990. Endotoxin and lysosomal protease activity in acute and chronic otitis media with effusion. *Ann. Otol. Rhinol. Laryngol.* **99**: 379–385.

1018. **Sano, M., Y. Wada, K. Ii, E. Kominami, N. Katunuma, and H. Tsukagoshi.** 1988. Immunolocalization of cathepsins B, H and L in skeletal muscle of X-linked muscular dystrophy (mdx) mouse. *Acta Neuropathol. Berl.* **75**: 217–225.

1019. **Schaefer, R. M., L. Paczek, S. Huang, M. Teschner, L. Schaefer, and A. Heidland.** 1992. Role of glomerular proteinases in the evolution of glomerulosclerosis. *Eur. J. Clin. Chem. Clin. Biochem.* **30**: 641–646.

1020. **Schive, K. and G. Volden.** 1982. Cathepsin B, cathepsin C and arylamidase in rabbit cornea. *Acta Ophthalmol. Copenh.* **60**: 765–772.

1021. **Schmidt, M., J. Heinrich, and M. Pfeifer.** 1990. [Proteolytic activities in bronchoalveolar lavage fluid in pneumonia and chronic bronchitis]. *Pneumologie* **44**(Suppl. 1): 308–309.

1022. **Seetharam, B., K. Y. Yeh, and D. H. Alpers.** 1980. Turnover of intestinal brush-border proteins during postnatal development in rat. *Am. J. Physiol.* **239**: G524–G531.

1023. **Shuto, K., H. Nakagawa, and S. Tsurufuji.** 1981. Cellular origin of cathepsin B in carrageenin-induced granuloma tissues in rats. *J. Pharmacobiodyn.* **4**: 513–519.

1024. **Sinha, A. A., D. F. Gleason, O. F. Deleon, M. J. Wilson, and B. F. Sloane.** 1993. Localization of a biotinylated cathepsin B oligonucleotide probe in human prostate including invasive cells and invasive edges by *in situ* hybridization. *Anat. Rec.* **235**: 233–240.

1025. **Sinha, A. A., D. F. Gleason, C. Limas, P. K. Reddy, M. R. Wick, K. A. Hagen, and M. J. Wilson.** 1989. Localization of cathepsin B in normal and hyperplastic human prostate by immunoperoxidase and protein A-gold techniques. *Anat. Rec.* **223**: 266–275.

1026. **Skrzydlewski, Z., A. Jasiewicz, S. Michalak, J. Tomaszewski, and K. Worowski.** 1984. [Lysosomal protease activity and protein degradation in the amniotic fluid]. *Ginekol. Pol.* **55**: 253–256.

1027. **Smith, S. M. and M. M. Gottesman.** 1989. Activity and deletion analysis of recombinant human cathepsin L expressed in *Escherichia coli*. *J. Biol. Chem.* **264**: 20487–20495.

1028. **Smith, S. M., S. E. Kane, S. Gal, R. W. Mason, and M. M. Gottesman.** 1989. Glycosylation of procathepsin L does not account for species molecular-mass differences and is not required for proteolytic activity. *Biochem. J.* **262**: 931–938.

1029. **Sohar, I., A. Laszlo, K. Gaal, and F. Mechler.** 1988. Cysteine and metalloproteinase activities in serum of Duchenne muscular dystrophic genotypes. *Biol. Chem. Hoppe Seyler* **369**(Suppl.): 277–279.

1030. **Stauber, W. T. and V. K. Fritz.** 1985. Decreased lysosomal protease content of skeletal muscles from streptozotocin-induced diabetic rats: a biochemical and histochemical study. *Histochem. J.* **17**: 613–622.

1031. **Stauber, W. T. and S. H. Ong.** 1981. Fluorescence demonstration of cathepsin B activity in skeletal, cardiac, and vascular smooth muscle. *J. Histochem. Cytochem.* **29**: 866–869.

1032. **Stauber, W. T. and S. H. Ong.** 1982. Fluorescence demonstration of dipeptidyl peptidase I (cathepsin C) in skeletal, cardiac, and vascular smooth muscles. *J. Histochem. Cytochem.* **30**: 162–164.

1033. **Stauber, W. T. and S. H. Ong.** 1982. Fluorescence demonstration of a cathepsin H-like protease in cardiac, skeletal and vascular smooth muscles. *Histochem. J.* **14**: 585–591.

1034. **Suleiman, S. A., G. L. Jones, H. Singh, and D. R. Labrecque.** 1980. Changes in lysosomal cathepsins during liver regeneration. *Biochim. Biophys. Acta* **627**: 17–22.

1035. **Suominen, J. and V. K. Hopsu Havu.** 1971. Cathepsin B′ in the thyroid gland. *Acta Chem. Scand.* **25**: 2531–2540.

1036. **Sutherland, J. H. and L. M. Greenbaum.** 1983. Paradoxical effect of leupeptin *in vivo* on cathepsin B activity. *Biochem. Biophys. Res. Commun.* **110**: 332–338.

1037. **Takahashi, H., K. Ishidoh, D. Muno, A. Ohwada, T. Nukiwa, E. Kominami, and S. Kira.** 1993. Cathepsin L activity is increased in alveolar macrophages and bronchoalveolar lavage fluid of smokers. *Am. Rev. Respir. Dis.* **147**: 1562–1568.

1038. **Takeda, E., Y. Kuroda, K. Toshima, E. Naito, M. Ito, M. Miyao, E. Kominami, and N. Katunuma.** 1986. Involvement of thiol proteases in galactosialidosis. *Clin. Chim. Acta* **155**: 109–115.

1039. **Tanabe, H., N. Kumagai, T. Tsukahara, S. Ishiura, E. Kominami, H. Nishina, and H. Sugita.** 1991. Changes of lysosomal proteinase activities and their expression in rat cultured keratinocytes during differentiation. *Biochim. Biophys. Acta* **1094**: 281–287.

1040. **Taniguchi, K., M. Tomita, E. Kominami, and Y. Uchiyama.** 1993. Cysteine proteinases in rat dorsal root ganglion and spinal cord, with special reference to the co-localization of these enzymes with calcitonin gene-related peptide in lysosomes. *Brain Res.* **601**: 143–153.

1041. **Tawa, N. E. J., I. C. Kettelhut, and A. L. Goldberg.** 1992. Dietary protein deficiency reduces lysosomal and nonlysosomal ATP-dependent proteolysis in muscle. *Am. J. Physiol.* **263**: E326–E334.

1042. **Taylor, M. A., R. E. Almond, and D. J. Etherington.** 1987. The immunohistochemical location of cathepsin L in rabbit skeletal muscle. Evidence for a fibre type dependent distribution. *Histochemistry* **86**: 379–383.

1043. **Teschner, M., L. Paczek, R. M. Schaefer, and A. Heidland.** 1991. Obese Zucker rat: potential role of intraglomerular proteolytic enzymes in the development of glomerulosclerosis. *Res. Exp. Med. Berl.* **191**: 129–135.

1044. **Trabandt, A., W. K. Aicher, R. E. Gay, V. P. Sukhatme, M. Nilson Hamilton, R. T. Hamilton, J. R. McGhee, H. G. Fassbender, and S. Gay.** 1990. Expression of the collagenolytic and Ras-induced cysteine proteinase cathepsin L and proliferation-associated oncogenes in synovial cells of MRL/l mice and patients with rheumatoid arthritis. *Matrix* **10**: 349–361.

1045. **Trabandt, A., R. E. Gay, H. G. Fassbender, and S. Gay.** 1991. Cathepsin B in synovial cells at the site of joint destruction in rheumatoid arthritis. *Arthritis Rheum.* **34**: 1444–1451.

1046. **Tsuchida, K., R. Yamazaki, K. Kaneko, and H. Aihara.** 1986. Effects of propranolol on tissue necrosis in experimental myocardial infarction in dogs. *J. Pharmacobiodyn.* **9**: 836–841.

1047. **Tsukahara, T., S. Ishiura, E. Kominami, and H. Sugita.** 1990. Changes in proteinase activities during the differentiation of murine erythroleukemia cells. *Exp. Cell Res.* **188**: 111–116.

1048. **Tsung, P. K., J. B. Lombardini, and F. J. Holly.** 1984. Intracellular distribution and properties of cathepsin B in the rat retina and its inhibition by a cytosolic fraction. *Exp. Eye Res.* **38**: 73–79.

1049. **Tuck, A. B., S. M. Wilson, R. Khokha, and A. F. Chambers.** 1991. Different patterns of gene expression in ras-resistant and ras-sensitive cells. *J. Natl Cancer Inst.* **83**: 485–491.

1050. **Turyna, B. and K. Szuba.** 1988. The comparison of lysosomal enzymes activities in alveolar and peritoneal macrophages of rat. *Biochem. Int.* **17**: 433–440.

1051. **Uchiyama, Y., M. Nakajima, D. Muno, T. Watanabe, Y. Ishii, S. Waguri, N. Sato, and E. Kominami.** 1990. Immunocytochemical localization of cathepsins B and H in corticotrophs and melanotrophs of rat pituitary gland. *J. Histochem. Cytochem.* **38**: 633–639.

1052. **Uchiyama, Y., M. Watanabe, T. Watanabe, Y. Ishii, H. Matsuba, S. Waguri, and E. Kominami.** 1989. Variations in immunocytochemical localization of cathepsin B and thyroxine in follicular cells of the rat thyroid gland and plasma TSH concentrations over 24 hours. *Cell Tissue Res.* **256**: 355–360.

1053. **van Noorden, C. J., R. E. Smith, and D. Rasnick.** 1988. Cysteine proteinase activity in arthritic rat knee joints and the effects of a selective systemic inhibitor, Z-Phe-AlaCH$_2$F. *J. Rheumatol.* **15**: 1525–1535.

1054. **van Noorden, C. J., I. M. Vogels, V. Everts, and W. Beertsen.** 1987. Localization of cathepsin B activity in fibroblasts and chondrocytes by continuous monitoring of the formation of a final fluorescent reaction product using 5-nitrosalicylaldehyde. *Histochem. J.* **19**: 483–487.

1055. **van Noorden, C. J., I. M. Vogels, and R. E. Smith.** 1989. Localization and cytophotometric analysis of cathepsin B activity in unfixed and undecalcified cryostat sections of whole rat knee joints. *J. Histochem. Cytochem.* **37**: 617–624.

1056. **Vasil'ev, A. V., N. S. Bregvadze, L. F. Poriadkov, and V. A. Tutel'ian.** 1990. [Proteolytic activity of lysosomes in various rat organs during total parenteral nutrition]. *Vopr. Med. Khim.* **36**: 51–53.

1057. **Vasil'ev, A. V., N. S. Bregvadze, and V. A. Tutel'ian.** 1990. [Lysosomal hydrolases of various rat organs in the process of cell nutrition]. *Vopr. Med. Khim.* **36**: 45–47.

1058. **Vasil'ev, A. V., L. S. Konovalova, K. D. Pletsitnyi, L. G. Ponomareva, and V. A. Tutel'ian.** 1988. [Lysosomal hydrolase activity of the rat liver, spleen and thymus during antigenic stimulation and vitamin B$_6$ deficiency]. *Vopr. Med. Khim.* **34**: 97–99.

1059. **Vasil'ev, A. V., L. S. Konovalova, K. D. Pletsityi, T. V. Davydova, and V. A. Tutel'ian.** 1988. [Lysosome proteinases and protein turnover in the liver, spleen and thymus of rats undergoing antigenic stimulation and fasting]. *Biull. Eksp. Biol. Med.* **105**: 181–184.

1060. **Vasil'ev, A. V., L. S. Konovalova, K. D. Pletsityi, L. G. Ponomareva, and V. A. Tutel'ian.** 1987. [Activity of lysosomal hydrolases in the rat liver, spleen and thymus during antigenic stimulation and vitamin A administration]. *Vopr. Med. Khim.* **33**: 81–83.

1061. **Vasil'ev, A. V., L. K. Ren, A. V. Pogozheva, L. G. Ponomareva, and N. P. Shimanovskaia.** 1986. [Effect of neonatal nutrition on the enzyme activity of liver lysosomes, adipocytes and thrombocytes in young and old rats]. *Vopr. Pitan.* 43–47.

1062. **Vasil'ev, A. V., N. P. Shimanovskaia, A. V. Pogozheva, M. A. Samsonov, and V. A. Tutel'ian.** 1987. [Thrombocyte lysosomal hydrolase activity in patients with ischemic heart disease, hyperlipidemia and obesity against a background of different diets]. *Vopr. Pitan.* 14–17.

1063. **Vasil'ev, A. V., T. A. Vorobeichik, D. V. Gutkin, A. B. Levitskaia, and V. A. Tutel'ian.** 1991. [The effect of actinomycin D on lysosomal proteinase activity in murine macrophages and spleen during immune response]. *Vopr. Med. Khim.* **37**: 33–35.

1064. **Vasil'ev, A. V., T. A. Vorobeichik, K. D. Pletsityi, and V. A. Tutel'ian.** 1989. [Activity of lysosomal proteinases in the liver, spleen, thymus and peritoneal macrophages after immunization of mice with thymus-dependent and thymus-independent antigens]. *Vopr. Med. Khim.* **35**: 121–123.

1065. **Vasishta, A., P. R. Baker, P. E. Preece, R. A. Wood, and A. Cuschieri.** 1988. Inhibition of proteinase-like peptidase activities in serum and tissue from breast cancer patients. *Anticancer Res.* **8**: 785–789.

1066. **Villanova, M., M. Kawai, U. Lubke, S. J. Oh, G. Perry, C. Six, C. Ceuterick, J. J. Martin, and P. Cras.** 1993. Rimmed vacuoles of inclusion body myositis and oculopharyngeal muscular dystrophy contain amyloid precursor protein and lysosomal markers. *Brain Res.* **603**: 343–347.

1067. **Volden, G.** 1978. Acid hydrolases in blister fluid. 3. Characterization and quantification of peptide hydrolases. *Br. J. Dermatol.* **99**: 49–52.

1068. **Waguri, S., T. Watanabe, E. Kominami, and Y. Uchiyama.** 1990. Variations in immunoreactivity of angiotensinogen and cathepsins B and H in rat hepatocytes over 24 hours. *Am. J. Anat.* **187**: 175–182.

1069. **Ward, C. J., J. Crocker, S. J. Chan, R. A. Stockley, and D. Burnett.** 1990. Changes in the expression of elastase and cathepsin B with differentiation of U937 promonocytes by GMCSF. *Biochem. Biophys. Res. Commun.* **167**: 659–664.

1070. **Warfel, A. H., C. Cardozo, O. H. Yoo, and D. Zucker Franklin.** 1991. Cystatin C and cathepsin B production by alveolar macrophages from smokers and nonsmokers. *J. Leukoc. Biol.* **49**: 41–47.

1071. **Warner, A. H. and M. A. Ryan.** 1990. Thiol protease-thiol protease inhibitor imbalance in cardiac tissue of ageing cardiomyopathic hamsters. *J. Mol. Cell Cardiol.* **22**: 577–586.

1072. **Warwas, M.** 1981. Isozymes of cathepsin B1 in developing human placenta. *Experientia.* **37**: 966–967.

1073. **Warwas, M. and W. Dobryszycka.** 1976. Cathepsins B$_1$ from human fetal membranes. *Biochim. Biophys. Acta* **429**: 573–580.

1074. **Watanabe, M., T. Watanabe, Y. Ishii, H. Matsuba, S. Kimura, T. Fujita, E. Kominami, N. Katunuma, and Y. Uchiyama.** 1988. Immunocytochemical localization of cathepsins B, H, and their endogenous inhibitor, cystatin β, in islet endocrine cells of rat pancreas. *J. Histochem. Cytochem.* **36**: 783–791.

1075. **Watanabe, T., S. Waguri, M. Watanabe, Y. Ishii, E. Kominami, and Y. Uchiyama.** 1989. Immunocytochemical localization of angiotensinogen and cathepsins B, H, and L in rat hepatocytes, with special reference to degradation of angiotensinogen in lysosomes after colchicine. *J. Histochem. Cytochem.* **37**: 1899–1911.

1076. **Watanabe, T., M. Watanabe, Y. Ishii, H. Matsuba, S. Kimura, T. Fujita, E. Kominami, N. Katunuma, and Y. Uchiyama.** 1989. An immunocytochemical study on co-localization of cathepsin B and atrial natriuretic peptides in secretory granules of atrial myoendocrine cells of rat heart. *J. Histochem. Cytochem.* **37**: 347–351.

1077. **Wattiaux, R., F. Gentinne, M. Jadot, F. Dubois, and S. Wattiaux De Coninck.** 1993. Chloroquine allows to distinguish between hepatocyte lysosomes and sinusoidal cell lysosomes. *Biochem. Biophys. Res. Commun.* **190**: 808–813.

1078. **Waydhas, C., D. Nast Kolb, M. Kick, M. Richter Turtur, A. Trupka, W. Machleidt, M. Jochum, and L. Schweiberer.** 1993. [Operative injury in spinal surgery in the management of polytrauma patients]. *Unfallchirurg.* **96**: 62–65.

1079. **Wheeler, T. L., J. W. Savell, H. R. Cross, D. K. Lunt, and S. B. Smith.** 1990. Mechanisms associated with the variation in tenderness of meat from Brahman and Hereford cattle. *J. Anim. Sci.* **68**: 4206–4220.

1080. **Whipple, G., M. Koohmaraie, M. E. Dikeman, and J. D. Crouse.** 1990. Predicting beef-longissimus tenderness from various biochemical and histological muscle traits. *J. Anim. Sci.* **68**: 4193–4199.

1081. **Whipple, G., M. Koohmaraie, M. E. Dikeman, and J. D. Crouse.** 1990. Effects of high-temperature conditioning on enzymatic activity and tenderness of Bos indicus longissimus muscle. *J. Anim. Sci.* **68**: 3654–3662.

1082. **Whipple, G., M. Koohmaraie, M. E. Dikeman, J. D. Crouse, M. C. Hunt, and R. D. Klemm.** 1990. Evaluation of attributes that affect longissimus muscle tenderness in *Bos taurus* and *Bos indicus* cattle. *J. Anim. Sci.* **68**: 2716–2728.

1083. **Wiederanders, B., D. Bromme, H. Kirschke, K. von Figura, B. Schmidt, and C. Peters.** 1992. Phylogenetic conservation of cysteine proteinases. Cloning and expression of a cDNA coding for human cathepsin S. *J. Biol. Chem.* **267**: 13708–13713.

1084. **Wiener, E. and Z. Curelaru.** 1975. The intracellular distribution of cathepsins and other acid hydrolases in mouse peritoneal macrophages. *J. Reticuloendothel. Soc.* **17**: 319–332.

1085. **Wilson, P. D., D. Hreniuk, and J. Lenard.** 1989. Reduced cytotoxicity of the lysosomotropic detergent N-dodecylimidazole after differentiation of HL60 promyelocytes. *Cancer Res.* **49**: 507–510.

1086. **Wolinsky, H., L. Capron, S. Goldfischer, F. Capron, B. Coltoff Schiller, and L. E. Kasak.** 1978. Hydrolase activities in the rat aorta. II. Effects of hypertension alone and in combination with diabetes mellitus. *Circ. Res.* **42**: 831–839.

1087. **Wolinsky, H., S. Goldfischer, L. Capron, F. Capron, B. Coltoff Schiller, and L. Kasak.** 1978. Hydrolase activities in the rat aorta. I. Effects of diabetes mellitus and insulin treatment. *Circ. Res.* **42**: 821–831.

1088. **Wolinsky, H., S. Goldfischer, D. Katz, R. Markle, L. Gidez, S. Wassertheil Smoller, and B. Coltoff Schiller.** 1979. Hydrolase activities in the rat aorta. III. Effects of regular swimming activity and its cessation. *Circ. Res.* **45**: 546–553.

1089. **Wright, W. W., C. Fiore, and B. R. Zirkin.** 1993. The effect of aging on the seminiferous epithelium of the brown Norway rat. *J. Androl.* **14**: 110–117.

1090. **Yamashita, M. and S. Konagaya.** 1990. High activities of cathepsins B, D, H, and L in the white muscle of chum salmon in spawning migration. *Comp. Biochem. Physiol. B* **95**: 149–152.

1091. **Yamashita, M. and S. Konagaya.** 1991. Increase in catheptic activity and appearance of phagocytes in the white muscle of chum salmon during spawning migration. *Biomed. Biochim. Acta* **50**: 565–567.

1092. **Yokota, S. and K. Kato.** 1987. Immunocytochemical localization of cathepsins B and H in rat liver. *Histochemistry* **88**: 97–103.

1093. **Yokota, S. and K. Kato.** 1988. Involvement of cathepsins B and H in lysosomal degradation of horseradish peroxidase endocytosed by the proximal tubule cells of the rat kidney: II. Immunocytochemical studies using protein A-gold technique applied to conventional and serial sections. *Anat. Rec.* **221**: 791–801.

1094. **Yokota, S. and K. Kato.** 1988. Involvement of cathepsins B and H in lysosomal degradation of horseradish peroxidase endocytosed by the proximal tubule cells of the rat kidney: I. Histochemical and immunohistochemical studies. *Anat. Rec.* **221**: 783–790.

1095. **Yokota, S. and K. Kato.** 1988. The heterogeneity of rat kidney lysosomes revealed by immunoelectron microscopic staining for cathepsins B and H. *Histochemistry* **89**: 499–504.

1096. **Yokota, S., Y. Nishimura, and K. Kato.** 1988. Localization of cathepsin L in rat kidney revealed by immunoenzyme and immunogold techniques. *Histochemistry* **90**: 277–283.

1097. **Yokota, S., H. Tsuji, and K. Kato.** 1986. Immunocytochemical localization of cathepsin B in rat kidney. II. Electron microscopic study using the protein A-gold technique. *J. Histochem. Cytochem.* **34**: 899–907.

1098. **Yokota, S., H. Tsuji, and K. Kato.** 1986. Immunocytochemical localization of cathepsin B in rat kidney. I. Light microscopic study using the indirect immunoenzyme technique. *J. Histochem. Cytochem.* **34**: 891–897.

1099. **Zdolsek, J. M.** 1993. Acridine orange-mediated photodamage to cultured cells. *APMIS* **101**: 127–132.

Processing

1100. **Burge, V., F. Mainferme, and R. Wattiaux.** 1991. Transient membrane association of the precursors of cathepsin C during their transfer into lysosomes. *Biochem. J.* **275**: 797–800.

1101. **Chi, L. M., M. X. Zhu, and N. Marks.** 1990. High affinity binding of rat brain cathepsin B to purified rat/bovine cerebral mannose 6-phosphate receptors. *Biol. Chem. Hoppe Seyler* **371**(Suppl.): 259–264.

1102. **Dalet Fumeron, V., N. Guinec, and M. Pagano.** 1993. *In vitro* activation of pro-cathepsin B by three serine proteinases: leucocyte elastase, cathepsin G, and the urokinase-type plasminogen activator. *FEBS Lett.* **332**: 251–254.

1103. **Dong, J. M. and G. G. Sahagian.** 1990. Basis for low affinity binding of a lysosomal cysteine protease to the cation-independent mannose 6-phosphate receptor. *J. Biol. Chem.* **265**: 4210–4217.

1104. **Hara, K., E. Kominami, and N. Katunuma.** 1988. Effect of proteinase inhibitors on intracellular processing of cathepsin B, H and L in rat macrophages. *FEBS Lett.* **231**: 229–231.

1105. **Hasnain, S., T. Hirama, C. P. Huber, P. Mason, and J. S. Mort.** 1993. Characterization of cathepsin B specificity by site-directed mutagenesis. Importance of Glu245 in the S_2-P_2 specificity for arginine and its role in transition state stabilization. *J. Biol. Chem.* **268**: 235–240.

1106. **Kane, S. E.** 1993. Mouse procathepsin L lacking a functional glycosylation site is properly folded, stable, and secreted by NIH 3T3 cells. *J. Biol. Chem.* **268**: 11456–11462.

1107. **Kawabata, T., Y. Nishimura, M. Higaki, and K. Kato.** 1993. Purification and processing of rat liver procathepsin B. *J. Biochem. Tokyo* **113**: 389–394.

1108. **Kominami, E., T. Tsukahara, K. Hara, and N. Katunuma.** 1988. Biosyntheses and processing of lysosomal cysteine proteinases in rat macrophages. *FEBS Lett.* **231**: 225–228.

1109. **Lazzarino, D. and C. A. Gabel.** 1990. Protein determinants impair recognition of procathepsin L phosphorylated oligosaccharides by the cation-independent mannose 6-phosphate receptor. *J. Biol. Chem.* **265**: 11864–11871.

1110. **Mach, L., J. S. Mort, and J. Glössl.** 1994. Noncovalent complexes between the lysosomal proteinase cathepsin B and its propeptide account for stable, extracellular, high molecular mass forms of the enzyme. *J. Biol. Chem.* **269**: 13036–13040.

1111. **Mach, L., H. Schwihla, K. Stuwe, A. D. Rowan, J. S. Mort, and J. Glossl.** 1993. Activation of procathepsin B in human hepatoma cells: the conversion into the mature enzyme relies on the action of cathepsin B itself. *Biochem. J.* **293**: 437–442.

1112. **Mach, L., K. Stuwe, A. Hagen, C. Ballaun, and J. Glossl.** 1992. Proteolytic processing and glycosylation of cathepsin B. The role of the primary structure of the latent precursor and of the carbohydrate moiety for cell-type-specific molecular forms of the enzyme. *Biochem. J.* **282**: 577–582.

1113. **Mason, R. W. and S. D. Massey.** 1992. Surface activation of pro-cathepsin L. *Biochem. Biophys. Res. Commun.* **189**: 1659–1666.

1114. **McDonald, J. K. and S. Kadkhodayan.** 1988. Cathepsin L – a latent proteinase in guinea pig sperm. *Biochem. Biophys. Res. Commun.* **151**: 827–835.

1115. **McIntyre, G. F. and A. H. Erickson.** 1993. The lysosomal proenzyme receptor that binds procathepsin L to microsomal membranes at pH 5 is a 43-kDa integral membrane protein. *Proc. Natl Acad. Sci. U. S. A.* **90**: 10588–10592.

1116. **Muno, D., K. Ishidoh, T. Ueno, and E. Kominami.** 1993. Processing and transport of the precursor of cathepsin C during its transfer into lysosomes. *Arch. Biochem. Biophys.* **306**: 103–110.

1117. **Nishimura, Y., J. Amano, H. Sato, H. Tsuji, and K. Kato.** 1988. Biosynthesis of lysosomal cathepsins B and H in cultured rat hepatocytes. *Arch. Biochem. Biophys.* **262**: 159–170.

1118. **Nishimura, Y., K. Furuno, and K. Kato.** 1988. Biosynthesis and processing of lysosomal cathepsin L in primary cultures of rat hepatocytes. *Arch. Biochem. Biophys.* **263**: 107–116.

1119. **Nishimura, Y. and K. Kato.** 1988. Identification of latent procathepsin H in microsomal lumen: characterization of proteolytic processing and enzyme activation. *Arch. Biochem. Biophys.* **260**: 712–718.

1120. **Nishimura, Y., T. Kawabata, K. Furuno, and K. Kato.** 1989. Evidence that aspartic proteinase is involved in the proteolytic processing event of procathepsin L in lysosomes. *Arch. Biochem. Biophys.* **271**: 400–406.

1121. **Nishimura, Y., T. Kawabata, and K. Kato.** 1988. Identification of latent procathepsins B and L in microsomal lumen: characterization of enzymatic activation and proteolytic processing *in vitro*. *Arch. Biochem. Biophys.* **261**: 64–71.

1122. **Nishimura, Y., T. Kawabata, S. Yano, and K. Kato.** 1990. Inhibition of intracellular sorting and processing of lysosomal cathepsins H and L at reduced temperature in primary cultures of rat hepatocytes. *Arch. Biochem. Biophys.* **283**: 458–463.

1123. **Oda, K. and Y. Nishimura.** 1989. Brefeldin A inhibits the targeting of cathepsin D and cathepsin H to lysosomes in rat hepatocytes. *Biochem. Biophys. Res. Commun.* **163**: 220–225.

1124. **Prence, E. M., J. M. Dong, and G. G. Sahagian.** 1990. Modulation of the transport of a lysosomal enzyme by PDGF. *J. Cell Biol.* **110**: 319–326.

1125. **Rowan, A. D., R. Feng, Y. Konishi, and J. S. Mort.** 1993. Demonstration by electrospray mass spectrometry that the peptidyldipeptidase activity of cathepsin B is capable of rat cathepsin B C-terminal processing. *Biochem. J.* **294**: 923–927.

1126. **Rowan, A. D., P. Mason, L. Mach, and J. S. Mort.** 1992. Rat procathepsin B. Proteolytic processing to the mature form *in vitro*. *J. Biol. Chem.* **267**: 15993–15999.

1127. **Salminen, A. and M. M. Gottesman.** 1990. Inhibitor studies indicate that active cathepsin L is probably essential to its own processing in cultured fibroblasts. *Biochem. J.* **272**: 39–44.

1128. **Strous, G. J., P. van Kerkhof, G. van Meer, S. Rijnboutt, and W. Stoorvogel.** 1993. Differential effects of brefeldin A on transport of secretory and lysosomal proteins. *J. Biol. Chem.* **268**: 2341–2347.

1129. **Wiederanders, B. and H. Kirschke.** 1989. The processing of a cathepsin L precursor *in vitro*. *Arch. Biochem. Biophys.* **272**: 516–521.

Gene structure, nucleotide and amino acid sequences

1130. **Barrett, A. J., M. J. H. Nicklin, and N. D. Rawlings.** 1984. The papain superfamily of cysteine proteinases and their protein inhibitors. *Symp. Biol. Hung.* **25**: 203–217.

1131. **Baudys, M., B. Meloun, T. Gan Erdene, J. Pohl, and V. Kostka.** 1990. Disulfide bridges of bovine spleen cathepsin B. *Biol. Chem. Hoppe Seyler* **371**: 485–491.

1132. **Baudys, M., B. Meloun, J. Pohl, and V. Kostka.** 1988. Identification of the second (buried) cysteine residue and of the C-terminal disulfide bridge of bovine spleen cathepsin B. *Biol. Chem. Hoppe Seyler* **369**(Suppl.): 169–174.

1133. **Cejudo, F. J., G. Murphy, C. Chinoy, and D. C. Baulcombe.** 1992. A gibberellin-regulated gene from wheat with sequence homology to cathepsin B of mammalian cells. *Plant J.* **2**: 937–948.

1134. **Chan, S. J., B. San Segundo, M. B. McCormick, and D. F. Steiner.** 1986. Nucleotide and predicted amino acid sequences of cloned human and mouse preprocathepsin B cDNAs. *Proc. Natl Acad. Sci. U. S. A.* **83**: 7721–7725.

1135. **Chauhan, S. S., N. C. Popescu, D. Ray, R. Fleischmann, M. M. Gottesman, and B. R. Troen.** 1993. Cloning, genomic organization, and chromosomal localization of human cathepsin L. *J. Biol. Chem.* **268**: 1039–1045.

1136. **Cox, G. N., D. Pratt, R. Hageman, and R. J. Boisvenue.** 1990. Molecular cloning and primary sequence of a cysteine protease expressed by *Haemonchus contortus* adult worms. *Mol. Biochem. Parasitol.* **41**: 25–34.

1137. **Delaria, K., L. Fiorentino, L. Wallace, P. Tamburini, E. Brownell, and D. Muller.** 1994. Inhibition of cathepsin L-like cysteine proteases by cytotoxic T-lymphocyte antigen-2β. *J. Biol. Chem.* **269**: 25172–25177.

1138. **Denhardt, D. T., R. T. Hamilton, C. L. Parfett, D. R. Edwards, R. St.Pierre, P. Waterhouse, and M. Nilsen Hamilton.** 1986. Close relationship of the major excreted protein of transformed murine fibroblasts to thiol-dependent cathepsins. *Cancer Res.* **46**: 4590–4593.

1139. **Denizot, F., J. F. Brunet, P. Roustan, K. Harper, M. Suzan, M. F. Luciani, M. G. Mattei, and P. Golstein.** 1989. Novel structures CTLA-2α and CTLA-2β expressed in mouse activated T cells and mast cells and homologous to cysteine proteinase proregions. *Eur. J. Immunol.* **19**: 631–635.

1140. **Devereux, J., P. Haeberli, and O. Smithies.** 1984. A comprehensive set of sequence analysis programs for the VAX. *Nucleic Acids Res.* **12**: 387–395.

1141. **Dolenc, I., A. Ritonja, A. Colic, M. Podobnik, T. Ogrinc, and V. Turk.** 1992. Bovine cathepsins S and L: isolation and amino acid sequences. *Biol. Chem. Hoppe Seyler* **373**: 407–412.

1142. **Dufour, E.** 1988. Sequence homologies, hydrophobic profiles and secondary structures of cathepsins B, H and L: comparison with papain and actinidin. *Biochimie* **70**: 1335–1342.

1143. **Dufour, E., A. Obled, C. Valin, D. Bechet, B. Ribadeau Dumas, and J. C. Huet.** 1987. Purification and amino acid sequence of chicken liver cathepsin L. *Biochemistry* **26**: 5689–5695.

1144. **Fan, Y. S., M. G. Byers, R. L. Eddy, L. J. Joseph, V. P. Sukhatme, S. J. Chan, and T. B. Shows.** 1989. Cathepsin L (CTSL) is located in the chromosome 9q21–q22 region; a related sequence is located on chromosome 10. *Cytogenet. Cell Genet.* **51**: 996.

1145. **Ferrara, M., F. Wojcik, H. Rhaissi, S. Mordier, M. P. Roux, and D. Bechet.** 1990. Gene structure of mouse cathepsin B. *FEBS Lett.* **273**: 195–199.

1146. **Fong, D., D. H. Calhoun, W. T. Hsieh, B. Lee, and R. D. Wells.** 1986. Isolation of a cDNA clone for the human lysosomal proteinase cathepsin B. *Proc. Natl Acad. Sci. U. S. A.* **83**: 2909–2913.

1147. **Fong, D., M. M. Chan, and W. T. Hsieh.** 1991. Gene mapping of human cathepsins and cystatins. *Biomed. Biochim. Acta* **50**: 595–598.

1148. **Fong, D., M. M. Chan, W. T. Hsieh, J. C. Menninger, and D. C. Ward.** 1992. Confirmation of the human cathepsin B gene (CTSB) assignment to chromosome 8. *Hum. Genet.* **89**: 10–12.

1149. **Friemert, C., E. I. Closs, M. Silbermann, V. Erfle, and P. G. Strauss.** 1991. Isolation of a cathepsin B-encoding cDNA from murine osteogenic cells. *Gene* **103**: 259–261.

1150. **Fuchs, R. and H. G. Gassen.** 1989. Nucleotide sequence of human preprocathepsin H, a lysosomal cysteine proteinase. *Nucleic Acids Res.* **17**: 9471.

1151. **Fuchs, R., W. Machleidt, and H. G. Gassen.** 1988. Molecular cloning and sequencing of a cDNA coding for mature human kidney cathepsin H. *Biol. Chem. Hoppe Seyler* **369**: 469–475.

1152. **Gal, S. and M. M. Gottesman.** 1988. Isolation and sequence of a cDNA for human pro-(cathepsin L). *Biochem. J.* **253**: 303–306.

1153. **Gong, Q., S. J. Chan, A. S. Bajkowski, D. F. Steiner, and A. Frankfater.** 1993. Characterization of the cathepsin B gene and multiple mRNAs in human tissues: evidence for alternative splicing of cathepsin B pre-mRNA. *DNA Cell Biol.* **12**: 299–309.

1154. **Gotz, B. and M. O. Klinkert.** 1993. Expression and partial characterization of a cathepsin B-like enzyme (Sm31) and a proposed 'haemoglobinase' (Sm32) from *Schistosoma mansoni. Biochem. J.* **290**: 801–806.

1155. **Holwerda, B. C. and J. C. Rogers.** 1992. Purification and characterization of aleurain. A plant thiol protease functionally homologous to mammalian cathepsin H. *Plant Physiol.* **99**: 848–855.

1156. **Hsieh, W. T., P. E. Barker, M. Smith, and D. Fong.** 1990. A TaqI DNA polymorphism in the human cathepsin B gene (CTSB). *Nucleic Acids Res.* **18**: 3430.

1157. **Ishidoh, K., S. Imajoh, Y. Emori, S. Ohno, H. Kawasaki, Y. Minami, E. Kominami, N. Katunuma, and K. Suzuki.** 1987. Molecular cloning and sequencing of cDNA for rat cathepsin H. Homology in pro-peptide regions of cysteine proteinases. *FEBS Lett.* **226**: 33–37.

1158. **Ishidoh, K., E. Kominami, N. Katunuma, and K. Suzuki.** 1989. Gene structure of rat cathepsin H. *FEBS Lett.* **253**: 103–107.

1159. **Ishidoh, K., E. Kominami, K. Suzuki, and N. Katunuma.** 1989. Gene structure and 5′-upstream sequence of rat cathepsin L. *FEBS Lett.* **259**: 71–74.

1160. **Ishidoh, K., D. Muno, N. Sato, and E. Kominami.** 1991. Molecular cloning of cDNA for rat cathepsin C. Cathepsin C, a cysteine proteinase with an extremely long propeptide. *J. Biol. Chem.* **266**: 16312–16317.

1161. **Ishidoh, K., K. Suzuki, N. Katunuma, and E. Kominami.** 1991. Gene structures of rat cathepsins H and L. *Biomed. Biochim. Acta* **50**: 541–547.

1162. **Ishidoh, K., T. Towatari, S. Imajoh, H. Kawasaki, E. Kominami, N. Katunuma, and K. Suzuki.** 1987. Molecular cloning and sequencing of cDNA for rat cathepsin L. *FEBS Lett.* **223**: 69–73.

1163. **Jaffe, R. C., K. M. Donnelly, P. A. Mavrogianis, and H. G. Verhage.** 1989. Molecular cloning and characterization of a progesterone-dependent cat endometrial secretory protein complementary deoxyribonucleic acid. *Mol. Endocrinol.* **3**: 1807–1814.

1164. **Joseph, L., S. Lapid, and V. Sukhatme.** 1987. The major ras induced protein in NIH3T3 cells is cathepsin L. *Nucleic Acids Res.* **15**: 3186.

1165. **Joseph, L. J., L. C. Chang, D. Stamenkovich, and V. P. Sukhatme.** 1988. Complete nucleotide and deduced amino acid sequences of human and murine preprocathepsin L. An abundant transcript induced by transformation of fibroblasts. *J. Clin. Invest.* **81**: 1621–1629.

1166. **Karrer, K. M., S. L. Peiffer, and M. E. DiTomas.** 1993. Two distinct gene subfamilies within the family of cysteine protease genes. *Proc. Natl Acad. Sci. U. S. A.* **90**: 3063–3067.

1167. **Kirschke, H., A. A. Kembhavi, P. Bohley, and A. J. Barrett.** 1982. Action of rat liver cathepsin L on collagen and other substrates. *Biochem. J.* **201**: 367–372.

1168. **Klinkert, M. Q., D. Cioli, E. Shaw, V. Turk, W. Bode, and R. Butler.** 1994. Sequence and structure similarities of cathepsin B from the parasite *Schistosoma mansoni* and human liver. *FEBS Lett.* **351**: 397–400.

1169. **Klinkert, M. Q., R. Felleisen, G. Link, A. Ruppel, and E. Beck.** 1989. Primary structures of Sm31/32 diagnostic proteins of *Schistosoma mansoni* and their identification as proteases. *Mol. Biochem. Parasitol.* **33**: 113–122.

1170. **Kominami, E., K. Ishido, D. Muno, and N. Sato.** 1992. The primary structure and tissue distribution of cathepsin C. *Biol. Chem. Hoppe Seyler* **373**: 367–373.

1171. **Laycock, M. V., R. M. MacKay, M. Di Fruscio, and J. W. Gallant.** 1991. Molecular cloning of three cDNAs that encode cysteine proteinases in the digestive gland of the American lobster (*Homarus americanus*). *FEBS Lett.* **292**: 115–120.

1172. **Li, X. Y., S. W. Rogers, and J. C. Rogers.** 1991. A copy of exon 3–intron 3 from the barley aleurain gene is present on chromosome 2. *Plant. Mol. Biol.* **17**: 509–512.

1173. **Machleidt, W., A. Ritonja, T. Popovic, M. Kotnik, J. Brzin, V. Turk, I. Machleidt, and W. Muller Esterl.** 1986. Human cathepsins B, H and L: characterization by amino acid sequences and some kinetics of inhibition by the kininogens. In *Cysteine Proteinases and Their Inhibitors*, V. Turk (ed.), pp. 3–18. Walter de Gruyter, Berlin.

1174. **Mason, R. W., J. E. Walker, and F. D. Northrop.** 1986. The N-terminal amino acid sequences of the heavy and light chains of human cathepsin L. Relationship to a cDNA clone for a major cysteine proteinase from a mouse macrophage cell line. *Biochem. J.* **240**: 373–377.

1175. **Meloun, B., M. Baudys, J. Pohl, M. Pavlik, and V. Kostka.** 1988. Amino acid sequence of bovine spleen cathepsin B. *J. Biol. Chem.* **263**: 9087–9093.

1176. **Meloun, B., J. Pohl, and V. Kostka.** 1986. Tentative amino acid sequence of bovine spleen cathepsin B. In *Cysteine Proteinases and Their Inhibitors*, V. Turk (ed.), pp. 19–29. Walter de Gruyter, Berlin.

1177. **Merckelbach, A., S. Hasse, R. Dell, A. Eschibeck, and A. Ruppel.** 1994. CDNA sequences of *Schistosoma japonicum* coding for two cathepsin B-like proteins and Sj32. *Trop. Med. Parasitol.* **45**: 193–198.

1178. **Mordier, S., D. Bechet, M. P. Roux, A. Obled, and M. Ferrara.** 1993. Nucleotide sequence of bovine preprocathepsin B. A study of polymorphism in the protein coding region. *Biochim. Biophys. Acta* **1174**: 305–311.

1179. **Morrison, R. I., A. J. Barrett, J. T. Dingle, and D. Prior.** 1973. Cathepsins B₁ and D. Action on human cartilage proteoglycans. *Biochim. Biophys. Acta* **302**: 411–419.

1180. **Petanceska, S. and L. Devi.** 1992. Sequence analysis, tissue distribution, and expression of rat cathepsin S. *J. Biol. Chem.* **267**: 26038–26043.

1181. **Pohl, J., M. Baudys, V. Tomasek, and V. Kostka.** 1982. Identification of the active site cysteine and of the disulfide bonds in the *N*-terminal part of the molecule of bovine spleen cathepsin B. *FEBS Lett.* **142**: 23–26.

1182. **Portnoy, D. A., A. H. Erickson, J. Kochan, J. V. Ravetch, and J. C. Unkeless.** 1986. Cloning and characterization of a mouse cysteine proteinase. *J. Biol. Chem.* **261**: 14697–14703.

1183. **Pratt, D., L. G. Armes, R. Hageman, V. Reynolds, R. J. Boisvenue, and G. N. Cox.** 1992. Cloning and sequence comparisons of four distinct cysteine proteases expressed by *Haemonchus contortus* adult worms. *Mol. Biochem. Parasitol.* **51**: 209–218.

1184. **Pratt, D., R. J. Boisvenue, and G. N. Cox.** 1992. Isolation of putative cysteine protease genes of *Ostertagia ostertagi. Mol. Biochem. Parasitol.* **56**: 39–48.

1185. **Pratt, D., G. N. Cox, M. J. Milhausen, and R. J. Boisvenue.** 1990. A developmentally regulated cysteine protease gene family in *Haemonchus contortus. Mol. Biochem. Parasitol.* **43**: 181–191.

1186. **Qian, F., A. Frankfater, S. J. Chan, and D. F. Steiner.** 1991. The structure of the mouse cathepsin B gene and its putative promoter. *DNA Cell Biol.* **10**: 159–168.

1187. **Rawlings, N. D. and A. J. Barrett.** 1994. Families of cysteine peptidases. *Methods Enzymol.* **244**: 461–486.

1188. **Ray, C. and J. H. McKerrow.** 1992. Gut-specific and developmental expression of a *Caenorhabditis elegans* cysteine protease gene. *Mol. Biochem. Parasitol.* **51**: 239–249.

1189. **Rhaissi, H., D. Bechet, and M. Ferrara.** 1993. Multiple leader sequences for mouse cathepsin B mRNA. *Biochimie* **75**: 899–904.

1190. **Ritonja, A., A. Colic, I. Dolenc, T. Ogrinc, M. Podobnik, and V. Turk.** 1991. The complete amino acid sequence of bovine cathepsin S and a partial sequence of bovine cathepsin L. *FEBS Lett.* **283**: 329–331.

1191. **Ritonja, A., T. Popovic, M. Kotnik, W. Machleidt, and V. Turk.** 1988. Amino acid sequences of the human kidney cathepsins H and L. *FEBS Lett.* **228**: 341–345.

1192. **Ritonja, A., T. Popovic, V. Turk, K. Wiedenmann, and W. Machleidt.** 1985. Amino acid sequence of human liver cathepsin B. *FEBS Lett.* **181**: 169–172.

1193. **Salvesen, G., N. Gay, and J. Walker.** 1986. Cloning of a bovine protein homologous with cysteine proteinases and identification of the gene. In *Cysteine Proteinases and Their Inhibitors*, V. Turk (ed.), pp. 55–62. Walter de Gruyter, Berlin.

1194. **San Segundo, B., S. J. Chan, and D. F. Steiner.** 1985. Identification of cDNA clones encoding a precursor of rat liver cathepsin B. *Proc. Natl Acad. Sci. U. S. A.* **82**: 2320–2324.

1195. **Shi, G. P., J. S. Munger, J. P. Meara, D. H. Rich, and H. A. Chapman.** 1992. Molecular cloning and expression of human alveolar macrophage cathepsin S, an elastinolytic cysteine protease. *J. Biol. Chem.* **267**: 7258–7262.

1196. **Shi, G. P., A. C. Webb, K. E. Foster, J. H. M. Knoll, C. A. Lemere, J. S. Munger, and H. A. Chapman.** 1994. Human cathepsin S: chromosomal localization, gene structure, and tissue distribution. *J. Biol. Chem.* **269**: 11530–11536.

1197. **Takahashi, N., S. Kurata, and S. Natori.** 1993. Molecular cloning of cDNA for the 29 kDa proteinase participating in decomposition of the larval fat body during metamorphosis of *Sarcophaga peregrina* (flesh fly). *FEBS Lett.* **334**: 153–157.

1198. **Takio, K., T. Towatari, N. Katunuma, D. C. Teller, and K. Titani.** 1983. Homology of amino acid sequences of rat liver cathepsins B and H with that of papain. *Proc. Natl Acad. Sci. U. S. A.* **80**: 3666–3670.

1199. **Takio, K., T. Towatari, N. Katunuma, and K. Titani.** 1980. Primary structure study of rat liver cathepsin B – a striking resemblance to papain. *Biochem. Biophys. Res. Commun.* **97**: 340–346.

1200. **Taniguchi, T., T. Mizuochi, T. Towatari, N. Katunuma, and A. Kobata.** 1985. Structural studies on the carbohydrate moieties of rat liver cathepsins B and H. *J. Biochem. Tokyo* **97**: 973–976.

1201. **Tezuka, K., Y. Tezuka, A. Maejima, T. Sato, K. Nemoto, H. Kamioka, Y. Hakeda, and M. Kumegawa.** 1994. Molecular cloning of a possible cysteine proteinase predominantly expressed in osteoclasts. *J. Biol. Chem.* **269**: 1106–1109.

1202. **Towatari, T. and N. Katunuma.** 1988. Amino acid sequence of rat liver cathepsin L. *FEBS Lett.* **236**: 57–61.

1203. **Troen, B. R., D. Ascherman, D. Atlas, and M. M. Gottesman.** 1988. Cloning and expression of the gene for the major excreted protein of transformed mouse fibroblasts. A secreted lysosomal protease regulated by transformation. *J. Biol. Chem.* **263**: 254–261.

1204. **Troen, B. R., S. S. Chauhan, D. Ray, and M. M. Gottesman.** 1991. Downstream sequences mediate induction of the mouse cathepsin L promoter by phorbol esters. *Cell Growth Differ.* **2**: 2331.

1205. **Troen, B. R., S. Gal, and M. M. Gottesman.** 1987. Sequence and expression of the cDNA for MEP (major excreted protein), a transformation-regulated secreted cathepsin. *Biochem. J.* **246**: 731–735.

1206. **Turk, V. and W. Bode.** 1991. The cystatins: protein inhibitors of cysteine proteinases. *FEBS Lett.* **285**: 213–219.

1207. **Velasco, G., A. A. Ferrando, X. S. Puente, L. M. Sanchez, and C. Lopez Otin.** 1994. Human cathepsin O. Molecular cloning from a breast carcinoma, production of the active enzyme in *Escherichia coli*, and expression analysis in human tissues. *J. Biol. Chem.* **269**: 27136–27142.

1208. **Wada, K., T. Takai, and T. Tanabe.** 1987. Amino acid sequence of chicken liver cathepsin L. *Eur. J. Biochem.* **167**: 13–18.

1209. **Wada, K. and T. Tanabe.** 1986. N-terminal amino acid sequences of the heavy and light chains of chicken liver cathepsin L. *FEBS Lett.* **209**: 330–334.

1210. **Wang, X., S. J. Chan, R. L. Eddy, M. G. Byers, Y. Fukushima, W. M. Henry, L. L. Haley, D. F. Steiner, and T. B. Shows.** 1987. Chromosome assignment of cathepsin B (CTSB) to 8p22 and cathepsin H (CTSH) to 15q24–q25. *Cytogenet. Cell Genet.* **46**: 710–711.

1211. **Watanabe, H., K. Abe, Y. Emori, H. Hosoyama, and S. Arai.** 1991. Molecular cloning and gibberellin-induced expression of multiple cysteine proteinases of rice seeds (oryzains). *J. Biol. Chem.* **266**: 16897–16902.

1212. **Whittier, R. F., D. A. Dean, and J. C. Rogers.** 1987. Nucleotide sequence analysis of α-amylase and thiol protease genes that are hormonally regulated in barley aleurone cells. *Nucleic Acids Res.* **15**: 2515–2535.

1213. **Wiederanders, B., D. Broemme, H. Kirschke, N. Kalkkinen, A. Rinne, T. Paquette, and P. Toothman.** 1991. Primary structure of bovine cathepsin S. Comparison to cathepsins L, H, B and papain. *FEBS Lett.* **286**: 189–192.

Structure–function relationship

1214. **Akahane, K. and H. Umeyama.** 1986. Binding specificity of papain and cathepsin B. *Enzyme* **36**: 141–149.

1215. **Baudys, M., B. Meloun, T. Gan Erdene, M. Fusek, M. Mares, V. Kostka, J. Pohl, and C. C. Blake.** 1991. S–S bridges of cathepsin B and H from bovine spleen: a basis for cathepsin B model building and possible functional implications for discrimination between exo- and endopeptidase activities among cathepsins B, H and L. *Biomed. Biochim. Acta* **50**: 569–577.

1216. **Benenson, A. M. and A. D. Morozkin.** 1971. [Quaternary structure of cathepsin C from bovine spleen]. *Vopr. Med. Khim.* **17**: 54–57.

1217. **Bode, W., R. Engh, D. Musil, U. Thiele, R. Huber, A. Karshikov, J. Brzin, J. Kos, and V. Turk.** 1988. The 2.0 Å X-ray crystal structure of chicken egg white cystatin and its possible mode of interaction with cysteine proteinases. *EMBO J.* **7**: 2593–2599.

1218. **Bode, W. and R. Huber.** 1992. Natural protein proteinase inhibitors and their interaction with proteinases. *Eur. J. Biochem.* **204**: 433–451.

1219. **Bromme, D., P. R. Bonneau, P. Lachance, and A. C. Storer.** 1994. Engineering the S_2 subsite specificity of human cathepsin S to a cathepsin L- and cathepsin B-like specificity. *J. Biol. Chem.* **269**: 30238–30242.

1220. **Cheng, C. Y., I. Morris, and C. W. Bardin.** 1993. Testins are structurally related to the mouse cysteine proteinase precursor but devoid of any protease/anti-protease activity. *Biochem. Biophys. Res. Commun.* **191**: 224–231.

1221. **Dufour, E., V. Dive, and F. Toma.** 1988. Delineation of chicken cathepsin L secondary structure; relationship between pH dependence activity and helix content. *Biochim. Biophys. Acta* **955**: 58–64.

1222. **Hasnain, S., C. P. Huber, A. Muir, A. D. Rowan, and J. S. Mort.** 1992. Investigation of structure function relationships in cathepsin B. *Biol. Chem. Hoppe Seyler* **373**: 413–418.

1223. **Kamphuis, I. G., J. Drenth, and E. N. Baker.** 1985. Thiol proteases. Comparative studies based on the high-resolution structures of papain and actinidin, and on amino acid sequence information for cathepsins B and H, and stem bromelain. *J. Mol. Biol.* **182**: 317–329.

1224. **Kamphuis, I. G., K. H. Kalk, M. B. A. Swarte, and J. Drenth.** 1984. Structure of papain refined at 1.65 Å resolution. *J. Mol. Biol.* **179**: 233–256.

1225. **Kraulis, P. J.** 1991. MOLSCRIPT: a program to produce both detailed and schematic plots of protein structures. *J. Appl. Cryst.* **24**: 946–950.

1226. **Lee, X., F. R. Ahmed, T. Hirama, C. P. Huber, D. R. Rose, R. To, S. Hasnain, A. Tam, and J. S. Mort.** 1990. Crystallization of recombinant rat cathepsin B. *J. Biol. Chem.* **265**: 5950–5951.

1227. **Metrione, R. M., Y. Okuda, and Jr. Fairclough, G. F.** 1970. Subunit structure of dipeptidyl transferase. *Biochemistry* **9**: 2427–2432.

1228. **Musil, D., D. Zucic, D. Turk, R. A. Engh, I. Mayr, R. Huber, T. Popovic, V. Turk, T. Towatari, N. Katunuma, and et al.** 1991. The refined 2.15 Å X-ray crystal structure of human liver cathepsin B: the structural basis for its specificity. *EMBO J.* **10**: 2321–2330.

1229. **Rothe, M. and J. Dodt.** 1992. Studies on the aminopeptidase activity of rat cathepsin H. *Eur. J. Biochem.* **210**: 759–764.

1230. **Rothe, M., A. Zichner, E. A. Auerswald, and J. Dodt.** 1994. Structure/function implications for the aminopeptidase specificity of aleurain. *Eur. J. Biochem.* **224**: 559–565.

1231. **Shaw, E. and C. Kettner.** 1981. The specificity of cathepsin B. *Acta Biol. Med. Ger.* **40**: 1503–1511.

1232. **Sudarsanam, S., G. D. Virca, C. J. March, and S. Srinivasan.** 1992. An approach to computer-aided inhibitor design: application to cathepsin L. *J. Comput. Aided. Mol. Des.* **6**: 223–233.

1233. **Sumiya, S., T. Yoneda, K. Kitamura, M. Murata, C. Yokoo, M. Tamai, A. Yamamoto, M. Inoue, and T. Ishida.** 1992. Molecular design of potent inhibitor specific for cathepsin B based on the tertiary structure prediction. *Chem. Pharm. Bull. Tokyo* **40**: 299–303.

1234. **Takahashi, K., M. Isemura, T. Ono, and T. Ikenaka.** 1980. Location of the essential thiol of porcine liver cathepsin B. *J. Biochem. Tokyo* **87**: 347–350.

1235. **Towatari, T. and N. Katunuma.** 1978. Crystallization and amino acid composition of cathepsin B from rat liver lysosomes. *Biochem. Biophys. Res. Commun.* **83**: 513–520.

1236. **Towatari, T., Y. Kawabata, and N. Katunuma.** 1979. Crystallization and properties of cathepsin B from rat liver. *Eur. J. Biochem.* **102**: 279–289.

1237. **Yamamoto, A., T. Kaji, K. Tomoo, T. Ishida, M. Inoue, M. Murata, and K. Kitamura.** 1992. Crystallization and preliminary X-ray study of the cathepsin B complexed with CA074, a selective inhibitor. *J. Mol. Biol.* **227**: 942–944.

References for second edition

Books

1238. **Alberts, B., D. Bray, J. Lewis, M. Raff, K. Roberts, and J. D. Watson.** 1994. *Molecular Biology of the Cell*, pp. 614–618. Garland, New York.

1239. **Barrett, A. J. and J. K. McDonald.** 1980. *Mammalian Proteases: A Glossary and Bibliography. Volume 1. Endopeptidases*, pp. 267–268. Academic Press, London.

1240. **McDonald, J. K. and A. J. Barrett.** 1986. *Mammalian Proteases: A Glossary and Bibliography. Volume 2. Exopeptidases*, pp. 92–98, 111–119, 251–257. Academic Press, London.

Reviews

1241. **Andreu, A. L. and S. Schwartz.** 1995. Nutrition, lysosomal proteases, and protein breakdown. *Nutrition* **11**: 382–387.

1242. **Barrett, A. J. and N. D. Rawlings.** 1996. Families and clans of cysteine peptidases. *Perspect. Drug Discov. Design* **6**: 1–11.

1243. **Bernstein, H. G.** 1994. The many faces of lysosomal proteinases (cathepsins) in human neuropathology. A histochemical perspective. *Eur. J. Histochem.* **38**: 189–192.

1244. **Berquin, I. M. and B. F. Sloane.** 1996. Cathepsin B expression in human tumors. In *Intracellular Protein Catabolism*, K. Suzuki and J. S. Bond (eds), pp. 281–294. Plenum Press, New York.

1245. **Bohley, P.** 1968. [Intracellular proteolysis]. *Naturwissenschaften* **55**: 211–217.

1246. **Bohley, P.** 1987. Intracellular proteolysis. In *Hydrolytic Enzymes*, A. Neuberger and K. Brocklehurst (eds), pp. 307–332. Elsevier, Amsterdam.

1247. **Bohley, P.** 1995. The fates of proteins in cells. *Naturwissenschaften* **82**: 544–550.

1248. **Bohley, P., H. Kirschke, J. Langner, S. Ansorge, B. Wiederanders, and H. Hanson.** 1971. Intracellular protein breakdown. In *Tissue Proteinases*, A. J. Barrett and J. T. Dingle (eds), pp. 187–219. North-Holland, Amsterdam.

1249. **Bohley, P., H. Kirschke, J. Langner, M. Miehe, S. Riemann, Z. Salama, E. Schön, B. Wiederanders, and S. Ansorge.** 1979. Intracellular protein turnover. In *Biological Functions of Proteinases*, H. Holzer and H. Tschesche (eds), pp. 17–34. Springer, Berlin.

1250. **Boyer, M. J. and I. F. Tannock.** 1993. Lysosomes, lysosomal enzymes, and cancer. *Adv. Cancer Res.* **60**: 269–291.

1251. **Calkins, C. C. and B. F. Sloane.** 1995. Mammalian cysteine protease inhibitors: biochemical properties and possible roles in tumor progression. *Biol. Chem. Hoppe-Seyler* **376**: 71–80.

1251a. **Cataldo, A. M., D. J. Hamilton, J. L. Barnett, P. A. Paskevich, and R. A. Nixon.** 1996. Abnormalities of the endosomal–lysosomal system in Alzheimer's disease. Relationship to disease pathogenesis. In *Intracellular Protein Catabolism*, K. Suzuki and J. S. Bond (eds), pp. 271–280. Plenum Press, New York.

1252. **Delaissé, J. M. and G. Vaes.** 1992. Mechanism of mineral solubilization and matrix degradation in osteoclastic bone resorption. In *Biology and Physiology of the Osteoclast*, B. R. Rifkin and C. V. Gay (eds), pp. 289–314. CRC Press, Boca Raton.

1253. **Elliott, E. and B. F. Sloane.** 1996. The cysteine protease cathepsin B in cancer. *Perspect. Drug Discov. Design* **6**: 12–32.

1253a. **Ezaki, J., L. S. Wolfe, K. Ishidoh, D. Muno, T. Ueno, and E. Kominami.** 1996. Lysosomal proteinosis based on decreased degradation of a specific protein, mitochondrial ATP synthase subunit C. Batten disease. In *Intracellular Protein Catabolism*, K. Suzuki and J. S. Bond (eds), pp. 121–128. Plenum Press, New York.

1254. **Grisolía, S., E. Knecht, and J. Hernández-Yago.** 1988. Intracellular protein degradation: past, present and future. In *The Roots of Modern Biochemistry*, H. Kleinkauf, H. von Döhren, and A. Jaenicke (eds), pp. 251–260. Walter de Gruyter & Co., Berlin.

1255. **Henskens, Y. M. C., E. C. I. Veerman, and A. V. N. Amerongen.** 1996. Cystatins in health and disease. *Biol. Chem.* **377**: 71–86.

1256. **Hopsu-Havu, V. K., I. Joronen, H. Kirschke, M. Järvinen, B. Wiederanders, and A. Rinne.** 1985. Localization of cathepsins H and L and their inhibitors in stratified epithelia and lymphatic tissue. In *The Biological Role of Proteinases and Their Inhibitors in Skin*, H. Ogawa, G. S. Lazarus, and V. K. Hopsu-Havu (eds), pp. 27–37. University of Tokyo Press, Tokyo.

1257. **Ii, K.** 1995. The role of beta-amyloid in the development of Alzheimer's disease. *Drugs Aging* **7**: 97–109.

1258. **Ishidoh, K. and E. Kominami.** 1995. Cathepsin L, gene regulation and extracellular functions. In *Proteases Involved in Cancer*, M. Suzuki and T. Hiwasa (eds), pp. 27–33. Monduzzi Editore, Bologna.

1259. **Ishii, S.** 1994. Legumain: asparaginyl endopeptidase. *Methods Enzymol.* **244**: 604–615.

1260. **Järvinen, M., A. Rinne, and V. K. Hopsu-Havu.** 1987. Human cystatins in normal and diseased tissues – a review. *Acta Histochem.* **82**: 5–18.

1261. **Kalnitsky, G., R. Chatterjee, H. Singh, M. Lones, and A. Paszkowski.** 1983. Bifunctional activities and possible modes of regulation of some lysosomal cysteinyl proteases. In *Proteinase Inhibitors. Medical and Biological Aspects*, N. Katunuma, H. Umezawa, and H. Holzer (eds), pp. 263–273. Japan Scientific Societies Press, Tokyo; Springer, Berlin.

1262. **Katunuma, N. and E. Kominami.** 1995. Structure, properties, mechanisms, and assays of cysteine protease inhibitors: cystatins and E-64 derivatives. *Methods Enzymol.* **251**: 382–397.

1263. **Katunuma, N., E. Kominami, T. Noda, and K. Isogai.** 1983. Lysosomal thiol proteinases and muscular dystrophy. In *Muscular Dystrophy: Biomedical Aspects*, S. Ebashi and E. Ozawa (eds), pp. 247–256. Japan Scientific Societies Press, Tokyo; Springer, Berlin.

1264. **Keilová, H.** 1971. On the specificity and inhibition of cathepsins D and B. In *Tissue Proteinases*, A. J. Barrett and J. T. Dingle (eds), pp. 45–68. North-Holland, Amsterdam.

1265. **Kirschke, H.** 1997. Lysosomal cysteine peptidases and malignant tumours. In *Cellular Peptidases in Immune Functions and Diseases*, S. Ansorge and J. Langner (eds), pp. 253–257. Plenum Press, New York (In press).

1266. **Kirschke, H., A. J. Barrett, and N. D. Rawlings.** 1995. Proteinases 1: lysosomal cysteine proteinases. *Protein Profile* **2**: 1587–1643.

1266a. **Mayer, R. J., C. Tipler, J. Arnold, L. Laszlo, A. Al-Khedhairy, J. Lowe, and M. Landon.** 1996. Endosome-lysosomes, ubiquitin and neurodegeneration. In *Intracellular Protein Catabolism*, K. Suzuki and J. S. Bond (eds), pp. 261–269. Plenum Press, New York.

1267. **McDonald, J. K., P. X. Callahan, S. Ellis, and R. E. Smith.** 1971. Polypeptide degradation by dipeptidyl aminopeptidase I (cathepsin C) and related peptidases. In *Tissue Proteinases*, A. J. Barrett and J. T. Dingle (eds), pp. 69–107. North-Holland, Amsterdam.

1268. **McDonald, J. K. and C. Schwabe.** 1977. Intracellular exopeptidases. In *Proteinases in Mammalian Cells and Tissues*, A. J. Barrett (ed), pp. 311–391. North-Holland, Amsterdam.

1269. **Mosolov, V. V.** 1994. [Proteolytic enzyme inhibitors of protein nature]. *Bioorg. Khim.* **20**: 153–161.

1270. **Muller-Ladner, U., R. E. Gay, and S. Gay.** 1996. Cysteine proteinases in arthritis and inflammation. *Perspect. Drug Discov. Design* 6: 87–98.

1271. **Otawara Hamamoto, Y.** 1994. [Biochemistry of bone matrix]. *Nippon Rinsho* **52**: 2239–2245.

1272. **Otto, K.** 1971. Cathepsins B1 and B2. In *Tissue Proteinases*, A. J. Barrett and J. T. Dingle (eds), pp. 1–28. North-Holland, Amsterdam.

1273. **Ozawa, H. and N. Amizuka.** 1994. [Structure and function of bone cells]. *Nippon Rinsho* **52**: 2246–2254.

1274. **Pagano, M.** 1995. Overexpression of cathepsins B and L in malignant tissues and cells. In *Proteases Involved in Cancer*, M. Suzuki and T. Hiwasa (eds), pp. 3–9. Monduzzi Editore, Bologna.

1274a. **Palmer, D. N. and J. M. Hay.** 1996. The neuronal ceroid lipofuscinoses (Batten disease). A group of lysosomal proteinoses. In *Intracellular Protein Catabolism*, K. Suzuki and J. S. Bond (eds), pp. 129–136. Plenum Press, New York.

1275. **Paweletz, N. and H. J. Boxberger.** 1994. Defined tumor cell–host interactions are necessary for malignant growth. *Crit. Rev. Oncog.* **5**: 69–105.

1276. **Rasnick, D.** 1996. Small synthetic inhibitors of cysteine proteases. *Perspect. Drug Discov. Design* **6**: 47–63.

1277. **Storer, A. C. and R. Menard.** 1996. Recent insights into cysteine protease specificity – lessons for drug design. *Perspect. Drug Discov. Design* **6**: 33–46.

1278. **Turk, V., J. Brzin, M. Kopitar, I. Kregar, P. Locnicar, M. Longer, T. Popovic, A. Ritonja, L. Vitale, W. Machleidt, T. Giraldi, and G. Sava.** 1983. Lysosomal cysteine proteinases and their protein inhibitors – structural studies. In *Proteinase Inhibitors: Medical and Biological Aspects*, N. Katunuma, H. Umezawa, and H. Holzer (eds), pp. 125–134. Japan Scientific Societies Press, Tokyo; Springer, Berlin.

1279. **Uchiyama, Y., S. Waguri, N. Sato, T. Watanabe, K. Ishido, and E. Kominami.** 1994. Cell and tissue distribution of lysosomal cysteine proteinases, cathepsins B, H, and L, and their biological roles. *Acta Histochem. Cytochem.* **27**: 287–308.

1280. **Wilhelm, O., U. Reuning, F. Jänicke, M. Schmitt, and H. Graeff.** 1994. The role of proteases in tumor invasion and metastasis: prognostic impact and therapeutical challenge? *Onkologie* **17**: 358–366.

Catalytic properties

1281. **Abul Faiz, M., J. B. Harris, C. A. Maltin, and D. Mantle.** 1995. Comparison of structural protein and proteolytic enzyme levels in degenerating and regenerating rat muscle induced by *Notechis scutatus* venom. *Comp. Biochem. Physiol. B Biochem. Mol. Biol.* **110**: 241–253.

1282. **Aibe, K., H. Yazawa, K. Abe, K. Teramura, M. Kumegawa, H. Kawashima, and K. Honda.** 1996. Substrate specificity of recombinant osteoclast-specific cathepsin K from rabbits. *Biol. Pharm. Bull.* **19**: 1026–1031.

1282a. **Baumgrass, R., M. K. Williamson, and P. A. Price.** 1997. Identification of peptide fragments generated by digestion of bovine and human osteocalcin with the lysosomal proteinases cathepsins B, D, L, H, and S. *J. Bone Miner. Res.* **12**: 447–455.

1283. **Bilge, A., J. Howell-Clark, S. Ramakrishnan, and O. W. Press.** 1994. Degradation of ricin A chain by endosomal and lysosomal enzymes – the protective role of ricin B chain. *Ther. Immunol.* **1**: 197–204.

1284. **Bohley, P., H. Kirschke, J. Langner, S. Ansorge, B. Wiederanders, and H. Hanson.** 1974. Degradation of rat liver proteins. In *Intracellular Protein Catabolism*, H. Hanson and P. Bohley (eds), pp. 201–209. J. A. Barth, Leipzig.

1285. **Bohley, P. and S. Riemann.** 1977. Intracellular protein catabolism. IX. Hydrophobicity of substrate proteins is a molecular basis of selectivity. *Acta Biol. Med. Ger.* **36**: 1823–1827.

1286. **Bossard, M. J., T. A. Tomaszek, S. K. Thompson, B. Y. Amegadzie, C. R. Hanning, C. Jones, J. T. Kurdyla, D. E. McNulty, F. H. Drake, M. Gowen, and M. A. Levy.** 1996. Proteolytic activity of human osteoclast cathepsin K. Expression, Purification, activation, and substrate identification. *J. Biol. Chem.* **271**: 12517–12524.

1287. **Brömme, D. and K. Okamoto.** 1995. The baculovirus cysteine protease has a cathepsin B-like S2-subsite specificity. *Biol. Chem. Hoppe-Seyler* **376**: 611–615.

1288. **Brömme, D., K. Okamoto, B. B. Wang, and S. Biroc.** 1996. Human cathepsin O2, a matrix protein-degrading cysteine protease expressed in osteoclasts. Functional expression of human cathepsin O2 in *Spodoptera frugiperda* and characterization of the enzyme. *J. Biol. Chem.* **271**: 2126–2132.

1289. **Callahan, P. X., J. K. McDonald, and S. Ellis.** 1972. Sequencing of peptides with dipeptidyl aminopeptidase I. *Methods Enzymol.* **25**: 282–298.

1290. **Cheng, D. L., W. P. Shu, J. C. S. Choi, E. J. Margolis, M. J. Droller, and B. C. Liu.** 1994. Bacillus Calmette–Guérin interacts with the carboxyl-terminal heparin binding domain of fibronectin: implications for BCG-mediated antitumor activity. *J. Urol.* **152**: 1275–1280.

1291. **Conlon, J. M., A. Hoog, and L. Grimelius.** 1995. Intracellular degradation of the C-peptide of proinsulin, in a human insulinoma: identification of sites of cleavage and evidence for a role for cathepsin B. *Pancreas* **10**: 167–172.

1292. **Csoma, C. and L. Polgár.** 1984. Proteinase from germinating bean cotyledons. Evidence for involvement of a thiol group in catalysis. *Biochem. J.* **222**: 769–776.

1293. **Dehrmann, F. M., T. H. T. Coetzer, R. N. Pike, and C. Dennison.** 1995. Mature cathepsin L is substantially active in the ionic milieu of the extracellular medium. *Arch. Biochem. Biophys.* **324**: 93–98.

1294. **Dehrmann, F. M., E. Elliott, and C. Dennison.** 1996. Reductive activation markedly increases the stability of cathepsin B and cathepsin L to extracellular ionic conditions. *Biol. Chem.* **377**: 391–394.

1295. **Doran, J. D., P. J. Tonge, J. S. Mort, and P. R. Carey.** 1996. Deacylation and reacylation for a series of acyl cysteine proteases, including acyl groups derived from novel chromophoric substrates. *Biochemistry* **35**: 12487–12494.

1296. **Dufour, E., M. Dalgalarrondo, G. Hervé, R. Goutefongea, and T. Haertlé.** 1996. Proteolysis of type III collagen by collagenase and cathepsin B under high hydrostatic pressure. *Meat Sci.* **3**: 261–269.

1297. **Fox, T., P. Mason, A. C. Storer, and J. S. Mort.** 1995. Modification of S_1 subsite specificity in the cysteine protease cathepsin B. *Protein Eng.* **8**: 53–57.

1298. **Fruton, J. S., W. R. Hearn, V. M. Ingram, D. S. Wiggans, and M. Winitz.** 1953. Synthesis of polymeric peptides in proteinase-catalyzed transamidation reactions. *J. Biol. Chem.* **204**: 891–902.

1299. **Fruton, J. S. and M. H. Knappenberger.** 1962. Polymerization reactions catalyzed by intracellular proteinases. III. Action of cathepsin C on a tetrapeptide amide. *Biochemistry* **1**: 674–676.

1300. **Fruton, J. S. and M. J. Mycek.** 1956. Studies on beef spleen cathepsin C. *Arch. Biochem. Biophys.* **65**: 11–20.

1301. **Gittel, C. and F. P. Schmidtchen.** 1995. Directed N-terminal elongation of unprotected peptides catalyzed by cathepsin C in water. *Bioconjug. Chem.* **6**: 70–76.

1302. **Gordon, M. M., T. Howard, M. J. Becich, and D. H. Alpers.** 1995. Cathepsin L mediates intracellular ileal digestion of gastric intrinsic factor. *Am. J. Physiol.* **268**: G33–G40.

1303. **Gorter, J. and M. Gruber.** 1970. Cathepsin C: an allosteric enzyme. *Biochim. Biophys. Acta* **198**: 546–555.

1304. **Guinec, N., V. Dalet-Fumeron, and M. Pagano.** 1993. 'In vitro' study of basement membrane degradation by the cysteine proteinases, cathepsins B, B-like and L. Digestion of collagen IV, laminin, fibronectin, and release of gelatinase activities from basement membrane fibronectin. *Biol. Chem. Hoppe-Seyler* **374**: 1135–1146.

1305. **Hara-Nishimura, I., T. Shimada, N. Hiraiwa, and M. Nishimura.** 1995. Vacuolar processing enzyme responsible for maturation of seed proteins. *J. Plant Physiol.* **145**: 632–640.

1306. **Heinrich, C. P. and J. S. Fruton.** 1968. The action of dipeptidyl transferase as a polymerase. *Biochemistry* **7**: 3556–3565.

1307. **Hiwasa, T., J. Ma, Y. Ike, N. Katunuma, and S. Sakiyama.** 1995. Increase of cyclin B by overexpression of cystatin alpha. *Cell Biochem. Funct.* **13**: 293–296.

1308. **Homma, K. and S. Natori.** 1996. Identification of substrate proteins for cathepsin L that are selectively hydrolyzed during the differentiation of imaginal discs of *Sarcophaga peregrina*. *Eur. J. Biochem.* **240**: 443–447.

1309. **Hopsu-Havu, V. K. and P. Rintola.** 1968. A sulphydryl-dependent and chloride-activated peptidase (cathepsin C) that hydrolyses alanyl-alanine naphthylamide. *Histochem. J.* **1**: 1–17.

1310. **Inaoka, T., H. Togame, O. Ishibashi, M. Kagoshima-Maezono, M. Kumegawa, and T. Kokubo.** 1996. Functional expression of human cathepsin K by baculovirus expression system and characterization of the recombinant enzyme. *J. Bone Miner. Res.* **11**: S180.

1311. **Ishibashi, O., H. Togame, Y. Mori, M. Kumegawa, and T. Kokubo.** 1996. Cathepsin K expressed at a high level in osteoclasts shows strong collagenolytic activity. *J. Bone Miner. Res.* **11**: S181.

1312. **Ishidoh, K. and E. Kominami.** 1995. Procathepsin L degrades extracellular matrix proteins in the presence of glycosaminoglycans *in vitro*. *Biochem. Biophys. Res. Commun.* **217**: 624–631.

1313. **Izumiya, N. and J. S. Fruton.** 1956. Specificity of cathepsin C. *J. Biol. Chem.* **218**: 59–76.

1314. **Jiang, S. T., J. J. Lee, and H. C. Chen.** 1996. Proteolysis of actomyosin by cathepsin B, cathepsin L, cathepsin L-like, and cathepsin X from mackerel (*Scomber australasicus*). *J. Agricult. Food Chem.* **44**: 769–773.

1315. **Jones, M. E., W. R. Hearn, M. Fried, and J. S. Fruton.** 1952. Transamidation reactions catalyzed by cathepsin C. *J. Biol. Chem.* **195**: 645–656.

1316. **Kakiuchi, S. and H. H. Tomizawa.** 1964. Properties of a glucagon-degrading enzyme of beef liver. *J. Biol. Chem.* 239: 2160–2164.

1317. **Katunuma, N.** 1994. [Participation of cathepsin B in exogenous antigen processing and regulation of the antigen presentation by invariant chain]. *Seikagaku* **66**: 510–520.

1318. **Katunuma, N., Y. Matsunaga, and T. Saibara.** 1994. Mechanism and regulation of antigen processing by cathepsin B. *Adv. Enzyme Regul.* **34**: 145–158.

1319. **Kawada, A., K. Hara, K. Morimoto, M. Hiruma, and A. Ishibashi.** 1995. Rat epidermal cathepsin B: purification and characterization of proteolytic properties toward filaggrin and synthetic substrates. *Int. J. Biochem. Cell Biol.* **27**: 175–183.

1320. **Keilová, H. and J. Turková.** 1970. Analogy between active sites of cathepsin B_1 and papain. *FEBS Lett.* **11**: 287–288.

1321. **Kiess, W., C. Terry, W. H. Burgess, B. Linder, W. Lopaczynski, and P. Nissley.** 1994. Insulin-like growth factor-II is a substrate for dipeptidylpeptidase I (cathepsin C). Biological properties of the product. *Eur. J. Biochem.* **226**: 179–184.

1322. **Kirschke, H., P. Bohley, S. Fittkau, and B. Wiederanders.** 1984. Some properties of lysosomal cysteine proteinases and their complexes with antibodies. In *Proteinase Action*, P. Elödi (ed), pp. 147–153. Akadémiai Kiadó, Budapest.

1323. **Kong, Y., Y. B. Chung, S. Y. Cho, and S. Y. Kang.** 1994. Cleavage of immunoglobulin G by excretory-secretory cathepsin S-like protease of *Spirometra mansoni* plerocercoid. *Parasitology* **109**: 611–621.

1324. **Kummer, J. A., A. M. Kamp, F. Citarella, A. J. G. Horrevoets, and C. E. Hack.** 1996. Expression of human recombinant granzyme A zymogen and its activation by the cysteine proteinase cathepsin C. *J. Biol. Chem.* **271**: 9281–9286.

1325. **Lee, J. J., H. C. Chen, and S. T. Jiang.** 1996. Comparison of the kinetics of cathepsin B, cathepsin L, cathepsin L-like, and cathepsin X from the dorsal muscle of mackerel on the hydrolysis of methylcoumarylamide substrates. *J. Agricult. Food Chem.* **44**: 774–778.

1326. **Lynch, G. W. and S. L. Pfueller.** 1988. Thrombin-independent activation of platelet factor XIII by endogenous platelet acid protease. *Thromb. Haemost.* **59**: 372–377.

1327. **Machleidt, W., D. K. Nägler, I. Assfalg-Machleidt, M. T. Stubbs, H. Fritz, and E. A. Auerswald.** 1995. Temporary inhibition of papain by hairpin loop mutants of chicken cystatin. Distorted binding of the loops results in cleavage of the Gly^9-Ala^{10} bond. *FEBS Lett.* **361**: 185–190.

1328. **Maciewicz, R. A., D. J. Etherington, J. Kos, and V. Turk.** 1987. Collagenolytic cathepsins of rabbit spleen: a kinetic analysis of collagen degradation and inhibition by chicken cystatin. *Collagen Rel. Res.* **7**: 295–304.

1329. **Marks, N., M. J. Berg, V. S. Sapirstein, R. Durrie, J. Swistok, R. C. Makofske, and W. Danho.** 1995. Brain cathepsin B but not metalloendopeptidases degrade rAPP751 with production of amyloidogenic fragments. Comparison with synthetic peptides emulating beta- and gamma-secretase sites. *Int. J. Pept. Protein Res.* **46**: 306–313.

1330. **Matsukura, U., T. Matsumoto, Y. Tashiro, A. Okitani, and H. Kato.** 1984. Morphological changes in myofibrils and glycerinated muscle fibers on treatment with cathepsins D and L. *Int. J. Biochem.* **16**: 957–962.

1331. **McDonald, J. K., P. X. Callahan, and S. Ellis.** 1972. Preparation and specificity of dipeptidyl aminopeptidase I. *Methods Enzymol.* **25**: 272–281.

1332. **McDonald, J. K., S. Ellis, and T. J. Reilly.** 1966. Properties of dipeptidyl arylamidase I of the pituitary. *J. Biol. Chem.* **241**: 1494–1501.

1333. **McDonald, J. K., T. J. Reilly, and S. Ellis.** 1965. A chloride-activated dipeptidyl-β-naphthylamidase of the pituitary gland. *Life Sci.* **4**: 1665–1668.

1334. **Metrione, R. M. and N. L. MacGeorge.** 1975. The mechanism of action of dipeptidyl aminopeptidase. Inhibition by amino acid derivatives and amines; activation by aromatic compounds. *Biochemistry* **14**: 5249–5252.

1335. **Mizuochi, T., S. T. Yee, M. Kasai, T. Kakiuchi, D. Muno, and E. Kominami.** 1994. Both cathepsin B and cathepsin D are necessary for processing of ovalbumin as well as for degradation of class II MHC invariant chain. *Immunol. Lett.* **43**: 189–193.

1336. **Morton, P. A., M. L. Zacheis, K. S. Giacoletto, J. A. Manning, and B. D. Schwartz.** 1995. Delivery of nascent MHC class II-invariant chain complexes to lysosomal compartments and proteolysis of invariant chain by cysteine proteases precedes peptide binding in B-lymphoblastoid cells. *J. Immunol.* **154**: 137–150.

1337. **Munger, J. S., C. Haass, C. A. Lemere, G. P. Shi, W. S. F. Wong, D. B. Teplow, D. J. Selkoe, and H. A. Chapman.** 1995. Lysosomal processing of amyloid precursor protein to Aβ peptides: a distinct role for cathepsin S. *Biochem. J.* **311**: 299–305.

1338. **Nauland, U. and D. C. Rijken.** 1994. Activation of thrombin-inactivated single-chain urokinase-type plasminogen activator by dipeptidyl peptidase I (cathepsin C). *Eur. J. Biochem.* **223**: 497–501.

1339. **Neves, F. A., K. G. Duncan, and J. D. Baxter.** 1996. Cathepsin B is a prorenin processing enzyme. *Hypertension* **27**: 514–517.

1340. **Nguyen, Q. V., V. E. Reyes, and R. E. Humphreys.** 1995. Structural analysis of invariant chain subsets as a function of their association with MHC class II chains. *Arch. Biochem. Biophys.* **317**: 128–132.

1341. **Ohshita, T. and H. Kido.** 1994. Involvement of a cystatin-alpha-sensitive cysteine proteinase in the degradation of native L-lactate dehydrogenase and serum albumin by rat liver or kidney lysosomes. *Eur. J. Biochem.* **225**: 781–786.

1342. **Planta, R. J., J. Gorter, and M. Gruber.** 1964. The catalytic properties of cathepsin C. *Biochim. Biophys. Acta* **89**: 511–519.

1343. **Planta, R. J. and M. Gruber.** 1961. Specificity of cathepsin C. *Biochim. Biophys. Acta* **53**: 443–444.

1344. **Reyes, V. E., S. Lu, and R. E. Humphreys.** 1991. Cathepsin B cleavage of I$_i$ from class II MHC α- and β-chains. *J. Immunol.* **146**: 3877–3880.

1345. **Riese, R. J., P. R. Wolf, D. Brömme, L. R. Natkin, J. A. Villadangos, H. L. Ploegh, and H. A. Chapman.** 1996. Essential role for cathepsin S in MHC class II-associated invariant chain processing and peptide loading. *Immunity* **4**: 357–366.

1346. **Rodriguez, G. M. and S. Diment.** 1995. Destructive proteolysis by cysteine proteases in antigen presentation of ovalbumin. *Eur. J. Immunol.* **25**: 1823–1827.

1347. **Serveau, C., L. Juliano, P. Bernard, T. Moreau, R. Mayer, and F. Gauthier.** 1994. New substrates of papain, based on the conserved sequence of natural inhibitors of the cystatin family. *Biochimie* **76**: 153–158.

1348. **Serveau, C., G. Lalmanach, M. A. Juliano, J. Scharfstein, L. Juliano, and F. Gauthier.** 1996. Investigation of the substrate specificity of cruzipain, the major cysteine proteinase of *Trypanosoma cruzi*, through the use of cystatin-derived substrates and inhibitors. *Biochem. J.* **313**: 951–956.

1349. **Shinagawa, T., K. Nakayama, Y. Uchiyama, E. Kominami, Y. Doi, K. Hashiba, K. Yano, W. A. Hsueh, and K. Murakami.** 1995. Role of cathepsin B as prorenin processing enzyme in human kidney. *Hypertens. Res.* **18**: 131–136.

1350. **Singh, H., T. Kuo, and G. Kalnitsky.** 1978. Collagenolytic activity of lung BANA hydrolase and cathepsin B1. In *Protein Turnover and Lysosome Function*, H. L. Segal and D. J. Doyle (eds), pp. 315–331. Academic Press, New York.

1351. **Sires, U. I., T. M. Schmid, C. J. Fliszar, Z. Q. Wang, S. L. Gluck, and H. G. Welgus.** 1995. Complete degradation of type X collagen requires the combined action of interstitial collagenase and osteoclast-derived cathepsin-B. *J. Clin. Invest.* **95**: 2089–2095.

1352. **Smith, A. M., C. Carmona, A. J. Dowd, S. McGonigle, D. Acosta, and J. P. Dalton.** 1994. Neutralization of the activity of a *Fasciola hepatica* cathepsin L proteinase by anti-cathepsin L antibodies. *Parasite Immunol.* **16**: 325–328.

1353. **Taniguchi, K., N. Sato, and Y. Uchiyama.** 1995. Apoptosis and heterophagy of medial edge epithelial cells of the secondary palatine shelves during fusion. *Arch. Histol. Cytol.* **58**: 191–203.

1354. **Taralp, A., H. Kaplan, I. I. Sytwu, I. Vlattas, R. Bohacek, A. K. Knap, T. Hirama, C. P. Huber, and S. Hasnain.** 1995. Characterization of the S_3 subsite specificity of cathepsin B. *J. Biol. Chem.* **270**: 18036–18043.

1355. **Vanha-Perttula, T., V. K. Hopsu, and G. G. Glenner.** 1965. A dipeptide naphthylamidase from hog kidney. *Histochemie* **5**: 448–449.

1356. **Verma, N. K., H. K. Ziegler, B. A. Stocker, and G. K. Schoolnik.** 1995. Induction of a cellular immune response to a defined T-cell epitope as an insert in the flagellin of a live vaccine strain of Salmonella. *Vaccine* **13**: 235–244.

1357. **Voynick, I. M. and J. S. Fruton.** 1968. The specificity of dipeptidyl transferase. *Biochemistry* **7**: 40–44.

1358. **Wojcikiewicz, R. J. H. and J. A. Oberdorf.** 1996. Degradation of inositol 1,4,5-trisphosphate receptors during cell stimulation is a specific process mediated by cysteine protease activity. *J. Biol. Chem.* **271**: 16652–16655.

1359. **Würz, H., A. Tanaka, and J. S. Fruton.** 1962. Polymerization of dipeptide amides by cathepsin C. *Biochemistry* **1**: 19–29.

1360. **Xu, M., G. A. Capraro, M. Daibata, V. E. Reyes, and R. E. Humphreys.** 1994. Cathepsin B cleavage and release of invariant chain from MHC class II molecules follow a staged pattern. *Mol. Immunol.* **31**: 723–731.

1361. **Yamashita, N.** 1996. [Cathepsin B and cathepsin L]. *Nippon Suisan Gakkaishi* **62**: 145–146.

1362. **Zhloba, A. A. and A. Y. Dubikaitis.** 1995. Variations in the activity of renal and hepatic cysteine cathepsins. *Bull. Exp. Biol. Med.* **120**: 1202–1204.

1363. **Zvonar-Popovic, T., T. Lah, I. Kregar, and V. Turk.** 1980. Some characteristics of cathepsin B and α-N-Benzoylarginine-β-naphthylamide hydrolase from bovine lymph nodes. *Croat. Chem. Acta* **53**: 509–517.

Determination and purification

1364. **Alnemri, E. S., D. J. Livingston, D. W. Nicholson, G. Salvesen, N. A. Thornberry, W. W. Wong, and J. Y. Yuan.** 1996. Human ICE/CED-3 protease nomenclature. *Cell* **87**: 171.

1365. **Aranishi, F., K. Hara, K. Osatomi, and T. Ishihara.** 1996. Purification and immunological properties of cathepsin B in carp *Cyprinus carpio*. *Comp. Biochem. Physiol. B Biochem. Mol. Biol.* **114**: 371–376.

1366. **Berg, T. O., P. E. Stromhaug, T. Berg, and P. O. Seglen.** 1994. Separation of lysosomes and autophagosomes by means of glycyl-phenylalanine-naphthylamide, a lysosome-disrupting cathepsin-C substrate. *Eur. J. Biochem.* **221**: 595–602.

1367. **Bouma, J. M. W. and M. Gruber.** 1964. The distribution of cathepsins B and C in rat tissue. *Biochim. Biophys. Acta* **89**: 545–547.

1368. **Brömme, D. and M. E. McGrath.** 1996. High level expression and crystallization of recombinant human cathepsin S. *Protein Sci.* **5**: 789–791.

1369. **Clark, I. D., J. P. MacManus, and A. G. Szabo.** 1995. A protease assay using time-resolved lanthanide luminescence from an engineered calcium binding protein substrate. *Clin. Biochem.* **28**: 131–135.

1370. **Coetzer, T. H. T., K. M. Dennehy, R. N. Pike, and C. Dennison.** 1995. Baboon (*Papio ursinus*) cathepsin L: purification, characterization and comparison with human and sheep cathepsin L. *Comp. Biochem. Physiol. B Biochem. Mol. Biol.* **112**: 429–439.

1371. **Coetzer, T. H. T., E. Elliott, P. H. Fortgens, R. N. Pike, and C. Dennison.** 1991. Anti-peptide antibodies to cathepsins B, L and D and type IV collagenase. Specific recognition and inhibition of the enzymes. *J. Immunol. Methods* **136**: 199–210.

1372. **Davidson, E. and B. Poole.** 1975. Fractionation of the rat liver enzymes that hydrolyze benzoyl-arginine-2-naphthylamide. *Biochim. Biophys. Acta* **397**: 437–442.

1373. de Lumen, B. O. and A. L. Tappel. 1972. α-N-Benzoylarginine-β-naphthylamide amidohydrolase of rat liver lysosomes. *J. Biol. Chem.* **247**: 3552–3557.

1374. Dolenc, I., B. Turk, G. Pungercic, A. Ritonja, and V. Turk. 1995. Oligomeric structure and substrate induced inhibition of human cathepsin C. *J. Biol. Chem.* **270**: 21626–21631.

1375. Dolinar, M., D. B. Maganja, and V. Turk. 1995. Expression of full-length human procathepsin L cDNA in *Escherichia coli* and refolding of the expression product. *Biol. Chem. Hoppe-Seyler* **376**: 385–388.

1376. Franklin, S. G. and R. M. Metrione. 1972. Chromatographic evidence for the existence of multiple forms of cathepsin B1. *Biochem. J.* **127**: 207–213.

1377. Fruton, J. S., G. W. Irving, Jr., and M. Bergmann. 1941. On the proteolytic enzymes of animal tissues. III. The proteolytic enzymes of beef spleen, beef kidney, and swine kidney. Classification of the cathepsins. *J. Biol. Chem.* **141**: 763–774.

1378. Gazi, M. I., S. W. Cox, D. T. Clark, and B. M. Eley. 1996. A comparison of cysteine and serine proteinases in human gingival crevicular fluid with tissue, saliva and bacterial enzymes by analytical isoelectric focusing. *Arch. Oral Biol.* **41**: 393–400.

1379. Greenbaum, L. M. and J. S. Fruton. 1957. Purification and properties of beef spleen cathepsin B. *J. Biol. Chem.* **226**: 173–180.

1380. Gutmann, H. R. and J. S. Fruton. 1948. On the proteolytic enzymes of animal tissues. VIII. An intracellular enzyme related to chymotrypsin. *J. Biol. Chem.* **174**: 851–858.

1381. Hara-Nishimura, I., Y. Takeuchi, and M. Nishimura. 1993. Molecular characterization of a vacuolar processing enzyme related to a putative cysteine proteinase of *Schistosoma mansoni*. *Plant Cell* **5**: 1651–1659.

1382. Ikeda, T., Y. Oikawa, and T. Nishiyama. 1996. Enzyme linked immunosorbent assay using cysteine proteinase antigens for immunodiagnosis of human paragonimiasis. *Am. J. Trop. Med. Hyg.* **55**: 434–437.

1383. Inubushi, T., H. Kakegawa, Y. Kishino, and N. Katunuma. 1994. Specific assay method for the activities of cathepsin L-type cysteine proteinases. *J. Biochem. (Tokyo)* **116**: 282–284.

1384. Ishibashi, O., T. Inaoka, H. Togame, G. Bilbe, H. Nakamura, Y. Mori, Y. Honda, Y. Hakeda, H. Ozawa, M. Kumegawa, and T. Kokubo. 1995. A novel cysteine proteinase localized at ruffled border: cathepsin K. *J. Bone Miner. Res.* **10**: S426.

1385. Järvinen, M. and V. K. Hopsu-Havu. 1975. α-N-Benzoylarginine-2-naphthylamide hydrolase (cathepsin B1?) from rat skin. II. Purification of the enzyme and demonstration of two inhibitors in the skin. *Acta Chem. Scand. B* **29**: 772–780.

1386. Karlsrud, T. S., L. Buo, A. O. Aasen, and H. T. Johansen. 1996. Quantification of kininogens in plasma – a functional method based on the cysteine proteinase inhibitor activity. *Thromb. Res.* **82**: 265–273.

1387. Kawada, A., K. Hara, M. Hiruma, H. Noguchi, and A. Ishibashi. 1995. Rat epidermal cathepsin L-like proteinase: purification and some hydrolytic properties toward filaggrin and synthetic substrates. *J. Biochem. (Tokyo)* **118**: 332–337.

1388. Keilová, H. and V. Tomásek. 1973. On the isozymes of cathepsin B_1. *FEBS Lett.* **29**: 335–338.

1389. Kembhavi, A. A., D. J. Buttle, C. G. Knight, and A. J. Barrett. 1993. The two cysteine endopeptidases of legume seeds: purification and characterization by use of specific fluorometric assays. *Arch. Biochem. Biophys.* **303**: 208–213.

1390. Kopitar, G., M. Dolinar, B. Strukelj, J. Pungercar, and V. Turk. 1996. Folding and activation of human procathepsin S from inclusion bodies produced in *Escherichia coli*. *Eur. J. Biochem.* **236**: 558–562.

1391. Kuhelj, R., M. Dolinar, J. Pungercar, and V. Turk. 1995. The preparation of catalytically active human cathepsin B from its precursor expressed in *Escherichia coli* in the form of inclusion bodies. *Eur. J. Biochem.* **229**: 533–539.

1392. Lipps, G., R. Füllkrug, and E. Beck. 1996. Cathepsin B of *Schistosoma mansoni*. Purification and activation of the recombinant proenzyme secreted by *Saccharomyces cerevisiae*. *J. Biol. Chem.* **271**: 1717–1725.

1393. McDonald, J. K. and J. M. C. Emerick. 1995. Purification and characterization of procathepsin L, a self-processing zymogen of guinea pig spermatozoa that acts on a cathepsin D assay substrate. *Arch. Biochem. Biophys.* **323**: 409–422.

1394. Metrione, R. M. 1978. Chromatography of dipeptidyl aminopeptidase I on inhibitor–sepharose columns. *Biochim. Biophys. Acta* **526**: 531–536.

1395. Planta, R. J. and M. Gruber. 1963. A simple estimation of cathepsin C using a new chromogenic substrate. *Anal. Biochem.* **5**: 360–367.

1396. Planta, R. J. and M. Gruber. 1964. Chromatographic purification of the thiol enzyme cathepsin C. *Biochim. Biophys. Acta* **89**: 503–510.

1397. Popovic, T., V. Puizdar, A. Ritonja, and J. Brzin. 1996. Simultaneous isolation of human kidney cathepsin B, cathepsin H, cathepsin L and cathepsin C and their characterization. *J. Chromatogr. B Biomed. Appl.* **681**: 251–262.

1398. Ryan, R. E., B. F. Sloane, M. Sameni, and P. L. Wood. 1995. Microglial cathepsin B: an immunological examination of cellular and secreted species. *J. Neurochem.* **65**: 1035–1045.

1399. **Scharfstein, J., M. Abrahamson, C. B. Palatnik de Souza, A. Barral, and I. V. Silva.** 1995. Antigenicity of cystatin-binding proteins from parasitic protozoan. Detection by a proteinase inhibitor based capture immunoassay (PINC-ELISA). *J. Immunol. Methods* **182**: 63–72.

1400. **Strewler, G. J. and V. C. Manganiello.** 1979. Purification and characterization of phosphodiesterase activator from kidney. *J. Biol. Chem.* **254**: 11891–11898.

1401. **Tallan, H. H., M. E. Jones, and J. S. Fruton.** 1952. On the proteolytic enzymes of animal tissues. X. Beef spleen cathepsin C. *J. Biol. Chem.* **194**: 793–805.

1402. **Trabandt, A., R.E. Gay, V.P. Sikhatme, and S. Gay.** 1995. Enzymatic detection systems for non-isotopic in situ hybridization using biotinylated cDNA probes. *Histochem. J.* **27**: 280–290.

1403. **Ulbricht, B., E. Spiess, R. Schwartz-Albiez, and W. Ebert.** 1995. Quantification of intracellular cathepsin activities in human lung tumor cell lines by flow cytometry. *Biol. Chem. Hoppe-Seyler* **376**: 407–414.

Reaction with inhibitors

1404. **Abe, M., K. Abe, C. Domoto, and S. Arai.** 1995. Two distinct species of corn cystatin in corn kernels. *Biosci. Biotechnol. Biochem.* **59**: 756–758.

1405. **Abe, M., K. Abe, K. Iwabuchi, C. Domoto, and S. Arai.** 1994. Corn cystatin I expressed in *Escherichia coli*: investigation of its inhibitory profile and occurrence in corn kernels. *J. Biochem. (Tokyo)* **116**: 488–492.

1406. **Abrahamson, M. and A. Grubb.** 1994. Increased body temperature accelerates aggregation of the Leu-68→Gln mutant cystatin C, the amyloid-forming protein in hereditary cystatin C amyloid angiopathy. *Proc. Natl Acad. Sci. U. S. A.* **91**: 1416–1420.

1407. **Abrahamson, M., A. Ritonja, M. A. Brown, A. Grubb, W. Machleidt, and A. J. Barrett.** 1987. Identification of the probable inhibitory reactive sites of the cysteine proteinase inhibitors human cystatin C and chicken cystatin. *J. Biol. Chem.* **262**: 9688–9694.

1408. **Alves, J. B., M. S. Alves, and Y. Naito.** 1994. Induction of synthesis of the rat cystatin S protein by the submandibular gland during the acute phase of experimental Chagas disease. *Mem. Inst. Oswaldo Cruz* **89**: 81–85.

1409. **Ambroso, J. L. and C. Harris.** 1994. In vitro embryotoxicity of the cysteine proteinase inhibitors benzyloxycarbonyl-phenylalanine-alanine-diazomethane (Z-Phe-Ala-CHN$_2$) and benzyloxycarbonyl-phenylalanine-phenylalanine-diazomethane (Z-Phe-Phe-CHN$_2$). *Teratology* **50**: 214–228.

1410. **Auerswald, E. A., D. K. Nägler, I. Assfalg-Machleidt, M. T. Stubbs, W. Machleidt, and H. Fritz.** 1995. Hairpin loop mutations of chicken cystatin have different effects on the inhibition of cathepsin B, cathepsin L and papain. *FEBS Lett.* **361**: 179–184.

1411. **Auerswald, E. A., D. K. Nagler, S. Gross, I. Assfalg-Machleidt, M. T. Stubbs, C. Eckerskorn, W. Machleidt, and H. Fritz.** 1996. Hybrids of chicken cystatin with human kininogen domain 2 sequences exhibit novel inhibition of calpain, improved inhibition of actinidin and impaired inhibition of papain, cathepsin L and cathepsin B. *Eur. J. Biochem.* **235**: 534–542.

1412. **Auerswald, E. A., D. K. Nagler, A. J. Schulze, R. A. Engh, G. Genenger, W. Machleidt, and H. Fritz.** 1994. Production, inhibitory activity, folding and conformational analysis of an N-terminal and an internal deletion variant of chicken cystatin. *Eur. J. Biochem.* **224**: 407–415.

1413. **Balbín, M., A. Hall, A. Grubb, R. W. Mason, C. López-Otín, and M. Abrahamson.** 1994. Structural and functional characterization of two allelic variants of human cystatin D sharing a characteristic inhibition spectrum against mammalian cysteine proteinases. *J. Biol. Chem.* **269**: 23156–23162.

1414. **Bano, B., S. P. Kunapuli, H. N. Bradford, and R. W. Colman.** 1996. Structural requirements for cathepsin B and cathepsin H inhibition by kininogens. *J. Protein Chem.* **15**: 519–525.

1415. **Barna, J. B. and R. R. Kew.** 1995. Inhibition of neutrophil chemotaxis by protease inhibitors. Differential effect of inhibitors of serine and thiol proteases. *Inflammation* **19**: 561–574.

1416. **Barrett, A. J. and P. M. Starkey.** 1973. The interaction of α_2-macroglobulin with proteinases. Characteristics and specificity of the reaction, and a hypothesis concerning its molecular mechanism. *Biochem. J.* **133**: 709–724.

1417. **Berg, K. and J. Moan.** 1994. Lysosomes as photochemical targets. *Int. J. Cancer* **59**: 814–822.

1418. **Bevec, T., V. Stoka, G. Pungercic, I. Dolenc, and V. Turk.** 1996. Major histocompatibility complex class II-associated p41 invariant chain fragment is a strong inhibitor of lysosomal cathepsin L. *J. Exp. Med.* **183**: 1331–1338.

1419. **Bihovsky, R., J. C. Powers, C. M. Kam, R. Walton, and R. C. Loewi.** 1993. Further evidence for the importance of free carboxylate in epoxysuccinate inhibitors of thiol proteases. *J. Enzyme Inhib.* **7**: 15–25.

1420. **Björk, I., I. Brieditis, and M. Abrahamson.** 1995. Probing the functional role of the N-terminal region of cystatins by equilibrium and kinetic studies of the binding of Gly-11 variants of recombinant human cystatin C to target proteinases. *Biochem. J.* **306**: 513–518.

1421. Björk, I., I. Brieditis, E. Raub-Segall, E. Pol, K. Håkansson, and M. Abrahamson. 1996. The importance of the second hairpin loop of cystatin C for proteinase binding, characterization of the interaction of Trp-106 variants of the inhibitor with cysteine proteinases. *Biochemistry* **35**: 10720–10726.

1422. Björk, I., E. Pol, E. Raub-Segall, M. Abrahamson, A. D. Rowan, and J. S. Mort. 1994. Differential changes in the association and dissociation rate constants for binding of cystatins to target proteinases occurring on N-terminal truncation of the inhibitors indicate that the interaction mechanism varies with different enzymes. *Biochem. J.* **299**: 219–225.

1423. Bobek, L. A., N. Ramasubbu, X. Wang, T. R. Weaver, and M. J. Levine. 1994. Biological activities and secondary structures of variant forms of human salivary cystatin SN produced in *Escherichia coli*. *Gene* **151**: 303–308.

1424. Brömme, D., J. L. Klaus, K. Okamoto, D. Rasnick, and J. T. Palmer. 1996. Peptidyl vinyl sulphones: a new class of potent and selective cysteine protease inhibitors. S_2P_2 specificity of human cathepsin O2 in comparison with cathepsins S and L. *Biochem. J.* **315**: 85–89.

1425. Brömme, D., R. A. Smith, P. J. Coles, H. Kirschke, A. C. Storer, and A. Krantz. 1994. Potent inactivation of cathepsins S and L by peptidyl (acyloxy)methyl ketones. *Biol. Chem. Hoppe-Seyler* **375**: 343–347.

1426. Brown, W. M., N. R. Saunders, K. Møllgård, and K. M. Dziegielewska. 1992. Fetuin–an old friend revisited. *Bioessays* **14**: 749–755.

1427. Cambra, O. and L. Berrens. 1996. Monoclonal antibodies against *Dermatophagoides* group I allergens as *pseudo*-cystatins blocking the catalytic site of cysteine proteinases. *Immunol. Lett.* **50**: 173–177.

1428. Carmona, E., É. Dufour, C. Plouffe, S. Takebe, P. Mason, J. S. Mort, and R. Ménard. 1996. Potency and selectivity of the cathepsin L propeptide as an inhibitor of cysteine proteases. *Biochemistry* **35**: 8149–8157.

1429. Chang, T. E. and C. R. Abraham. 1996. A novel brain cysteine protease forms an SDS stable complex with the β-amyloid precursor protein. *Ann. N. Y. Acad. Sci.* **777**: 183–188.

1430. Cimerman, N., M. D. Kosorok, B. D. Korant, B. Turk, and V. Turk. 1996. Characterization of cystatin C from bovine parotid glands – cysteine proteinase inhibition and antiviral properties. *Biol. Chem.* **377**: 19–23.

1431. Colella, R., I. Kaplan, and G. D. Mower. 1994. Localization of cystatin mRNA in chicken brain by in situ hybridization. *J. Histochem. Cytochem.* **42**: 1487–1491.

1432. Delaisse, J. M., A. Boyde, E. Maconnachie, N. N. Ali, C. H. J. Sear, Y. Eeckhout, G. Vaes, and S. J. Jones. 1987. The effects of inhibitors of cysteine-proteinases and collagenase on the resorptive activity of isolated osteoclasts. *Bone* **8**: 305–313.

1433. Demuth, H. U., A. Schierhorn, P. Bryan, R. Höfke, H. Kirschke, and D. Brömme. 1996. N-peptidyl, O-acyl hydroxamates: comparison of the selective inhibition of serine and cysteine proteinases. *Biochim. Biophys. Acta* **1295**: 179–186.

1434. Dolenc, I., B. Turk, J. Kos, and V. Turk. 1996. Interaction of human cathepsin C with chicken cystatin. *FEBS Lett.* **392**: 277–280.

1435. Dolle, R. E., J. Singh, D. Whipple, I. K. Osifo, G. Speier, T. L. Graybill, J. S. Gregory, A. L. Harris, C. T. Helaszek, R. E. Miller, *et al.* 1995. Aspartyl alpha-((diphenylphosphinyl)oxy)methyl ketones as novel inhibitors of interleukin-1 beta converting enzyme. Utility of the diphenylphosphinic acid leaving group for the inhibition of cysteine proteases. *J. Med. Chem.* **38**: 220–222.

1436. Drake, F., B. J. Votta, A. Badger, R. A. Dodds, I. James, M. A. Levy, S. Thompson, M. J. Bossard, T. Carr, J. R. Connor, T. A. Tomaszek, L. Szewczuk, D. Veber, and M. Gowen. 1996. Peptide aldehyde inhibitors of cathepsin K inhibit bone resorption both *in vitro* and *in vivo*. *J. Bone Miner. Res.* **11**: P246.

1437. Dufour, É., A. C. Storer, and R. Ménard. 1995. Peptide aldehydes and nitriles as transition state analog inhibitors of cysteine proteases. *Biochemistry* **34**: 9136–9143.

1438. Ekiel, I. and M. Abrahamson. 1996. Folding-related dimerization of human cystatin C. *J. Biol. Chem.* **271**: 1314–1321.

1439. Emilsson, V., L. Thorsteinsson, O. Jensson, and G. Gudmundsson. 1996. Human cystatin C expression and regulation by TGF-β-1 – implications for the pathogenesis of hereditary cystatin C amyloid angiopathy causing brain hemorrhage. *Amyloid – Int. J. Exp. Clin. Invest.* **3**: 110–118.

1440. Esser, R. E., R. A. Angelo, M. D. Murphey, L. M. Watts, L. P. Thornburg, J. T. Palmer, J. W. Talhouk, and R. E. Smith. 1994. Cysteine proteinase inhibitors decrease articular cartilage and bone destruction in chronic inflammatory arthritis. *Arthritis Rheum.* **37**: 236–247.

1441. Ganz, T. 1994. Biosynthesis of defensins and other antimicrobial peptides. *Ciba Found. Symp.* **186**: 62–71 [Discussion 71–76].

1442. Gao, X., L. D. Devoe, and K. S. Given. 1994. Effects of amniotic fluid on proteases: a possible role of amniotic fluid in fetal wound healing. *Ann. Plast. Surg.* **33**: 128–134.

1443. Gour-Salin, B. J., P. Lachance, P. R. Bonneau, A. C. Storer, H. Kirschke, and D. Broemme. 1994. E-64 analogs as inhibitors of cathepsin L and cathepsin S: importance of the S_2–P_2 interactions for potency and selectivity. *Bioorg. Chem.* **22**: 227–241.

1444. **Gour-Salin, B. J., P. Lachance, M. C. Magny, C. Plouffe, R. Ménard, and A. C. Storer.** 1994. E64 [*trans*-epoxysuccinyl-L-leucylamido-(4-guanidino)butane] analogues as inhibitors of cysteine proteinases: investigation of S_2 subsite interactions. *Biochem. J.* **299**: 389–392.

1445. **Grubb, A., O. Jensson, G. Gudmundsson, A. Arnason, H. Löfberg, and J. Malm.** 1984. Abnormal metabolism of γ-trace alkaline microprotein: the basic defect in hereditary cerebral hemorrhage with amyloidosis. *New Engl. J. Med.* **311**: 1547–1549.

1446. **Hall, A., K. Håkansson, R. W. Mason, A. Grubb, and M. Abrahamson.** 1995. Structural basis for the biological specificity of cystatin C. Identification of leucine 9 in the N-terminal binding region as a selectivity-conferring residue in the inhibition of mammalian cysteine peptidases. *J. Biol. Chem.* **270**: 5115–5121.

1447. **Harada, M., A. Kozakai, A. Sugure, S. Kajita, A. Ikeda, K. Shinyashiki, T. Takamura, S. Higuchi, M. Murata, E. Kominami, and Y. Uchiyama.** 1996. Effects of a cysteine proteinase inhibitor, Ca-315, on experimental models of bone resorption and rat osteopenia. *J. Bone Miner. Res.* **11**: M665.

1448. **Hibino, T., T. Takahashi, A. Takeda, D. Leuthardt, P. Baciu, and P. F. Goetinck.** 1995. Cloning of psoriastatin, a cathepsin L-specific inhibitor, from psoriatic epidermis. *Mol. Biol. Cell* **6**: 345a.

1449. **Hiwasa T. and E. Kominami.** 1995. Physical association of Ras and cathepsins B and L in the conditioned medium of v-Ha-*ras*-transformed NIH3T3 cells. *Biochem. Biophys. Res. Commun.* **216**: 828–834.

1450. **Hiwasa, T., T. Sawada, and S. Sakiyama.** 1996. Synergistic induction of anchorage-independent growth of NIH3T3 mouse fibroblasts by cysteine proteinase inhibitors and a tumor promoter. *J. Biol. Chem.* **271**: 9181–9184.

1451. **Hiwasa, T., T. Sawada, S. Sakiyama, E. Kominami, and N. Katunuma.** 1989. Cysteine proteinase-inhibitory activity of *ras* gene product is not affected by mutations at GTPase-activating protein-binding sites. *Biol. Chem. Hoppe-Seyler* **370**: 1215–1220.

1452. **Hu, B., L. Coulson, B. Moyer, and P. A. Price.** 1995. Isolation and molecular cloning of a novel bone phosphoprotein related in sequence to the cystatin family of thiol protease inhibitors. *J. Biol. Chem.* **270**: 431–436.

1453. **Irie, K., H. Hosoyama, T. Takeuchi, K. Iwabuchi, H. Watanabe, M. Abe, K. Abe, and S. Arai.** 1996. Transgenic rice established to express corn cystatin exhibits strong inhibitory activity against insect gut proteinases. *Plant Mol. Biol.* **30**: 149–157.

1454. **Isemura, S. and E. Saitoh.** 1994. Inhibitory activities of partially degraded salivary cystatins. *Int. J. Biochem.* **26**: 825–831.

1455. **Jerala, R., L. Kroon-Zitko, T. Popovic, and V. Turk.** 1994. Elongation on the amino-terminal part of stefin B decreases inhibition of cathepsin H. *Eur. J. Biochem.* **224**: 797–802.

1456. **Kastelic, L., B. Turk, N. Kopitar Jerala, A. Stolfa, S. Rainer, V. Turk, and T. T. Lah.** 1994. Stefin B, the major low molecular weight inhibitor in ovarian carcinoma. *Cancer Lett.* **82**: 81–88.

1457. **Katunuma, N., H. Kakegawa, Y. Matsunaga, and T. Saibara.** 1994. Immunological significances of invariant chain from the aspect of its structural homology with the cystatin family. *FEBS Lett.* **349**: 265–269.

1458. **Kominami, E., T. Tsukahara, K. Ii, K. Hizawa, and N. Katunuma.** 1984. Studies on thiol proteinase inhibitors in rat peripheral blood cells. *Biochem. Biophys. Res. Commun.* **123**: 816–821.

1459. **Lenarcic, B., I. Krizaj, P. Zunec, and V. Turk.** 1996. Differences in specificity for the interactions of stefins A, B and D with cysteine proteinases. *FEBS Lett.* **395**: 113–118.

1460. **Leonardi, A., B. Turk, and V. Turk.** 1996. Inhibition of bovine cathepsin L and cathepsin S by stefins and cystatins. *Biol. Chem.* **377**: 319–321.

1461. **Li, R., G. L. Kenyon, F. E. Cohen, X. Chen, B. Gong, J. N. Dominguez, E. Davidson, G. Kurzban, R. E. Miller, E. O. Nuzum, et al.** 1995. In vitro antimalarial activity of chalcones and their derivatives. *J. Med. Chem.* **38**: 5031–5037.

1462. **Li, Z. Z., A. C. Ortega-Vilain, G. S. Patil, D. L. Chu, J. E. Foreman, D. D. Eveleth, and J. C. Powers.** 1996. Novel peptidyl α-keto amide inhibitors of calpains and other cysteine proteases. *J. Med. Chem.* **39**: 4089–4098.

1463. **Lindahl, P., D. Ripoll, M. Abrahamson, J. S. Mort, and A. C. Storer.** 1994. Evidence for the interaction of valine-10 in cystatin C with the S_2 subsite of cathepsin B. *Biochemistry* **33**: 4384–4392.

1464. **Martichonok, V., C. Plouffe, A. C. Storer, R. Menard, and J. B. Jones.** 1995. Aziridine analogs of [[trans-(epoxysuccinyl)-L-leucyl]amino]-4-guanidinobutane (E-64) as inhibitors of cysteine proteases. *J. Med. Chem.* **38**: 3078–3085.

1465. **Martin, J. R., R. Jerala, L. Kroon Zitko, E. Zerovnik, V. Turk, and J. P. Waltho.** 1994. Structural characterisation of human stefin A in solution and implications for binding to cysteine proteinases. *Eur. J. Biochem.* **225**: 1181–1194.

1466. **Matsumoto, K., D. Yamamoto, H. Ohishi, K. Tomoo, T. Ishida, M. Inoue, T. Sadatome, K. Kitamura, and H. Mizuno.** 1989. Mode of binding of E-64-c, a potent thiol protease inhibitor, to papain as determined by X-ray crystal analysis of the complex. *FEBS Lett.* **245**: 177–180.

1467. **Moin, K., L. T. Emmert, and B. F. Sloane.** 1992. A membrane-associated cysteine protease inhibitor from murine hepatoma. *FEBS Lett.* **309**: 279–282.

1468. **Morishita, A., S. Ishikawa, Y. Ito, K. Ogawa, M. Takada, S. Yaginuma, and S. Yamamoto.** 1994. AM4299 A and B, novel thiol protease inhibitors. *J. Antibiot. (Tokyo)* **47**: 1065–1068.

1469. **Okada, Y., Y. Tsuda, Y. Mu, K. Hirano, H. Okamoto, Y. Okamoto, H. Kakegawa, H. Matsumoto, and T. Sato.** 1995. Amino acids and peptides. XXXIX. Synthesis of iNoc-Gln-Val-Val-Ala-Ala-pNA and its action on thiol proteinases. *Chem. Pharm. Bull. (Tokyo)* **43**: 96–99.

1470. **Paczek, L., M. Teschner, R. M. Schaefer, M. Lao, L. Gradowska, and A. Heidland.** 1996. Low-density lipoprotein suppresses cathepsins B and L activity in rat mesangial cells. *Miner. Electrolyte Metab.* **22**: 51–53.

1471. **Pennacchio, L. A., A. E. Lehesjoki, N. E. Stone, V. L. Willour, K. Virtaneva, J. Miao, E. D'Amato, L. Ramirez, M. Faham, M. Koskiniemi, J. A. Warrington, R. Norio, A. de la Chapelle, D. R. Cox, and R. M. Myers.** 1996. Mutations in the gene encoding cystatin B in progressive myoclonus epilepsy (EPM1). *Science* **271**: 1731–1734.

1472. **Pol, E., S. L. Olsson, S. Estrada, T. W. Prasthofer, and I. Björk.** 1995. Characterization by spectroscopic, kinetic and equilibrium methods of the interaction between recombinant human cystatin A (stefin A) and cysteine proteinases. *Biochem. J.* **311**: 275–282.

1473. **Rauber, P., B. Walker, S. Stone, and E. Shaw.** 1988. Synthesis of lysine-containing sulphonium salts and their properties as proteinase inhibitors. *Biochem. J.* **250**: 871–876.

1474. **Rosenblum, W. I.** 1996. Cystatin C – Icelandic-like mutation in an animal model of cerebrovascular β-amyloidosis – Editorial Comment. *Stroke* **27**: 2085.

1475. **Saitoh, E. and S. Isemura.** 1994. Production of human salivary type cysteine proteinase inhibitors (cystatins) by an *Escherichia coli* system and partial characterization of recombinant cystatin S and its mutant (117 arginine → tryptophan). *J. Biochem. (Tokyo)* **116**: 399–405.

1476. **Sato, N., T. Horiuchi, M. Hamano, H. Sekine, S. Chiba, H. Yamamoto, T. Yoshioka, I. Kimura, M. Satake, and Y. Ida.** 1996. Kojistatin A, a new cysteine protease inhibitor produced by *Aspergillus oryzae*. *Biosci. Biotech. Biochem.* **60**: 1747–1748.

1477. **Shibuya, K., H. Kaji, T. Itoh, Y. Ohyama, A. Tsujikami, S. Tate, A. Takeda, I. Kumagai, I. Hirao, K. Miura, F. Inagaki, and T. Samejima.** 1995. Human cystatin A is inactivated by engineered truncation. The NH_2-terminal region of the cysteine proteinase inhibitor is essential for expression of its inhibitory activity. *Biochemistry* **34**: 12185–12192.

1478. **Sivaparvathi, M., I. McCutcheon, R. Sawaya, G. L. Nicolson, and J. S. Rao.** 1996. Expression of cysteine protease inhibitors in human gliomas and meningiomas. *Clin. Exp. Metastasis* **14**: 344–350.

1479. **Starkey, P. M. and A. J. Barrett.** 1973. Human cathepsin B1. Inhibition by α_2-macroglobulin and other serum proteins. *Biochem. J.* **131**: 823–831.

1480. **Takahashi, M., T. Tezuka, H. Kakegawa, and N. Katunuma.** 1994. Linkage between phosphorylated cystatin alpha and filaggrin by epidermal transglutaminase as a model of cornified envelope and inhibition of cathepsin L activity by cornified envelope and the conjugated cystatin alpha. *FEBS Lett.* **340**: 173–176.

1481. **Takeda, A., T. Yamamoto, Y. Nakamura, T. Takahashi, and T. Hibino.** 1995. Squamous cell carcinoma antigen is a potent inhibitor of cysteine proteinase cathepsin L. *FEBS Lett.* **359**: 78–80.

1482. **Teramura, K., M. Orita, H. Matsumoto, K. Yasumuro, and K. Abe.** 1996. Effects of Ym-51084 and Ym-51085, new inhibitors produced by *Streptomyces* sp. Q21705, on cathepsin L. *J. Enzyme Inhib.* **11**: 115–121.

1483. **Tsuji, A., T. Akamatsu, H. Nagamune, and Y. Matsuda.** 1994. Identification of targeting proteinase for rat alpha 1-macroglobulin *in vivo*. Mast-cell tryptase is a major component of the alpha 1-macroglobulin-proteinase complex endocytosed into rat liver lysosomes. *Biochem. J.* **298**: 79–85.

1484. **Turk, B., J. G. Bieth, I. Bjork, I. Dolenc, D. Turk, N. Cimerman, J. Kos, A. Colic, V. Stoka, and V. Turk.** 1995. Regulation of the activity of lysosomal cysteine proteinases by pH-induced inactivation and/or endogenous protein inhibitors, cystatins. *Biol. Chem. Hoppe-Seyler* **376**: 225–230.

1485. **Turk, B., A. Colić, V. Stoka, and V. Turk.** 1994. Kinetics of inhibition of bovine cathepsin S by bovine stefin B. *FEBS Lett.* **339**: 155–159.

1486. **Turk, B., A. Ritonja, I. Björk, V. Stoka, I. Dolenc, and V. Turk.** 1995. Identification of bovine stefin A, a novel protein inhibitor of cysteine proteinases. *FEBS Lett.* **360**: 101–105.

1487. **Turk, B., V. Stoka, I. Björk, C. Boudier, G. Johansson, I. Dolenc, A. Colic, J. G. Bieth, and V. Turk.** 1995. High-affinity binding of two molecules of cysteine proteinases to low-molecularweight kininogen. *Protein Sci.* **4**: 1874–1880.

1488. **Turk, B., V. Stoka, V. Turk, G. Johansson, J. J. Cazzulo, and I. Björk.** 1996. High-molecular-weight kininogen binds two molecules of cysteine proteinases with different rate constants. *FEBS Lett.* **391**: 109–112.

1489. **Wagner, B. M., R. A. Smith, P. J. Coles, L. J. Copp, M. J. Ernest, and A. Krantz.** 1994. In vivo inhibition of cathepsin B by peptidyl (acyloxy)methyl ketones. *J. Med. Chem.* **37**: 1833–1840.

1490. **Wei, L. H., L. C. Walker, and E. Levy.** 1996. Cystatin C – Icelandic-like mutation in an animal model of cerebrovascular β-amyloidosis. *Stroke* **27**: 2080–2085.

1491. **Woo, J. T., S. Sigeizumi, K. Yamaguchi, K. Sugimoto, T. Kobori, T. Tsuji, and K. Kondo.** 1995. Peptidyl aldehyde derivatives as potent and selective inhibitors of cathepsin L. *Bioorg. Med. Chem. Lett.* **5**: 1501–1504.

1492. **Woo, J. T., K. Yamaguchi, T. Hayama, T. Kobori, S. Sigeizumi, K. Sugimoto, K. Kondo, T. Tsuji, Y. Ohba, K. Tagami, and K. Sumitani.** 1996. Suppressive effect of N-(benzyloxycarbonyl)-L-phenylalanyl-L-tyrosinal on bone resorption in vitro and in vivo. *Eur. J. Pharmacol.* **300**: 131–135.

1493. **Ylinenjärvi, K., T. W. Prasthofer, N. C. Martin, and I. Björk.** 1995. Interaction of cysteine proteinases with recombinant kininogen domain 2, expressed in *Escherichia coli. FEBS Lett.* **357**: 309–311.

1494. **Zimacheva, A. V. and V. V. Mosolov.** 1996. A new inhibitor of cysteine proteinases from soybean. *Biochemistry (Moscow)* **61**: 1193–1200.

Expression of the enzymes in cancer

1495. **Abe, S., N. Sukoh, S. Ogura, and H. Isobe.** 1994. [Clinical indicators of malignancy of lung cancer]. *Hokkaido Igaku Zasshi* **69**: 391–395.

1496. **Achkar, C., Q. Gong, S. Mehtani, A. S. Bajkowski, and A. Frankfater.** 1995. Expression of cathepsin B and cathepsin B-like enzymes in cancer. In *Proteases Involved in Cancer*, M. Suzuki and T. Hiwasa (eds), pp. 35–44. Monduzzi Editore, Bologna.

1497. **Adenis, A., G. Huet, F. Zerimech, B. Hecquet, M. Balduyck, and J. P. Peyrat.** 1995. Cathepsin B, L, and D activities in colorectal carcinomas: relationship with clinico-pathological parameters. *Cancer Lett.* **96**: 267–275.

1498. **Arkona, C. and B. Wiederanders.** 1996. Expression, subcellular distribution and plasma membrane binding of cathepsin B and gelatinases in bone metastatic tissue. *Biol. Chem.* **377**: 695–702.

1499. **Baracos, V. E., C. DeVivo, D. H. R. Hoyle, and A. L. Goldberg.** 1995. Activation of the ATP–ubiquitin–proteasome pathway in skeletal muscle of cachectic rats bearing a hepatoma. *Am. J. Physiol.* **268**: E996-E1006.

1500. **Bellelli, A., M. Mattioni, V. Rusconi, M. L. Sezzi, and L. Bellelli.** 1990. Inhibition of tumor growth, invasion and metastasis in papain-immunized mice. *Invasion Metastasis* **10**: 142–169.

1501. **Berg, K. and J. Moan.** 1995. The influence of the cysteine protease inhibitor L-trans-epoxysuccinyl-leucyl amido(4-guanidio)butane (E64) on photobiological effects of tetra(4-sulfonatophenyl)porphine. *Cancer Lett.* **88**: 227–236.

1502. **Bhuvarahamurthy, V. and S. Govindasamy.** 1995. Extracellular matrix components and proteolytic enzymes in uterine cervical carcinoma. *Mol. Cell. Biochem.* **144**: 35–43.

1503. **Bjornland, K., L. Buø, I. Kjønniksen, M. Larsen, Ø. Fodstad, H. T. Johansen, and A. O. Aasen.** 1996. Cysteine proteinase inhibitors reduce malignant melanoma cell invasion in vitro. *Anticancer Res.* **16**: 1627–1631.

1504. **Boike, G., T. Lah, B. F. Sloane, J. Rozhin, K. Honn, R. Guirguis, M. L. Stracke, L. A. Liotta, and E. Schiffmann.** 1991. A possible role for cysteine proteinase and its inhibitors in motility of malignant melanoma and other tumour cells. *Melanoma Res.* **1**: 333–340.

1505. **Bongers, V., C. H. Konings, A. M. Grijpma, I. Steen, B. J. M. Braakhuis, and G. B. Snow.** 1995. Serum proteinase activities in head and neck squamous cell carcinoma patients. *Anticancer Res.* **15**: 2763–2766.

1506. **Brown, V., A. Courtney, A. Trimble, and D. McCormick.** 1996. Cathepsin B in monolayer and spheroid cultures of human gliomas. *Neuropath. Appl. Neurobiol.* **22**: 454.

1507. **Budihna, M., J. Skrk, B. Zakotnik, D. Gabrijelcic, and J. Lindtner.** 1995. Prognostic value of total cathepsin B in invasive ductal carcinoma of the breast. *Eur. J. Cancer* **31A**: 661–664.

1508. **Budihna, M., P. Strojan, L. Smid, J. Skrk, I. Vrhovec, A. Zupevc, Z. Rudolf, M. Zargi, M. Krasovec, B. Svetic, N. Kopitar-Jerala, and J. Kos.** 1996. Prognostic value of cathepsin B, cathepsin H, cathepsin L, cathepsin D and their endogenous inhibitors stefin A and stefin B in head and neck carcinoma. *Biol. Chem.* **377**: 385–390.

1509. **Campo, E., J. Muñoz, R. Miquel, A. Palacin, A. Cardesa, B. F. Sloane, and M. R. Emmert Buck.** 1994. Cathepsin B expression in colorectal carcinomas correlates with tumor progression and shortened patient survival. *Am. J. Pathol.* **145**: 301–309.

1510. **Castiglioni, T., M. J. Merino, B. Elsner, T. T. Lah, B. F. Sloane, and M. R. Emmert-Buck.** 1994. Immunohistochemical analysis of cathepsins D, B, and L in human breast cancer. *Hum. Pathol.* **25**: 857–862.

1511. **Chung, S. M.** 1990. Variant cathepsin L activity from gastric cancer tissue. *Japan. J. Cancer Res.* **81**: 813–819.

1512. **Conese, M. and F. Blasi.** 1995. The urokinase/urokinase-receptor system and cancer invasion. *Baillière's Clin. Haematol.* **8**: 365–389.

1513. Conway, J. G., S. J. Trexler, J. A. Wakefield, B. E. Marron, D. L. Emerson, D. M. Bickett, D. N. Deaton, D. Garrison, M. Elder, A. McElroy, N. Willmott, A. J. Dockerty, and G. M. McGeehan. 1996. Effect of matrix metalloproteinase inhibitors on tumor growth and spontaneous metastasis. *Clin. Exp. Metastasis* **14**: 115–124.

1514. David, L., F. Gartner, and M. Sobrinh-Simões. 1996. Do cathepsins play a role in the biological behavior of gastric carcinoma? *Hum. Pathol.* **27**: 997–998.

1515. Ebert, W., H. Knoch, B. Werle, G. Trefz, T. Muley, and E. Spiess. 1994. Prognostic value of increased lung tumor tissue cathepsin B. *Anticancer Res.* **14**: 895–900.

1516. Emmert-Buck, M. R., M. J. Roth, Z. Zhuang, E. Campo, J. Rozhin, B. F. Sloane, L. A. Liotta, and W. G. Stetler-Stevenson. 1994. Increased gelatinase A (MMP-2) and cathepsin B activity in invasive tumor regions of human colon cancer samples. *Am. J. Pathol.* **145**: 1285–1290.

1517. Feng, B., E. E. Rollo, and D. T. Denhardt. 1995. Osteopontin (OPN) may facilitate metastasis by protecting cells from macrophage NO-mediated cytotoxicity: evidence from cell lines down-regulated for OPN expression by a targeted ribozyme. *Clin. Exp. Metastasis* **13**: 453–462.

1518. Fröhlich, E., G. Schaumburg-Lever, and C. Klessen. 1995. Immunocytochemical and immunoelectron microscopic demonstration of cathepsin B in human malignant melanoma. *Br. J. Dermatol.* **132**: 867–875.

1519. Gong, Q. and A. Frankfater. 1993. Alternative splicing of cathepsin B in human tumors produces a novel isoform which can be folded into an active enzyme *in vitro*. *FASEB J.* **7**: A1208.

1520. Hadman, M., L. Gabos, M. Loo, A. Sehgal, and T. J. Bos. 1996. Isolation and cloning of JTAP1: a cathepsin like gene upregulated in response to V-Jun induced cell transformation. *Oncogene* **12**: 135–142.

1521. Haczynska, H. and M. Warwas. 1994. [Cathepsin B – role in cancer invasion and diagnosis]. *Postepy Hig. Med. Dosw.* **48**: 729–743.

1522. Heidtmann, H. H., U. Salge, K. Havemann, H. Kirschke, and B. Wiederanders. 1993. Secretion of a latent, acid activatable cathepsin L precursor by human non-small cell lung cancer cell lines. *Oncol. Res.* **5**: 441–451.

1523. Herszenyi, L., M. Plebani, P. Carraro, M. De Paoli, G. Roveroni, M. Rugge, R. Cardin, R. Naccaarato, and F. Farinati. 1995. [Role and behavior of cathepsin B and cathepsin L in gastric cancer]. *Orv. Hetil.* **136**: 1315–1318.

1524. Honn, K. V., J. Timár, J. Rozhin, R. Bazaz, M. Sameni, G. Ziegler, and B. F. Sloane. 1994. A lipoxygenase metabolite, 12-(*S*)-HETE, stimulates protein kinase C-mediated release of cathepsin B from malignant cells. *Exp. Cell Res.* **214**: 120–130.

1525. Inoue, T., T. Ishida, K. Sugio, and K. Sugimachi. 1994. Cathepsin B expression and laminin degradation as factors influencing prognosis of surgically treated patients with lung adenocarcinoma. *Cancer Res.* **54**: 6133–6136.

1526. Jean, D., M. Bar-Eli, S. Huang, K. Xie, F. Rodrigues-Lima, J. Hermann, and R. Frade. 1996. A cysteine proteinase, which cleaves human C3, the third component of complement, is involved in tumorigenicity and metastasis of human melanoma. *Cancer Res.* **56**: 254–258.

1527. Jean, D., J. Hermann, F. Rodrigues-Lima, M. Barel, M. Balbo, and R. Frade. 1995. Identification on melanoma cells of p39, a cysteine proteinase that cleaves C3, the third component of complement: amino-acid-sequence identities with procathepsin L. *Biochem. J.* **312**: 961–969.

1528. Jessup, J. M. 1994. Cathepsin B and other proteases in human colorectal carcinoma [comment]. *Am. J. Pathol.* **145**: 253–262.

1529. Jiang, L. W., V. M. Maher, J. J. McCormick, and M. Schindler. 1990. Alkalinization of the lysosomes is correlated with *ras* transformation of murine and human fibroblasts. *J. Biol. Chem.* **265**: 4775–4777.

1530. Kageshita, T., T. Ono, M. Himeno, and Y. Nishimura. 1995. Biochemical and immunohistochemical analysis of cathepsins B, H, L and D in human melanocytic tumors. In *Proteases Involved in Cancer*, M. Suzuki and T. Hiwasa (eds), pp. 65–69. Monduzzi Editore, Bologna.

1531. Kageshita, T., A. Yoshii, T. Kimura, K. Maruo, T. Ono, M. Himeno, and Y. Nishimura. 1995. Biochemical and immunohistochemical analysis of cathepsins B, H, L and D in human melanocytic tumours. *Arch. Dermatol. Res.* **287**: 266–272.

1532. Kawada, A., K. Hara, E. Kominami, T. Kobayashi, M. Hiruma, and A. Ishibashi. 1996. Cathepsins B and D expression in squamous cell carcinoma. *Br. J. Dermatol.* **135**: 905–910.

1533. Keppler, D., M. Abrahamson, and B. Sordat. 1994. Secretion of cathepsin B and tumour invasion. *Biochem. Soc. Trans.* **22**: 43–49.

1534. Keppler, D., P. Waridel, M. Abrahamson, D. Bachmann, J. Berdoz, and B. Sordat. 1994. Latency of cathepsin B secreted by human colon carcinoma cells is not linked to secretion of cystatin C and is relieved by neutrophil elastase. *Biochim. Biophys. Acta* **1226**: 117–125.

1535. Kirschke, H., T. Clausen, B. Göhring, D. Günther, E. Heucke, F. Laube, E. Löwe, H. Neef, H. Papesch, S. Peinze, G. Plehn, U. Rebmann, A. Rinne, R. Rüdrich, and E. Weber. 1997. Concentrations of lysosomal cysteine proteases are decreased in renal cell carcinoma compared with normal kidney. *J. Cancer Res. Clin. Oncol.* **123**: 402–406.

1536. **Kirschke, H., I. Joronen, A. Rinne, M. Järvinen, and V. K. Hopsu-Havu.** 1997. Overexpression of cathepsins B, H, L and S in malignant tumours of the lymph node. In *Proteolysis in Cell Functions*, V. K. Hopsu-Havu, M. Järvinen, and H. Kirschke (eds), pp. 479–484. IOS Press, Amsterdam..

1537. **Kirschke, H., S. Peinze, F. Laube, and E. Weber.** 1995. Processing and secretion of cathepsin S by normal and transformed cells. In *Proteases Involved in Cancer*, M. Suzuki and T. Hiwasa (eds), pp. 45–49. Monduzzi Editore, Bologna.

1538. **Kos, J., A. Smid, M. Krasovec, B. Svetic, B. Lenarcic, I. Vrhovec, J. Skrk, and V. Turk.** 1995. Lysosomal proteases cathepsins D, B, H, L and their inhibitors stefins A and B in head and neck cancer. *Biol. Chem. Hoppe-Seyler* **376**: 401–405.

1539. **Kozyreva, E. A., K. I. Zhordanina, L. S. Basalyk, and A. V. Vasil'ev.** 1994. [Prognostic significance of determining cathepsin B activity in malignant ovarian tumors]. *Vopr. Med. Khim.* **40**: 25–27.

1540. **Krepela, E., J. Prochazka, B. Karova, J. Cermak, and H. Roubkova.** 1996. Cathepsin B and cathepsin C and gamma-glutamyl-transferase activities and levels of reduced glutathione and cysteine in human lung carcinomas and lungs. *Chem. Listy* **90**: 654–655.

1541. **Krepela, E., J. Procházka, H. Mynaríková, B. Kárová, J. Polák, J. Cermák, and H. Roubková.** 1995. Multiple forms of cathepsin B in human lung cancer. *Int. J. Cancer.* **61**: 44–53.

1542. **Kusunoki, T., S. Nishida, K. Kabashi, K. Murata, and T. Tomura.** 1995. Study on cathepsin B and L in human thyroid tumors. In *Proteases Involved in Cancer*, M. Suzuki and T. Hiwasa (eds), pp. 71–74. Monduzzi Editore, Bologna.

1543. **Kusunoki, T., S. Nishida, T. Nakano, K. Funasaka, S. Kimoto, K. Murata, and T. Tomura.** 1995. Study on cathepsin B activity in human thyroid tumors. *Auris Nasus Larynx* **22**: 43–48.

1544. **Lah, T. T., G. Calaf, E. Kalman, B. G. Shinde, J. Russo, D. Jarosz, J. Zabrecky, R. Somers, and I. Daskal.** 1995. Cathepsin D, cathepsin B and cathepsin L in breast carcinoma and in transformed human breast epithelial cells (Hbec). *Biol. Chem. Hoppe-Seyler* **376**: 357–363.

1545. **Lah, T. T., G. Calaf, E. Kalman, B. G. Shinde, R. Somers, S. Estrada, E. Salero, J. Russo, and I. Daskal.** 1996. Cathepsin D, cathepsin B, and cathepsin L in transformed human breast epithelial cells. *Breast Cancer Res. Treat.* **39**: 221–233.

1546. **Leto, G., F. M. Tumminello, A. Russo, G. Pizzolanti, V. Bazan, and N. Gebbia.** 1994. Cathepsin D activity levels in colorectal cancer – correlation with cathepsin B and cathepsin L and other biological and clinical parameters. *Int. J. Oncol.* **5**: 509–515.

1547. **Llovera, M., C. García-Martínez, N. Agell, M. Marzábal, F. J. López-Soriano, and J. M. Argilés.** 1994. Ubiquitin gene expression is increased in skeletal muscle of tumour-bearing rats. *FEBS. Lett.* **338**: 311–318.

1548. **Makarewicz, R., G. Drewa, W. Szymanski, and I. Skonieczna-Makarewicz.** 1995. Cathepsin B in predicting the extent of the cervix carcinoma. *Neoplasma* **42**: 21–24.

1549. **McCormick, D., A. Trimble, I. Allen, and P. Winter.** 1996. Cathepsin S-like activity in human gliomas. *Neuropathol. Appl. Neurobiol.* **22**: 461.

1550. **Mikkelsen, T., P. S. Yan, K. L. Ho, M. Sameni, B. F. Sloane, and M. L. Rosenblum.** 1995. Immunolocalization of cathepsin B in human glioma: implications for tumor invasion and angiogenesis. *J. Neurosurg.* **83**: 285–290.

1551. **Moan, J., K. Berg, H. Anholt, and K. Madslien.** 1994. Sulfonated aluminium phthalocyanines as sensitizers for photochemotherapy. Effects of small light doses on localization, dye fluorescence and photosensitivity in V79 cells. *Int. J. Cancer* **58**: 865–870.

1552. **Moin, K., M. Sameni, E. Elliott, G. Ziegler, and B. F. Sloane.** 1995. *Ras* oncogene alters localization of cathepsins. In *Proteases Involved in Cancer*, M. Suzuki and T. Hiwasa (eds), pp. 51–58. Monduzzi Editore, Bologna.

1553. **Morris, V. L., S. Koop, I. C. MacDonald, E. E. Schmidt, M. Grattan, D. Percy, A. F. Chambers, and A. C. Groom.** 1994. Mammary carcinoma cell lines of high and low metastatic potential differ not in extravasation but in subsequent migration and growth. *Clin. Exp. Metastasis* **12**: 357–367.

1554. **Murnane, M. J., J. Cai, S. Shuja, L. Coté, E. Del Re, C. Iacobuzio-Donahue, K. Kim, and K. Sheahan.** 1995. Changing patterns of proteolytic expression with colorectal tumor progression. In *Proteases Involved in Cancer*, M. Suzuki and T. Hiwasa (eds), pp. 11–26. Monduzzi Editore, Bologna.

1555. **Narvaez, C. J., K. Vanweelden, I. Byrne, and J. Welsh.** 1996. Characterization of a vitamin D_3-resistant MCF-7 cell line. *Endocrinology* **137**: 400–409.

1556. **Nishida, Y., K. Kohno, T. Kawamata, K. Morimitsu, M. Kuwano, and I. Miyakawa.** 1995. Increased cathepsin L levels in serum in some patients with ovarian cancer: comparison with CA125 and CA72–3. *Gynecol. Oncol.* **56**: 357–361.

1557. **Oberhuber, H., B. Seliger, amd R. Schafer.** 1995. Partial restoration of pre-transformation levels of lysyl oxidase and transin mRNAs in phenotypic ras revertants. *Mol. Carcinog.* **12**: 198–204.

1558. **Ohta, T., T. Terada, T. Nagakawa, H. Tajima, H. Itoh, L. Fonseca, and I. Miyazaki.** 1994. Pancreatic trypsinogen and cathepsin B in human pancreatic carcinomas and associated metastatic lesions. *Br. J. Cancer* **69**: 152–156.

1559. Ostrowski, L. E., A. Ahsan, B. P. Suthar, P. Pagast, D. L. Bain, C. Wong, A. Patel, and R. M. Schultz. 1986. Selective inhibition of proteolytic enzymes in an *in vivo* mouse model for experimental metastasis. *Cancer Res.* **46**: 4121–4128.

1560. Oursler, M. J., B. L. Riggs, and T. C. Spelsberg. 1993. Glucocorticoid-induced activation of latent transforming growth factor-β by normal human osteoblast-like cells. *Endocrinology* **133**: 2187–2196.

1561. Paciucci, R., G. Berrozpe, M. Torá, E. Navarro, A. G. de Herreros, and F. X. Real. 1996. Isolation of tissue-type plasminogen activator, cathepsin H, and nonspecific cross-reacting antigen from SK-PC-1 pancreas cancer cells using subtractive hybridization. *FEBS Lett.* **385**: 72–76.

1562. Plebani, M., L. Herszènyi, R. Cardin, G. Roveroni, P. Carraro, M. D. Paoli, M. Rugge, W. F. Grigioni, D. Nitti, R. Naccarato, and F. Farinati. 1995. Cysteine and serine proteases in gastric cancer. *Cancer* **76**: 367–375.

1563. Recklies, A. D., J. S. Mort, and A. R. Poole. 1982. Secretion of a thiol proteinase from mouse mammary carcinomas and its characterization. *Cancer Res.* **42**: 1026–1032.

1564. Rempel, S. A., M. L. Rosenblum, T. Mikkelsen, P. S. Yan, K. D. Ellis, W. A. Golembieski, M. Sameni, J. Rozhin, G. Ziegler, and B. F. Sloane. 1994. Cathepsin B expression and localization in glioma progression and invasion. *Cancer Res.* **54**: 6027–6031.

1565. Rozhin, J., M. Sameni, G. Ziegler, and B. F. Sloane. 1994. Pericellular pH affects distribution and secretion of cathepsin B in malignant cells. *Cancer Res.* **54**: 6517–6525.

1566. Russo, A., V. Bazan, N. Gebbia, G. Pizzolanti, F. M. Tumminello, G. Dardanoni, F. Ingria, S. Restivo, R. M. Tomasino, and G. Leto. 1995. Flow cytometric DNA analysis and lysosomal cathepsins B and L in locally advanced laryngeal cancer. Relationship with clinicopathologic parameters and prognostic significance. *Cancer* **76**: 1757–1764.

1567. Sakai, H. 1996. Do cathepsins play a role in the biological behavior of gastric carcinoma? – reply. *Hum. Pathol.* **27**: 998.

1568. Scaddan, P. B. and M. J. Dufresne. 1993. Characterization of cysteine proteases and their endogenous inhibitors in MCF-7 and adriamycin-resistant MCF-7 human breast cancer cells. *Invasion Metastasis* **13**: 301–313.

1569. Schwartz, G. K., H. Wang, N. Lampen, N. Altorki, D. Kelsen, and A. P. Albino. 1994. Defining the invasive phenotype of proximal gastric cancer cells. *Cancer* **73**: 22–27.

1570. Shuja, S. and M. J. Murnane. 1996. Marked increases in cathepsin B and L activities distinguish papillary carcinoma of the thyroid from normal thyroid or thyroid with non-neoplastic disease. *Int. J. Cancer* **66**: 420–426.

1571. Sinha, A. A., D. F. Gleason, N. A. Staley, M. J. Wilson, M. Sameni, and B. F. Sloane. 1995. Cathepsin B in angiogenesis of human prostate: an immunohistochemical and immunoelectron microscopic analysis. *Anat. Rec.* **241**: 353–362.

1572. Sinha, A. A., M. J. Wilson, D. F. Gleason, P. K. Reddy, M. Sameni, and B. F. Sloane. 1995. Immunohistochemical localization of cathepsin B in neoplastic human prostate. *Prostate* **26**: 171–178.

1573. Sivaparvathi, M., R. Sawaya, Z. L. Gokaslan, K. S. Chintala, and J. S. Rao. 1996. Expression and the role of cathepsin H in human glioma progression and invasion. *Cancer Lett.* **104**: 121–126.

1574. Sivaparvathi, M., R. Sawaya, S. W. Wang, A. Rayford, M. Yamamoto, L. A. Liotta, G. L. Nicolson, and J. S. Rao. 1995. Overexpression and localization of cathepsin B during the progression of human gliomas. *Clin. Exp. Metastasis* **13**: 49–56.

1575. Sivaparvathi, M., M. Yamamoto, G. L. Nicolson, Z. L. Gokaslan, G. N. Fuller, L. A. Liotta, R. Sawaya, and J. S. Rao. 1996. Expression and immunohistochemical localization of cathepsin L during the progression of human gliomas. *Clin. Exp. Metastasis* **14**: 27–34.

1576. Sloane, B. F., K. Moin, M. Sameni, L. R. Tait, J. Rozhin, and G. Ziegler. 1994. Membrane association of cathepsin B can be induced by transfection of human breast epithelial cells with c-Ha-*ras* oncogene. *J. Cell Sci.* **107**: 373–384.

1577. Solovyeva, N. I., T. O. Balayevskaya, E. A. Dilakyan, T. A. Zakamaldina-Zama, V. F. Pozdnev, L. Z. Topol, and F. L. Kisseljov. 1995. Proteolytic enzymes at various stages of oncogenic transformation of rat fibroblasts. I. Aspartyl and cysteine proteinases. *Int. J. Cancer* **60**: 495–500.

1578. Spiess, E., A. Bruning, S. Gack, B. Ulbricht, H. Spring, G. Trefz, and W. Ebert. 1994. Cathepsin B activity in human lung tumor cell lines: ultrastructural localization, pH sensitivity, and inhibitor status at the cellular level. *J. Histochem. Cytochem.* **42**: 917–929.

1580. Sukoh, N., S. Abe, I. Nakajima, S. Ogura, H. Isobe, K. Inoue, and Y. Kawakami. 1994. Immunohistochemical distributions of cathepsin B and basement membrane antigens in human lung adenocarcinoma: association with invasion and metastasis. *Virchows Arch.* **424**: 33–38.

1581. Sukoh, N., S. Abe, S. Ogura, H. Isobe, H. Takekawa, K. Inoue, and Y. Kawakami. 1994. Immunohistochemical study of cathepsin B. Prognostic significance in human lung cancer. *Cancer* **74**: 46–51.

1582. Tang, D. G. and K. V. Honn. 1994. 12-Lipoxygenase, 12(*S*)-HETE, and cancer metastasis. *Ann. N. Y. Acad. Sci.* **744**: 199–215.

1583. Terada, T., T. Ohta, H. Minato, and Y. Nakanuma. 1995. Expression of pancreatic trypsinogen/trypsin and cathepsin B in human cholangiocarcinomas and hepatocellular carcinomas. *Hum. Pathol.* **26**: 746–752.

1584. Tomasino, R. M., V. Bazan, E. Daniele, R. Nuara, V. Morello, V. Tralongo, G. Leto, and A. Russo. 1996. Biological characterization of laryngeal squamous-cell carcinoma. *Anticancer Res.* **16**: 2257–2267.

1585. Tumminello, F. M., G. Leto, G. Pizzolanti, V. Candiloro, M. Crescimanno, L. Crosta, C. Flandina, G. Montalto, M. Soresi, A. Carroccio, F. Bascone, I. Ruggeri, S. Ippolito, and N. Gebbia. 1996. Cathepsins D, B and L circulating levels as prognostic markers of malignant progression. *Anticancer Res.* **16**: 2315–2319.

1586. Tumminello, F. M., G. Pizzolanti, N. Gebbia, and G. Leto. 1995. Stefin A and cathepsin B and cathepsin L circulating serum levels in patients with malignant or nonmalignant liver diseases – a preliminary report. *Med. Sci. Res.* **23**: 741–742.

1587. Turowski, G. A., Z. Rashid, F. Hong, J. A. Madri, and M. D. Basson. 1994. Glutamine modulates phenotype and stimulates proliferation in human colon cancer cell lines. *Cancer Res.* **54**: 5974–5980.

1588. Ueda, M. 1995. [A study on cathepsin B-like substance in cancer of the urinary tract]. *Nippon Hinyokika Gakkai Zasshi* **86**: 1429–1434.

1589. van der Stappen, J. W. J., A. C. Williams, R. A. Maciewicz, and C. Paraskeva. 1996. Activation of cathepsin B, secreted by a colorectal cancer cell line requires low pH and is mediated by cathepsin D. *Int. J. Cancer* **67**: 547–554.

1590. Vasishta, A., P. R. Baker, D. Hopwood, P. M. Holley, and A. Cuschieri. 1985. Proteinase-like peptidase activities in malignant and non-malignant gastric tissue. *Br. J. Surg.* **72**: 386–388.

1591. Visscher, D. W., B. F. Sloane, M. Sameni, J. W. Babiarz, J. Jacobson, and J. D. Crissman. 1994. Clinicopathologic significance of cathepsin B immunostaining in transitional neoplasia. *Mod. Pathol.* **7**: 76–81.

1592. Waguri, S., N. Sato, T. Watanabe, K. Ishidoh, E. Kominami, K. Sato, and Y. Uchiyama. 1995. Cysteine proteinases in GH4C1 cells, a rat pituitary tumor cell line, are secreted by the constitutive and regulated secretory pathways. *Eur. J. Cell Biol.* **67**: 308–318.

1593. Weber, E., D. Günther, F. Laube, B. Wiederanders, and H. Kirschke. 1994. Hybridoma cells producing antibodies to cathepsin L have greatly reduced potential for tumour growth. *J. Cancer Res. Clin. Oncol.* **120**: 564–567.

1594. Weber, E., D. Günther, F. Laube, B. Wiederanders, and H. Kirschke. 1995. Suppression of malignancy of myeloma cells by their production of antibodies to cathepsin L. In *Proteases Involved in Cancer*, M. Suzuki and T. Hiwasa (eds), pp. 75–79. Monduzzi Editore, Bologna.

1595. Werle, B., W. Ebert, W. Klein, and E. Spiess. 1994. Cathepsin B in tumors, normal tissue and isolated cells from the human lung. *Anticancer Res.* **14**: 1169–1176.

1596. Werle, B., W. Ebert, W. Klein, and E. Spiess. 1995. Assessment of cathepsin L activity by use of the inhibitor CA-074 compared to cathepsin B activity in human lung tumor tissue. *Biol. Chem. Hoppe-Seyler* **376**: 157–164.

1597. Yoshii, A., T. Kageshita, H. Tsushima, and T. Ono. 1995. Clinical relevance of cathepsin B-like enzyme activity and cysteine proteinase inhibitor in melanocytic tumours. *Arch. Dermatol. Res.* **287**: 209–213.

Expression of the enzymes in pancreatitis

1598. Abdo, E. E., A. M. Coelho, A. L. Montagnini, M. S. Kubrusly, K. R. Leite, S. N. Sampietre, N. A. Molan, M. C. Machado, and H. W. Pinotti. 1994. [Simplified model of induction of experimental acute pancreatitis with a supra-maximal dose of cerulein]. *Rev. Hosp. Clin. Fac. Med. Sao Paulo* **49**: 204–207.

1599. Apte, M. V., J. S. Wilson, M. A. Korsten, G. W. McCaughan, P. S. Haber, and R. C. Pirola. 1995. Effects of ethanol and protein deficiency on pancreatic digestive and lysosomal enzymes. *Gut* **36**: 287–293.

1600. Apte, M. V., J. S. Wilson, G. W. McCaughan, M. A. Korsten, P. S. Haber, I. D. Norton, and R. C. Pirola. 1995. Ethanol-induced alterations in messenger RNA levels correlate with glandular content of pancreatic enzymes. *J. Lab. Clin. Med.* **125**: 634–640.

1601. Baniukiewicz, A. A., J. W. Dlugosz, and A. Gabryelewicz. 1994. The lysosomal hydrolases in the rat pancreas after maximal or supramaximal stimulation with cerulein. *Int. J. Pancreatol.* **16**: 7179.

1602. Dalet-Fumeron, V., L. Boudjennah, and M. Pagano. 1996. Competition between plasminogen and procathepsin B as a probe to demonstrate the *in vitro* activation of procathepsin B by the tissue plasminogen activator. *Arch. Biochem. Biophys.* **335**: 351–357.

1603. **Gumaste, V. V.** 1994. Role of cathepsin B in the pathogenesis of acute pancreatitis. *Gastroenterology* **106**: 1123–1125.

1604. **Hirano, T.** 1994. Lysosomal enzyme secretion into pancreatic juice in rats injected with pancreatic secretagogues and augmented secretion after short-term pancreatic duct obstruction. *Nippon Geka Hokan* **63**: 21–35.

1605. **Hirano, T., H. Furuyama, Y. Kawakami, K. Ando, and T. Tsuchitani.** 1995. Protective effects of prophylaxis with a protease inhibitor and a free radical scavenger against a temporary ischemia model of pancreatitis. *Can. J. Surg.* **38**: 241–248.

1606. **Hirano, T. and S. Takeuchi.** 1994. Anti-ulcer agent, cetraxate hydrochloride (Neuer®), prevents subcellular redistribution of lysosomal enzyme in caerulein-induced pancreatitis in the rat. *J. Int. Med. Res.* **22**: 107–112.

1607. **Kharbanda, K. K., D. L. McVicker, R. K. Zetterman, and T. M. Donohue.** 1996. Ethanol consumption alters trafficking of lysosomal enzymes and affects the processing of procathepsin L in rat liver. *Biochim. Biophys. Acta* **1291**: 45–52.

1608. **Korsten, M. A., P. S. Haber, J. S. Wilson, and C. S. Lieber.** 1995. The effect of chronic alcohol administration on cerulein-induced pancreatitis. *Int. J. Pancreatol.* **18**: 25–31.

1609. **Lerch, M. M., H. Weidenbach, T. M. Gress, and G. Adler.** 1995. Effect of kinin inhibition in experimental acute pancreatitis. *Am. J. Physiol.* **269**: G490-G499.

1610. **Lüthen, R. E., C. Niederau, and J. H. Grendell.** 1994. Glutathione and ATP levels, subcellular distribution of enzymes, and permeability of duct system in rabbit pancreas following intravenous administration of alcohol and cerulein. *Dig. Dis. Sci.* **39**: 871–879.

1611. **Lüthen, R., C. Niederau, M. Niederau, L. D. Ferrell, and J. H. Grendell.** 1995. Influence of ductal pressure and infusates on activity and subcellular distribution of lysosomal enzymes in the rat pancreas. *Gastroenterology* **109**: 573–581.

1612. **Machado, M. C., A. M. Coelho, M. S. Kubrusly, A. Bonizzia, and H. W. Pinotti.** 1995. Reduction of pancreatic enzyme content and mortality in experimental acute pancreatitis in rats. *Braz. J. Med. Biol. Res.* **28**: 471–475.

1613. **Samuel, I., Y. Toriumi, D. P. Wilcockson, C. M. Turkelson, T. E. Solomon, and R. J. Joehl.** 1995. Bile and pancreatic juice replacement ameliorates early ligation-induced acute pancreatitis in rats. *Am. J. Surg.* **169**: 391–399.

1614. **Samuel, I., D. P. Wilcockson, J. P. Regan, and R. J. Joehl.** 1995. Ligation-induced acute pancreatitis in opossums: acinar cell necrosis in the absence of colocalization. *J. Surg. Res.* **58**: 69–74.

1615. **Wang, Z. H., T. Manabe, G. Ohshio, T. Hirano, N. Okada, T. Imamura, Y. Kawaguchi, F. Yotsumoto, K. Yamaki, and M. Imamura.** 1995. Effect of the oral administration of sepimostat mesilate on cerulein induced acute pancreatitis in rats. *Arzneimittelforschung* **45**: 1082–1086.

Expression of the enzymes in normal tissues and disorders

1616. **Amano, T., H. Nakanishi, T. Kondo, T. Tanaka, M. Oka, and K. Yamamoto.** 1995. Age-related changes in cellular localization and enzymatic activities of cathepsins B, L and D in the rat trigeminal ganglion neuron. *Mech. Ageing Dev.* **83**: 133–141.

1617. **Ambroso, J. L. and C. Harris.** 1994. Chloroquine accumulation and alterations of proteolysis and pinocytosis in the rat conceptus in vitro. *Biochem. Pharmacol.* **47**: 679–688.

1618. **An, H., M. Y. Peters, T. A. Seymour, and M. T. Morrissey.** 1995. Isolation and activation of cathepsin L-inhibitor complex from pacific whiting (*Merluccius productus*). *J. Agric. Food Chem.* **43**: 327–330.

1619. **Andreu, A. L., S. Schwartz, J. Asin, J. Lopez, E. Garcia, and M. A. Arbos.** 1995. Thiolprotease activity in skeletal and myocardial muscle and liver of fasted rats. *Nutrition* **11**: 289–291.

1620. **Asari, A., S. Kuriyama, E. Kominami, and Y. Uchiyama.** 1995. Cytochemical localization of hyaluronic acid in human synovium with special reference to its possible process of degradation. *Arch. Histol. Cytol.* **58**: 65–76.

1621. **Baici, A., D. Hörler, A. Lang, C. Merlin, and R. Kissling.** 1995. Cathepsin B in osteoarthritis: zonal variation of enzyme activity in human femoral head cartilage. *Ann. Rheum. Dis.* **54**: 281–288.

1622. **Baici, A., A. Lang, D. Hörler, R. Kissling, and C. Merlin.** 1995. Cathepsin B in osteoarthritis: cytochemical and histochemical analysis of human femoral head cartilage. *Ann. Rheum. Dis.* **54**: 289–297.

1623. **Bailey, A. J., J. F. Tarlton, J. Van der Stappen, T. J. Sims, and A. Messina.** 1994. The continuous elongation technique for severe Dupuytren's disease. A biochemical mechanism. *J. Hand Surg. [Br.]* **19**: 522–527.

1624. **Bechet, D. M., C. Deval, J. Robelin, M. J. Ferrara, and A. Obled.** 1996. Developmental control of cathepsin B expression in bovine fetal muscles. *Arch. Biochem. Biophys.* **334**: 362–368.

1625. **Bednarski, E. and G. Lynch.** 1996. Cytosolic proteolysis of tau by cathepsin D in hippocampus following suppression of cathepsins B and L. *J. Neurochem.* **67**: 1846–1855.

1626. **Bernstein, H. G., A. Reichenbach, H. Kirschke, and B. Wiederanders.** 1989. Cell type-specific distribution of cathepsin B and D immunoreactivity within the rabbit retina. *Neurosci. Lett.* **98**: 135–138.

1627. **Bernstein, H. G., T. Roskoden, H. Kirschke, H. J. Hahn, B. Wiederanders, and V. Novakova.** 1994. Cathepsin L immunoreactivity in the hypothalamus of normal, streptozotocin-diabetic and vasopressin-deficient Brattleboro rats. *Cell. Mol. Biol. (Noisy-le-grand)* **40**: 159–164.

1628. **Bever, Jr., C. T. and D. W. Garver.** 1995. Increased cathepsin B activity in multiple sclerosis brain. *J. Neurol. Sci.* **131**: 71–73.

1629. **Bever, Jr., C. T., H. S. Panitch, and K. P. Johnson.** 1994. Increased cathepsin B activity in peripheral blood mononuclear cells of multiple sclerosis patients. *Neurology* **44**: 745–748.

1630. **Boujrad, N., S. O. Ogwuegbu, M. Garnier, C. H. Lee, B. M. Martin, and V. Papadopoulos.** 1995. Identification of a stimulator of steroid hormone synthesis isolated from testis. *Science* **268**: 1609–1612.

1631. **Brix, K., P. Lemansky, and V. Herzog.** 1996. Evidence for extracellularly acting cathepsins mediating thyroid hormone liberation in thyroid epithelial cells. *Endocrinology* **137**: 1963–1974.

1632. **Burnett, D., M. Abrahamson, J. L. Devalia, R. J. Sapsford, R. J. Davies, and D. J. Buttle.** 1995. Synthesis and secretion of procathepsin B and cystatin C by human bronchial epithelial cells *in vitro*: modulation of cathepsin B activity by neutrophil elastase. *Arch. Biochem. Biophys.* **317**: 305–310.

1633. **Butor, C., G. Griffiths, N. A. Aronson, Jr., and A. Varki.** 1995. Co-localization of hydrolytic enzymes with widely disparate pH optima: implications for the regulation of lysosomal pH. *J. Cell Sci.* **108**: 2213–2219.

1634. **Carpintero, A., M. M. Sanchez-Martin, M. J. Cabezas-Delamare, and J. A. Cabezas.** 1996. Variation in serum arylesterase, β-glucuronidase, cathepsin L and plasminogen activators during pregnancy. *Clin. Chim. Acta* **255**: 153–164.

1635. **Chapman, Jr., H. A., J. S. Munger, and G. P. Shi.** 1994. The role of thiol proteases in tissue injury and remodeling. *Am. J. Respir. Crit. Care Med.* **150**: S155–S159.

1636. **Chugunova, L. G. and I. I. Dubinina.** 1994. [Lipid peroxidation parameters and activities of lysosomal enzymes in patients with diabetes mellitus]. *Probl. Endokrinol. (Mosk.)* **40**: 9–11.

1637. **Dalton, J. P., K. A. Clough, M. K. Jones, and P. J. Brindley.** 1996. Characterization of the cathepsin-like cysteine proteinases of *Schistosoma mansoni*. *Infect. Immun.* **64**: 1328–1334.

1638. **Dalton, J. P., L. Hola-Jamriska, and P. J. Brindley.** 1995. Asparaginyl endopeptidase activity in adult *Schistosoma mansoni*. *Parasitology* **111**: 575–580.

1639. **Dalton, J. P., S. McGonigle, T. P. Rolph, and S. J. Andrews.** 1996. Induction of protective immunity in cattle against infection with *Fasciola hepatica* by vaccination with cathepsin L proteinases and with hemoglobin. *Infect. Immun.* **64**: 5066–5074.

1640. **Debari, K., T. Sasaki, N. Udagawa, and B. R. Rifkin.** 1995. An ultrastructural evaluation of the effects of cysteine-proteinase inhibitors on osteoclastic resorptive functions. *Calcif. Tissue Int.* **56**: 566–570.

1641. **Diment, S., M. Eidelman, G. M. Rodriguez, and S. J. Orlow.** 1995. Lysosomal hydrolases are present in melanosomes and are elevated in melanizing cells. *J. Biol. Chem.* **270**: 4213–4215.

1642. **Dodds, R. A., J. R. Connor, F. Drake, J. Feild, and M. Gowen.** 1996. Cathepsin K mRNA expression is restricted to osteoclasts during fetal and neonatal mouse development. *J. Bone Miner. Res.* **11**: S180.

1643. **Dodds, R. A., I. James, J. Connor, E. Lee-Rykaczewski, S. Richardson, C. Debouck, D. Rieman, F. Drake, and M. Gowen.** 1995. Human osteoclasts express abundant cathepsin O (osteoclast-specific cysteine proteinase), but not cathepsins B, L, or S. *J. Bone Miner. Res.* **10**: S228.

1644. **Donohue, T. M. J., D. L. McVicker, K. K. Kharbanda, M. L. Chaisson, and R. K. Zetterman.** 1994. Ethanol administration alters the proteolytic activity of hepatic lysosomes. *Alcohol. Clin. Exp. Res.* **18**: 536–541.

1645. **Drake, F. H., R. A. Dodds, I. E. James, J. R. Connor, C. Debouck, S. Richardson, E. Lee-Rykaczewski, L. Coleman, D. Rieman, R. Barthlow, G. Hastings, and M. Gowen.** 1996. Cathepsin K, but not cathepsins B, L, or S, is abundantly expressed in human osteoclasts. *J. Biol. Chem.* **271**: 12511–12516.

1646. **Ebisui, C., T. Tsujinaka, T. Morimoto, J. Fujita, A. Ogawa, K. Ishidoh, E. Kominami, K. Tanaka, and M. Monden.** 1995. Changes of proteasomes and cathepsins activities and their expression during differentiation of C2C12 myoblasts. *J. Biochem. (Tokyo)* **117**: 1088–1094.

1647. **Ebisui, C., T. Tsujinaka, T. Morimoto, K. Kan, S. Iijima, M. Yano, E. Kominami, K. Tanaka, and M. Monden.** 1995. Interleukin-6 induces proteolysis by activating intracellular proteases (cathepsins B and L, proteasome) in C2C12 myotubes. *Clin. Sci. (Colch.)* **89**: 431–439.

1648. **Edelstein, C. L., E. D. Wieder, M. M. Yaqoob, P. E. Gengaro, T. J. Burke, R. A. Nemenoff, and R. W. Schrier.** 1995. The role of cysteine proteases in hypoxia-induced rat renal proximal tubular injury. *Proc. Natl Acad. Sci. U. S. A.* **92**: 7662–7666.

1649. **Eisenberger, U., L. M. Fels, C. J. Olbricht, and H. Stolte.** 1995. Cathepsin B and L in isolated proximal tubular segments during acute and chronic proteinuria. *Ren. Physiol. Biochem.* **18**: 89–96.

1650. **Eley, B. M. and S. W. Cox.** 1996. The relationship between gingival crevicular fluid cathepsin B activity and periodontal attachment loss in chronic periodontitis patients – a 2 year longitudinal study. *J. Periodontal Res.* **31**: 381–392.

1651. **Fujita, J., T. Tsujinaka, C. Ebisui, M. Yano, H. Shiozaki, A. Katsume, Y. Ohsugi, and M. Monden.** 1996. Role of interleukin 6 in skeletal muscle protein breakdown and cathepsin activity in vivo. *Eur. Surg. Res.* **28**: 361–366.

1652. **Gelb, B. D., G. P. Shi, H. A. Chapman, and R. J. Desnick.** 1996. Pycnodysostosis, a lysosomal disease caused by cathepsin K deficiency. *Science* **273**: 1236–1238.

1653. **Gerbaux, C., F. Van Bambeke, J. P. Montenez, J. Piret, G. Morlighem, and P. M. Tulkens.** 1996. Hyperactivity of cathepsin B and other lysosomal enzymes in fibroblasts exposed to azithromycin, a dicationic macrolide antibiotic with exceptional tissue accumulation. *FEBS Lett.* **394**: 307–310.

1654. **Goldspink, D. F., V. M. Cox, S. K. Smith, L. A. Eaves, N. J. Osbaldeston, D. M. Lee, and D. Mantle.** 1995. Muscle growth in response to mechanical stimuli. *Am. J. Physiol.* **268**: E288-E297.

1655. **Gordon, R. D., J. A. Young, S. Rayner, R. W. A. Luke, M. L. Crowther, P. Wordsworth, J. Bell, G. Hassall, J. Evans, S. A. Hinchliffe, E. J. Culbert, and M. D. Davison.** 1995. Purification and characterization of endogenous peptides extracted from HLA-DR isolated from the spleen of a patient with rheumatoid arthritis. *Eur. J. Immunol.* **25**: 1473–1476.

1656. **Guenette, R. S., M. Mooibroek, K. Wong, P. Wong, and M. Tenniswood.** 1994. Cathepsin B, a cysteine protease implicated in metastatic progression, is also expressed during regression of the rat prostate and mammary glands. *Eur. J. Biochem.* **226**: 311–321.

1657. **Gupta-Bansal, R., R. C. A. Frederickson, and K. R. Brunden.** 1995. Proteoglycan-mediated inhibition of A beta proteolysis. A potential cause of senile plaque accumulation. *J. Biol. Chem.* **270**: 18666–18671.

1658. **Hill, P. A., D. J. Buttle, S. J. Jones, A. Boyde, M. Murata, J. J. Reynolds, and M. C. Meikle.** 1994. Inhibition of bone resorption by selective inactivators of cysteine proteinases. *J. Cell. Biochem.* **56**: 118–130.

1659. **Homma, K., S. Kurata, and S. Natori.** 1994. Purification, characterization, and cDNA cloning of procathepsin L from the culture medium of NIH-Sape-4, an embryonic cell line of *Sarcophaga peregrina* (flesh fly), and its involvement in the differentiation of imaginal discs. *J. Biol. Chem.* **269**: 15258–15264.

1660. **Hong, D. and N. E. Forsberg.** 1994. Effects of serum and insulin-like growth factor I on protein degradation and protease gene expression in rat L8 myotubes. *J. Anim. Sci.* **72**: 2279–2288.

1661. **Ichimaru, E., M. Tanoue, M. Tani, Y. Tani, T. Kaneko, Y. Iwasaki, K. Kunimatsu, and I. Kato.** 1996. Cathepsin B in gingival crevicular fluid of adult periodontitis patients – identification by immunological and enzymological methods. *Inflamm. Res.* **45**: 277–282.

1662. **Inubushi, T., M. Shikiji, K. Endo, H. Kakegawa, Y. Kishino, and N. Katunuma.** 1996. Hormonal and dietary regulation of lysosomal cysteine proteinases in liver under gluconeogenesis conditions. *Biol. Chem.* **377**: 539–542.

1663. **Jane, D. T. and M. J. Dufresne.** 1994. Expression and regulation of three lysosomal cysteine protease activities during growth of a differentiating L6 rat myoblast cell line and its nonfusing variant. *Biochem. Cell Biol.* **72**: 267–274.

1664. **Jochum, M., C. Gippner-Steppert, W. Machleidt, and H. Fritz.** 1994. The role of phagocyte proteinases and proteinase inhibitors in multiple organ failure. *Am. J. Respir. Crit. Care Med.* **150**: S123–S130.

1665. **Kakegawa, H., K. Tagami, Y. Ohba, K. Sumitani, T. Kawata, and N. Katunuma.** 1995. Secretion and processing mechanisms of procathepsin L in bone resorption. *FEBS Lett.* **370**: 78–82.

1666. **Kar, N. C. and C. M. Pearson.** 1978. Dipeptidyl peptidases in human muscle disease. *Clin. Chim. Acta* **82**: 185–192.

1667. **Kawas, S. A., N. Amizuka, J. M. Bergeron, and H. Warshawsky.** 1996. Immunolocalization of the cation independent mannose 6-phosphate receptor and cathepsin B in the enamel organ and alveolar bone of the rat incisor. *Calcif. Tissue Int.* **59**: 192–199.

1668. **Kennett, C. N., S. W. Cox, and B. M. Eley.** 1994. Comparative histochemical, biochemical and immunocytochemical studies of cathepsin B in human gingiva. *J. Periodontal Res.* **29**: 203–213.

1669. **Keyszer, G. M., A. H. Heer, J. Kriegsmann, T. Geiler, A. Trabandt, M. Keysser, R. E. Gay, and S. Gay.** 1995. Comparative analysis of cathepsin L, cathepsin D, and collagenase messenger RNA expression in synovial tissues of patients with rheumatoid arthritis and osteoarthritis, by in situ hybridization. *Arthritis Rheum.* **38**: 976–984.

1670. **Kharbanda, K. K., D. L. McVicker, R. K. Zetterman, and T. M. Donohue, Jr.** 1995. Ethanol consumption reduces the proteolytic capacity and protease activities of hepatic lysosomes. *Biochim. Biophys. Acta* **1245**: 421–429.

1671. **Kinbara, K., H. Kitagaki, T. Kinouchi, M. Okano, H. Sorimachi, S. Ishiura, and K. Suzuki.** 1994. Processing and secretion of Alzheimer's disease amyloid precursor protein. *Tohoku J. Exp. Med.* **174**: 209–216.

1672. **Kishibuchi, M., T. Tsujinaka, S. Iijima, M. Yano, C. Ebisui, K. Kan, T. Morimoto, and T. Mori.** 1995. Interrelation of intracellular proteases with total parenteral nutrition-induced gut mucosal atrophy and increase of mucosal macromolecular transmission in rats. *J. Parenter. Enteral Nutr.* **19**: 187–192.

1673. **Kiyoshima, T., M. A. Kido, Y. Nishimura, M. Himeno, T. Tsukuba, H. Tashiro, K. Yamamoto, and T. Tanaka.** 1994. Immunocytochemical localization of cathepsin L in the synovial lining cells of the rat temporomandibular joint. *Arch. Oral Biol.* **39**: 1049–1056.

1674. **Kiyoshima, T., M. A. Kido, T. Tsukuba, H. Sakai, K. Yamamoto, and T. Tanaka.** 1994. Localization of cathepsins B and D in the synovial lining cells of the normal rat temporo-mandibular joint by immuno-light and -electron microscopy. *Acta Histochem. Cytochem.* **5**: 441–450.

1675. **Kohda, Y., T. Yamashima, K. Sakuda, J. Yamashita, T. Ueno, E. Kominami, and T. Yoshioka.** 1996. Dynamic changes of cathepsins B and L expression in the monkey hippocampus after transient ischemia. *Biochem. Biophys. Res. Commun.* **228**: 616–622.

1676. **Kremer, M., J. Judd, B. Rifkin, J. Auszmann, and M. J. Oursler.** 1995. Estrogen modulation of osteoclast lysosomal enzyme secretion. *J. Cell. Biochem.* **57**: 271–279.

1677. **Kreutzberg, G. W.** 1995. Microglia, the first line of defence in brain pathologies. *Arzneimittelforschung* **45**: 357–360.

1678. **Lafuse, W. P., D. Brown, L. Castle, and B. S. Zwilling.** 1995. IFN-γ increases cathepsin H mRNA levels in mouse macrophages. *J. Leukoc. Biol.* **57**: 663–669.

1679. **Lah, T. T., M. Hawley, K. L. Rock, and A. L. Goldberg.** 1995. Gamma-interferon causes a selective induction of the lysosomal proteases, cathepsins B and L, in macrophages. *FEBS Lett.* **363**: 85–89.

1680. **Leboulleux, M., B. Lelongt, B. Mougenot, G. Touchard, R. Makdassi, A. Rocca, L. H. Noel, P. M. Ronco, and P. Aucouturier.** 1995. Protease resistance and binding of Ig light chains in myeloma-associated tubulopathies. *Kidney Int.* **48**: 72–79.

1681. **Lecomte, V., I. Knott, M. Burton, J. Remacle, and M. Raes.** 1994. Cathepsin B and N-acetyl beta-D-glucosaminidase in human synovial cells in culture: effects of interleukin-1. *Clin. Chim. Acta* **228**: 143–159.

1682. **Lee, E. R., L. Lamplugh, N. L. Shepard, and J. S. Mort.** 1995. The septoclast, a cathepsin B-rich cell involved in the resorption of growth plate cartilage. *J. Histochem. Cytochem.* **43**: 525–536.

1683. **Lemaire, R., R. M. Flipo, D. Monté, T. Dupressoir, B. Duquesnoy, J. Y. Cesbron, A. Janin, A. Capron, and R. Lafyatis.** 1994. Synovial fibroblast-like cell transfection with the SV40 large T antigen induces a transformed phenotype and permits transient tumor formation in immunodeficient mice. *J. Rheumatol.* **21**: 1409–1419.

1684. **Lemere, C. A., J. S. Munger, G. P. Shi, L. Natkin, C. Haass, H. A. Chapman, and D. J. Selkoe.** 1995. The lysosomal cysteine protease, cathepsin S, is increased in Alzheimer's disease and Down syndrome brain. An immunocytochemical study. *Am. J. Pathol.* **146**: 848–860.

1685. **Li, Q. D. and C. T. Bever.** 1996. γ-Interferon induced increases in intracellular cathepsin B activity in PMA primed Thp-1 cells are blocked by inhibitors of protein kinase C. *Immunopharmacol. Immunotoxicol.* **18**: 375–396.

1686. **Ling, H., S. Vamvakas, G. Busch, J. Dämmrich, L. Schramm, F. Lang, and A. Heidland.** 1995. Suppressing role of transforming growth factor-β1 on cathepsin activity in cultured kidney tubule cells. *Am. J. Physiol.* **269**: F911-F917.

1687. **Ling, H., S. Vamvakas, M. Gekle, L. Schaefer, M. Teschner, R. M. Schaefer, and A. Heidland.** 1996. Role of lysosomal cathepsin activities in cell hypertrophy induced by NH_4Cl in cultured renal proximal tubule cells. *J. Am. Soc. Nephrol.* **7**: 73–80.

1688. **Ling, H., S. Vamvakas, L. Schaefer, R. M. Schaefer, M. Teschner, L. Schramm, and A. Heidland.** 1995. Insulin-like growth factor I induced reduction in cysteine proteinase activity in freshly isolated proximal tubule cells of the rat. *Nephron* **69**: 83–85.

1689. **Lohrke, B., J. Wegner, T. Viergutz, G. Dietl, and K. Ender.** 1995. Flow-cytometric analysis of oxidative and proteolytical activities in tissue-associated phagocytes from normal and hypertrophic muscles. *Anal. Cell. Pathol.* **9**: 281–293.

1690. **Maciewicz, R. A. and D. J. Etherington.** 1988. Enzyme immunoassay for cathepsins B and L in synovial fluids from patients with arthritis. *Biochem. Soc. Trans.* **16**: 812–813.

1691. **Marrogi, A. J., M. K. Cheles, and M. A. Gerber.** 1995. Chronic hepatitis C. Analysis of host immune response by immunohistochemistry. *Arch. Pathol. Lab. Med.* **119**: 232–237.

1692. **McCulloch, C. A.** 1994. Host enzymes in gingival crevicular fluid as diagnostic indicators of periodontitis. *J. Clin. Periodontol.* **21**: 497–506.

1693. **McDonald, J. K. and C. Schwabe.** 1982. Relaxin-induced elevations of cathepsin B and dipeptidyl peptidase I in the mouse pubic symphysis, with localization by fluorescence enzyme histochemistry. *Ann. N. Y. Acad. Sci.* **380**: 178–186.

1694. **Miles, C. A., R. J. Wardale, H. L. Birch, and A. J. Bailey.** 1994. Differential scanning calorimetric studies of superficial digital flexor tendon degeneration in the horse [Comment in *Equine Vet. J.* (1994) **26**: 255–256]. *Equine Vet. J.* **26**: 291–296.

1695. **Montenez, J. P., J. M. Delaissé, P. M. Tulkens, and B. K. Kishore.** 1994. Increased activities of cathepsin B and other lysosomal hydrolases in fibroblasts and bone tissue cultured in the presence of cysteine proteinases inhibitors. *Life Sci.* **55**: 1199–1208.

1696. **Moroi, R., T. Yamaza, Y. Ayukawa, T. Kiyoshima, Y. Ohsaki, Y. Nishimura, Y. Terada, M. Himeno, and T. Tanaka.** 1995. The changes in the immunocytochemical localization of cathepsin L and type I collagen in rat osteoclasts treated with E-64. *Acta Histochem. Cytochem.* **28**: 523–531.

1697. **Mort, J. S., A. R. Poole, and R. S. Decker.** 1981. Immunofluorescent localization of cathepsins B and D in human fibroblasts. *J. Histochem. Cytochem.* **29**: 649–657.

1698. **Nagy, L., S. Kusstatscher, P. V. Hauschka, and S. Szabo.** 1996. Role of cysteine proteases and protease inhibitors in gastric mucosal damage induced by ethanol or ammonia in the rat. *J. Clin. Invest.* **98**: 1047–1054.

1699. **Nakanishi, H., K. Tominaga, T. Amano, I. Hirotsu, T. Inoue, and K. Yamamoto.** 1994. Age-related changes in activities and localizations of cathepsins D, E, B, and L in the rat brain tissues. *Exp. Neurol.* **126**: 119–128.

1699a. **Nitatori, T., N. Sato, E. Kominami, and Y. Uchiyama.** 1996. Participation of cathepsins B, H, and L in perikaryal condensation of CA1 pyramidal neurons undergoing apoptosis after brief ischemia. In *Intracellular Protein Catabolism*, K. Suzuki and J. S. Bond (eds), pp. 177–185. Plenum Press, New York.

1700. **Nitatori, T., N. Sato, S. Waguri, Y. Karasawa, H. Araki, K. Shibanai, E. Kominami, and Y. Uchiyama.** 1995. Delayed neuronal death in the CA1 pyramidal cell layer of the gerbil hippocampus following transient ischemia is apoptosis. *J. Neurosci.* **15**: 1001–1011.

1701. **Noda, T., K. Isogai, N. Katunuma, Y. Tarumoto, and M. Ohzeki.** 1981. Effects on cathepsin B, H, and D in pectoral muscle of dystrophic chickens (line 413) of *in vivo* administration of E-64-c (*N*-[*N*-(L-3-transcarboxyoxirane-2-carbonyl)-L-leucyl]-3-methyl-butylamine). *J. Biochem. (Tokyo)* **90**: 893–896.

1702. **North, M. J., K. Nicol, T. W. Sands, and D. A. Cotter.** 1996. Acid activatable cysteine proteinases in the cellular slime mold *Dictyostelium discoideum*. *J. Biol. Chem.* **271**: 14462–14467.

1703. **Oh, Y. K. and J. A. Swanson.** 1996. Different fates of phagocytosed particles after delivery into macrophage lysosomes. *J. Cell Biol.* **132**: 585–593.

1704. **Okamura, N., M. Tamba, Y. Uchiyama, Y. Sugita, F. Dacheux, P. Syntin, and J. L. Dacheux.** 1995. Direct evidence for the elevated synthesis and secretion of procathepsin L in the distal caput epididymis of boar. *Biochim. Biophys. Acta* **1245**: 221–226.

1705. **Olbricht, C. J., M. Steinker, W. Auch-Schwelk, C. Bossaller, J. Haas, and K. M. Koch.** 1994. Effect of cyclosporin on kidney proteolytic enzymes in men and rats. *Nephrol. Dial. Transplant.* **9**: 22–26.

1706. **Paczek, L., J. Pazik, L. Gradowska, G. Senatorski, M. Lao, I. Bartlomiejczyk, A. Heidland, W. Rowiński, and J. Szmidt.** 1995. Intraglomerular cathepsin B and L activity in chronic kidney allograft rejection. *Transplant. Proc.* **27**: 932–933.

1707. **Parreno, M., R. Cusso, M. Gil, and C. Sarraga.** 1994. Development of cathepsin B, cathepsin L and cathepsin H activities and cystatin-like activity during 2 different manufacturing processes for spanish dry-cured ham. *Food Chem.* **49**: 15–21.

1708. **Penttila, T. L., H. Hakovirta, P. Mali, W. W. Wright, and M. Parvinen.** 1995. Follicle stimulating hormone regulates the expression of cyclic protein 2 cathepsin L messenger ribonucleic acid in rat Sertoli cells in a stage specific manner. *Mol. Cell. Endocrinol.* **113**: 175–181.

1709. **Petanceska, S., S. Burke, S. J. Watson, and L. Devi.** 1994. Differential distribution of messenger RNAs for cathepsins B, L and S in adult rat brain: an *in situ* hybridization study. *Neuroscience* **59**: 729–738.

1710. **Petanceska, S., P. Canoll, and L. A. Devi.** 1996. Expression of rat cathepsin S in phagocytic cells. *J. Biol. Chem.* **271**: 4403–4409.

1712. **Piqueras, A. I., M. Somers, T. G. Hammond, K. Strange, H. W. Harris, Jr., M. Gawryl, and M. L. Zeidel.** 1994. Permeability properties of rat renal lysosomes. *Am. J. Physiol.* **266**: C121-C133.

1713. **Porter, R., B. Koury, and F. Stone.** 1996. Comparison of cathepsin B, D, H and L activity in four species of pacific fish. *J. Food Biochem.* **19**: 429–442.

1714. **Rakoczy, P. E., K. Mann, D. M. Cavaney, T. Robertson, J. Papadimitreou, and I. J. Constable.** 1994. Detection and possible functions of a cysteine protease involved in digestion of rod outer segments by retinal pigment epithelial cells. *Invest. Ophthalmol. Vis. Sci.* **35**: 4100–4108.

1715. **Reddy, V. Y., Q. Y. Zhang, and S. J. Weiss.** 1995. Pericellular mobilization of the tissue-destructive cysteine proteinases, cathepsins B, L, and S, by human monocyte-derived macrophages. *Proc. Natl Acad. Sci. U. S. A.* **92**: 3849–3853.

1716. **Ren, W. P., R. Fridman, J. R. Zabrecky, L. D. Morris, N. A. Day, and B. F. Sloane.** 1996. Expression of functional recombinant human procathepsin B in mammalian cells. *Biochem. J.* **319**: 793–800.

1717. **Saneshige, S., H. Mano, K. Tezuka, S. Kakudo, Y. Mori, Y. Honda, A. Itabashi, T. Yamada, K. Miyata, Y. Hakeda, J. Ishii, and M. Kumegawa.** 1995. Retinoic acid directly stimulates osteoclastic bone resorption and gene expression of cathepsin K/OC-2. *Biochem. J.* **309**: 721–724.

1718. **Sangueza, O. P., J. K. Salmon, C. R. White, Jr., and J. H. Beckstead.** 1995. Juvenile xanthogranuloma: a clinical, histopathologic and immunohistochemical study. *J. Cutan. Pathol.* **22**: A327–335.

1719. **San Segundo, B., S. J. Chan, and D. F. Steiner.** 1986. Differences in cathepsin B mRNA levels in rat tissues suggest specialized functions. *FEBS Lett.* **201**: 251–256.

1720. **Schaefer, L., U. Gilge, A. Heidland, and R. M. Schaefer.** 1994. Urinary excretion of cathepsin B and cystatins as parameters of tubular damage. *Kidney Int.* **47** (Suppl.): S64-S67.

1721. **Schaefer, L., X. Han, N. Gretz, and R. M. Schaefer.** 1996. Alterations of cathepsin B, H, and L in proximal tubules from polycystic kidneys of the Han:SPRD rat. *Kidney Int.* **50**: 424–431.

1722. **Schaefer, L., R. M. Schaefer, H. Ling, M. Teschner, and A. Heidland.** 1994. Renal proteinases and kidney hypertrophy in experimental diabetes. *Diabetologia* **37**: 567–571.

1723. **Schafer, L., T. Lorenz, J. Daemmrich, A. Heidland, and R. M. Schaefer.** 1995. Role of proteinases in renal hypertrophy and matrix accumulation. *Nephrol. Dial. Transplant.* **10**: 801–807.

1724. **Schreurs, F. J., D. van der Heide, F. R. Leenstra, and W. de Wit.** 1995. Endogenous proteolytic enzymes in chicken muscles. Differences among strains with different growth rates and protein efficiencies. *Poult. Sci.* **74**: 523–537.

1725. **Schutze, N., M. J. Oursler, J. Nolan, B. L. Riggs, and T. C. Spelsberg.** 1995. Zeolite A inhibits osteoclast-mediated bone resorption in vitro. *J. Cell. Biochem.* **58**: 39–46.

1726. **Shechter, P., J. D. Shi, and R. Rabkin.** 1994. Renal tubular cell protein breakdown in uninephrectomized and ammonium chloride-loaded rats. *J. Am. Soc. Nephrol.* **5**: 1201–1207.

1727. **Shi, L., M. Sawada, U. Sester, and J. C. Carlson.** 1994. Alterations in free radical activity in aging Drosophila. *Exp. Gerontol.* **29**: 575–584.

1728. **Siddiq, T., D. K. Shori, G. B. Proctor, C. Luckhaus, P. J. Richardson, and V. R. Preedy.** 1994. The activities of cathepsins B, D and H in the hearts of rats treated with ethanol. *Biochem. Soc. Trans.* **22**: 171S.

1729. **Siebeck, M., J. Kohl, S. Endres, M. Spannagl, and W. Machleidt.** 1994. Delayed treatment with platelet activating factor receptor antagonist WEB 2086 attenuates pulmonary dysfunction in porcine endotoxin shock. *J. Trauma* **37**: 745–751.

1730. **Singhal, P. C., S. Sagar, and N. Gibbons.** 1995. Morphine modulates cathepsin B and L activity in isolated glomeruli and mesangial cells. *Inflammation* **19**: 67–73.

1731. **Sleat, D. E., I. Sohar, H. Lackland, J. Majercak, and P. Lobel.** 1996. Rat brain contains high levels of mannose-6-phosphorylated glycoproteins including lysosomal enzymes and palmitoyl-protein thioesterase, an enzyme implicated in infantile neuronal lipofuscinosis. *J. Biol. Chem.* **271**: 19191–19198.

1732. **Struckhoff, G.** 1995. Transforming growth factor beta 1 and parathyroid hormone-related protein control the secretion of dipeptidyl peptidase II by rat astrocytes. *Neurosci. Lett.* **189**: 117–120.

1733. **Suzuki, H., L. Schaefer, H. Ling, R. M. Schaefer, J. Dämmrich, M. Teschner, and A. Heidland.** 1995. Prevention of cardiac hypertrophy in experimental chronic renal failure by long-term ACE inhibitor administration: potential role of lysosomal proteinases. *Am. J. Nephrol.* **15**: 129–136.

1734. **Tagami, K., H. Kakegawa, H. Kamioka, K. Sumitani, T. Kawata, B. Lenarcic, V. Turk, and N. Katunuma.** 1994. The mechanisms and regulation of procathepsin L secretion from osteoclasts in bone resorption. *FEBS Lett.* **342**: 308–312.

1735. **Terada, T., Y. Okada, and Y. Nakanuma.** 1995. Expression of matrix proteinases during human intrahepatic bile duct development. A possible role in biliary cell migration. *Am. J. Pathol.* **147**: 1207–1213.

1736. **Tomomasa, H., S. Waguri, T. Umeda, K. Koiso, E. Kominami, and Y. Uchiyama.** 1994. Lysosomal cysteine proteinases in rat epididymis. *J. Histochem. Cytochem.* **42**: 417–425.

1737. **Trabandt, A., U. Müller-Ladner, J. Kriegsmann, R. E. Gay, and S. Gay.** 1995. Expression of proteolytic cathepsins B, D, and L in periodontal gingival fibroblasts and tissues. *Lab. Invest.* **73**: 205–212.

1738. **Tsujinaka, T., C. Ebisui, J. Fujita, M. Kishibuchi, T. Morimoto, A. Ogawa, A. Katsume, Y. Ohsugi, E. Kominami, and M. Monden.** 1995. Muscle undergoes atrophy in association with increase of lysosomal cathepsin activity in interleukin-6 transgenic mouse. *Biochem. Biophys. Res. Commun.* **207**: 168–174.

1739. **Tsujinaka, T., C. Ebisui, J. Fujita, T. Morimoto, A. Ogawa, K. Ishidoh, E. Kominami, M. Yano, H. Shiozaki, and M. Monden.** 1995. Autocatalytic inactivation of lysosomal cathepsins is associated with inhibition of protein breakdown by insulin-like growth factor-1 (IGF-1) in myotubes. *Biochem. Biophys. Res. Commun.* **208**: 353–359.

1740. Tsujinaka, T., J. Fujita, C. Ebisui, M. Yano, E. Kominami, K. Suzuki, K. Tanaka, A. Katsume, Y. Ohsugi, H. Shiozaki, and M. Monden. 1996. Interleukin 6 receptor antibody inhibits muscle atrophy and modulates proteolytic systems in interleukin 6 transgenic mice. *J. Clin. Invest.* **97**: 244–249.

1741. Tsujinaka, T., T. Homma, C. Ebisui, J. Fujita, Y. Kido, M. Yano, H. Shibata, T. Tanaka, and T. Mori. 1995. Modulation of muscle protein metabolism in disseminated intravascular coagulation. *Eur. Surg. Res.* **27**: 227–233.

1742. Uchiyama, Y., T. Watanabe, M. Watanabe, Y. Ishii, H. Matsuba, S. Waguri, and E. Kominami. 1989. Immunocytochemical localization of cathepsins B, H, L, and T4 in follicular cells of rat thyroid gland. *J. Histochem. Cytochem.* **37**: 691–696.

1743. van den Hemel-Grooten, H. N., M. Koohmaraie, J. T. Yen, J. R. Arbona, J. A. Rathmacher, S. L. Nissen, M. L. Fiorotto, G. J. Garssen, and M. W. Verstegen. 1995. Comparison between 3-methylhistidine production and proteinase activity as measures of skeletal muscle breakdown in protein-deficient growing barrows. *J. Anim. Sci.* **73**: 2272–2281.

1744. Voisin, L., D. Breuillé, L. Combaret, C. Pouyet, D. Taillandier, E. Aurousseau, C. Obled, and D. Attaix. 1996. Muscle wasting in a rat model of long-lasting sepsis results from the activation of lysosomal, Ca^{2+}-activated, and ubiquitin-proteasome proteolytic pathways. *J. Clin. Invest.* **97**: 1610–1617.

1745. Werle, B., W. Ebert, W. Klein, and E. Spiess. 1996. Charge polymorphism in human lung cell pro-cathepsin B. *Anticancer Res.* **16**: 49–53.

1746. Wright, W. W., S. D. Zabludoff, T. L. Penttila, and M. Parvinen. 1995. Germ cell–Sertoli cell interactions: regulation by germ cells of the stage-specific expression of CP-2/cathepsin L mRNA by Sertoli cells. *Dev. Genet.* **16**: 104–113.

1747. Yamada, E., C. H. Chue, N. Yukioka, and F. Hazama. 1994. Causative role of lysosomal enzymes in the pathogenesis of cerebral lesions due to brain edema under chronic hypertension. *Acta Neurochir. Suppl. (Wien)* **60**: 83–85.

1748. Yamada, T., J. J. Liepnieks, B. Kluve-Beckerman, and M. D. Benson. 1995. Cathepsin B generates the most common form of amyloid A (76 residues) as a degradation product from serum amyloid A. *Scand. J. Immunol.* **41**: 94–97.

1749. Yamamoto, H., Y. Murawaki, and H. Kawasaki. 1995. Hepatic collagen synthesis and degradation during liver regeneration after partial hepatectomy. *Hepatology* **21**: 155–161.

1750. Yamamoto, K. 1995. [Studies on periodontal pathogenic proteinases from *Porphyromonas gingivalis* and host cells]. *Nippon Yakurigaku Zasshi* **105**: 345–355.

1751. Yano, T., N. Takahashi, S. Kurata, and S. Natori. 1995. Regulation of the expression of cathepsin B in *Sarcophaga peregrina* (flesh fly) at the translational level during metamorphosis. *Eur. J. Biochem.* **234**: 39–43.

1752. Yih, L. H. and T. C. Lee. 1994. A proteolytic activity enhanced by arsenite in Chinese hamster ovary cells: possible involvement in arsenite-induced cell killing. *Biochem. Biophys. Res. Commun.* **202**: 1015–1022.

Processing

1753. Authier, F., J. S. Mort, A. W. Bell, B. I. Posner, and J. J. M. Bergeron. 1995. Proteolysis of glucagon within hepatic endosomes by membrane-associated cathepsins B and D. *J. Biol. Chem.* **270**: 15798–15807.

1754. Chagas, J. R., M. Ferrer-Di Martino, F. Gauthier, and G. Lalmanach. 1996. Inhibition of cathepsin B by its propeptide: use of overlapping peptides to identify a critical segment. *FEBS Lett.* **392**: 233–236.

1755. Chen, Y. M., C. Plouffe, R. Ménard, and A. C. Storer. 1996. Delineating functionally important regions and residues in the cathepsin B propeptide for inhibitory activity. *FEBS Lett.* **393**: 24–26.

1756. Cuozzo, J. W. and G. G. Sahagian. 1994. Lysine is a common determinant for mannose phosphorylation of lysosomal proteins. *J. Biol. Chem.* **269**: 14490–14496.

1757. Cuozzo, J. W., K. Tao, Q. Wu, W. Young, and G. G. Sahagian. 1995. Lysine-based structure in the proregion of procathepsin L is the recognition site for mannose phosphorylation. *J. Biol. Chem.* **270**: 15611–15619.

1758. Cygler, M., J. Sivaraman, P. Grochulski, R. Coulombe, A. C. Storer, and J. S. Mort. 1996. Structure of rat procathepsin B: model for inhibition of cysteine protease activity by the proregion. *Structure* **4**: 405–416.

1759. Deceuninck, F., S. Poiraudeau, M. Pagano, L. Tsagris, O. Blanchard, J. Willeput, and M. Corvol. 1995. Inhibition of chondrocyte cathepsin B and cathepsin L activities by insulin-like growth factor II (IGF II) and its Ser (29) variant in vitro – possible role of the mannose 6-phosphate IGF II receptor. *Mol. Cell. Endocrinol.* **113**: 205–213.

1760. Drake, F., B. Amegadzie, C. Jones, J. Kurdyla, M. Bossard, T. Tomaszek, C. Debouck, S. Richardson, M. Gowen, and M. Levy. 1995. Heterologous expression and activation of cathepsin O. *J. Bone Miner. Res.* **10**: S227.

1761. Erickson, A. H. 1989. Biosynthesis of lysosomal endopeptidases. *J. Cell. Biochem.* **40**: 31–41.

1762. Hiwasa, T. and S. Sakiyama. 1996. Nuclear localization of procathepsin L/MEP in ras-transformed mouse fibroblasts. *Cancer Lett.* **99**: 87–91.

1763. **Igdoura, S. A., A. Rasky, and C. R. Morales.** 1996. Trafficking of sulfated glycoprotein-1 (prosaposin) to lysosomes or to the extracellular space in rat Sertoli cells. *Cell Tissue Res.* **283**: 385–394.

1764. **Ishidoh, K. and E. Kominami.** 1994. Multi-step processing of procathepsin L in vitro. *FEBS Lett.* **352**: 281–284.

1765. **Ludwig, T., H. Munier-Lehmann, U. Bauer, M. Hollinshead, C. Ovitt, P. Lobel, and B. Hoflack.** 1994. Differential sorting of lysosomal enzymes in mannose 6-phosphate receptor-deficient fibroblasts. *EMBO J.* **13**: 3430–3437.

1766. **Mach, L., J. S. Mort, and J. Glössl.** 1994. Maturation of human procathepsin B. Proenzyme activation and proteolytic processing of the precursor to the mature proteinase, *in vitro*, are primarily unimolecular processes. *J. Biol. Chem.* **269**: 13030–13035.

1767. **Mahuran, D. J., K. Neote, M. H. Klavins, A. Leung, and R. A. Gravel.** 1988. Proteolytic processing of pro-α and pro-β precursors from human β-hexosaminidase. Generation of the mature α and $\beta_a\beta_b$ subunits. *J. Biol. Chem.* **263**: 4612–3618.

1768. **Maubach, G., K. Schilling, I. Wenz, W. Rommerskirch, and B. Wiederanders.** 1997. The propeptide of human cathepsin S is a powerful and specific inhibitor of cathepsin S activity. In *Proteolysis in Cell Functions*, V. K. Hopsu-Havu, M. Järvinen, and H. Kirschke (eds), pp. 167–173. IOS Press, Amsterdam. In press.

1769. **McIntyre, G. F., G. D. Godbold, and A. H. Erickson.** 1994. The pH-dependent membrane association of procathepsin L is mediated by a 9-residue sequence within the propeptide. *J. Biol. Chem.* **269**: 567–572.

1770. **Nishimura, Y., K. Kato, K. Furuno, and M. Himeno.** 1995. Inhibitory effect of leupeptin on the intracellular maturation of lysosomal cathepsin L in primary cultures of rat hepatocytes. *Biol. Pharm. Bull.* **18**: 945–950.

1771. **Nishimura, Y., H. Tsuji, K. Kato, H. Sato, J. Amano, and M. Himeno.** 1995. Biochemical properties and intracellular processing of lysosomal cathepsins B and H. *Biol. Pharm. Bull.* **18**: 829–836.

1772. **Punnonen, E. L., V. S. Marjomäki, and H. Reunanen.** 1994. 3-Methyladenine inhibits transport from late endosomes to lysosomes in cultured rat and mouse fibroblasts. *Eur. J. Cell Biol.* **65**: 14–25.

1773. **Tao, K., N. A. Stearns, J. Dong, Q. Wu, and G. G. Sahagian.** 1994. The proregion of cathepsin L is required for proper folding, stability, and ER exit. *Arch. Biochem. Biophys.* **311**: 19–27.

1774. **Ulbricht, B., W. Hagmann, W. Ebert, and E. Spiess.** 1996. Differential secretion of cathepsins B and L from normal and tumor human lung cells stimulated by 12(*S*)-hydroxy-eicosatetraenoic acid. *Exp. Cell Res.* **226**: 255–263.

1775. **Warren, L.** 1989. Stimulated secretion of lysosomal enzymes by cells in culture. *J. Biol. Chem.* **264**: 8835–8842.

Gene structure, nucleotide and amino acid sequences

1776. **Atkins, K. B. and B. R. Troen.** 1995. Phorbol ester stimulated cathepsin L expression in U937 cells. *Cell Growth Differ.* **6**: 713–718.

1777. **Bart, G., G. H. Coombs, and J. C. Mottram.** 1995. Isolation of *lmcpc*, a gene encoding a *Leishmania mexicana* cathepsin-B-like cysteine proteinase. *Mol. Biochem. Parasitol.* **73**: 271–274.

1778. **Berquin, I. M., L. Cao, D. Fong, and B. F. Sloane.** 1995. Identification of two new exons and multiple transcription start points in the 5′-untranslated region of the human cathepsin-B-encoding gene. *Gene* **159**: 143–149.

1779. **Berti, P. J. and A. C. Storer.** 1995. Alignment/ phylogeny of the papain superfamily of cysteine proteases. *J. Mol. Biol.* **246**: 273–283.

1780. **Brömme, D. and K. Okamoto.** 1995. Human cathepsin O2, a novel cysteine protease highly expressed in osteoclastomas and ovary molecular cloning, sequencing and tissue distribution. *Biol. Chem. Hoppe-Seyler* **376**: 379–384.

1781. **Bryce, S. D., S. Lindsay, A. J. Gladstone, K. Braithwaite, C. Chapman, N. K. Spurr, and J. Lunec.** 1994. A novel family of cathepsin L-like (CTSLL) sequences on human chromosome 10q and related transcripts. *Genomics* **24**: 568–576.

1782. **Cao, L., R. T. Taggart, I. M. Berquin, K. Moin, D. Fong, and B. F. Sloane.** 1994. Human gastric adenocarcinoma cathepsin B: isolation and sequencing of full-length cDNAs and polymorphisms of the gene. *Gene* **139**: 163–169.

1783. **Chen, J. M., P. M. Dando, N. D. Rawlings, M. A. Brown, N. E. Young, R. A. Stevens, E. Hewitt, C. Watts, and A. J. Barrett.** 1997. Cloning, isolation, and characterization of mammalian legumain, an asparaginyl endopeptidase. *J. Biol. Chem.* **272**: 8030–8098.

1784. **Conliffe, P. R., S. Ogilvie, R. C. M. Simmen, F. J. Michel, P. Saunders, and K. T. Shiverick.** 1995. Cloning and expression of a rat placental cDNA encoding a novel cathepsin L-related protein. *Mol. Reprod. Dev.* **40**: 146–156.

1785. **Day, S. R., J. P. Dalton, K. A. Clough, L. Leonardo, W. U. Tiu, and P. J. Brindley.** 1995. Characterization and cloning of the cathepsin L proteinases of *Schistosoma japonicum*. *Biochem. Biophys. Res. Commun.* **217**: 1–9.

1786. **Dong, S. S., G. I. Stransky, C. H. Whitaker, S. E. Jordan, P. H. Schlesinger, J. C. Edwards, and H. C. Blair.** 1995. Avian cathepsin B cDNA: sequence and demonstration that mRNAs of two sizes are produced in cell types producing large quantities of the enzyme. *Biochim. Biophys. Acta* **1251**: 69–73.

1787. **Drake, F., I. James, J. Connor, D. Rieman, R. Dodds, F. McCabe, D. Bertolini, R. Barthlow, G. Hastings, and M. Gowen.** 1994. Identification of a novel osteoclast-selective human cysteine proteinase. *J. Bone Miner. Res.* **9**: S177.

1788. **Frick, K. K., P. J. Doherty, M. M. Gottesman, and C. D. Scher.** 1985. Regulation of the transcript for a lysosomal protein: evidence for a gene program modified by platelet-derived growth factor. *Mol. Cell. Biol.* **5**: 2582–2589.

1789. **Gogos, J. A., R. Thompson, W. Lowry, B. F. Sloane, H. Weintraub, and M. Horwitz.** 1996. Gene trapping in differentiating cell lines: regulation of the lysosomal protease cathepsin B in skeletal myoblast growth and fusion. *J. Cell Biol.* **134**: 837–847.

1790. **Hart, K. A., W. Potts, J. Bowyer, G. Slynn, P. Elston, W. Vernon, D. Tucker, K. Tolley, and D. Johnstone.** 1996. Transgenic mice lacking cathepsin L activity. *J. Bone Miner. Res.* **11**: S181.

1791. **Hughes, A. L.** 1994. Evolution of cysteine proteinases in eukaryotes. *Mol. Phylogenet. Evol.* **3**: 310–321.

1792. **Inaoka, T., G. Bilbe, O. Ishibashi, K. Tezuka, M. Kumegawa, and T. Kokubo.** 1995. Molecular cloning of human cDNA for cathepsin K: novel cysteine proteinase predominantly expressed in bone. *Biochem. Biophys. Res. Commun.* **206**: 89–96.

1793. **Inaoka, T., M. Kagoshima-Maezono, S. Kuwano, I. Kondo, M. Kumegawa, and T. Kokubo.** 1996. Human cathepsin K: gene isolation, genomic organization, and chromosomal localization. *J. Bone Miner. Res.* **11**: S180.

1794. **Keppler, D. and B. F. Sloane.** 1996. Cathepsin B – multiple enzyme forms from a single gene and their relation to cancer. *Enzyme Protein* **49**: 94–105.

1795. **Larminie, C. G. and I. L. Johnstone.** 1996. Isolation and characterization of four developmentally regulated cathepsin B-like cysteine protease genes from the nematode *Caenorhabditis elegans*. *DNA Cell Biol.* **15**: 75–82.

1796. **Le Boulay, C., A. Van Wormhoudt, and D. Sellos.** 1995. Molecular cloning and sequencing of two cDNAs encoding cathepsin L-related cysteine proteinases in the nervous system and in the stomach of the Norway lobster (*Nephrops norvegicus*). *Comp. Biochem. Physiol. B Biochem. Mol. Biol.* **111**: 353–359.

1797. **Li, Y. P., M. B. Alexander, A. L. Wucherpfennig, W. Chen, P. Yelick, and P. Stashenko.** 1994. Cloning and characterization of a novel cathepsin from human osteoclasts. *Mol. Biol. Cell* **5**: 335a.

1798. **Li, Y. P., M. Alexander, A. L. Wucherpfennig, P. Yelick, W. Chen, and P. Stashenko.** 1995. Cloning and complete coding sequence of a novel human cathepsin expressed in giant cells of osteoclastomas. *J. Bone Miner. Res.* **10**: 1197–1202.

1799. **Mallinson, D. J., B. C. Lockwood, G. H. Coombs, and M. J. North.** 1994. Identification and molecular cloning of four cysteine proteinase genes from the pathogenic protozoon *Trichomonas vaginalis*. *Microbiology* **140**: 2725–2735.

1800. **Merckelbach, A., S. Hasse, R. Dell, A. Eschlbeck, and A. Ruppel.** 1994. cDNA sequences of *Schistosoma japonicum* coding for two cathepsin B-like proteins and Sj32. *Trop. Med. Parasitol.* **45**: 193–198.

1801. **Michel, A., H. Ghoneim, M. Resto, M. Q. Klinkert, and W. Kunz.** 1995. Sequence, characterization and localization of a cysteine proteinase cathepsin L in *Schistosoma mansoni*. *Mol. Biochem. Parasitol.* **73**: 7–18.

1802. **Mordier, S. B., D. M. Bechet, M. P. Roux, A. Obled, and M. J. Ferrara.** 1995. The structure of the bovine cathepsin B gene. Genetic variability in the 3′ untranslated region. *Eur. J. Biochem.* **229**: 35–44.

1803. **North, M. J.** 1986. Homology within the *N*-terminal extension of cysteine proteinases. *Biochem. J.* **238**: 623–624.

1804. **Paris, A., B. Strukelj, J. Pungercar, M. Renko, I. Dolenc, and V. Turk.** 1995. Molecular cloning and sequence analysis of human preprocathepsin C. *FEBS Lett.* **369**: 326–330.

1805. **Qian, F., S. J. Chan, C. Achkar, D. F. Steiner, and A. Frankfater.** 1994. Transcriptional regulation of cathepsin B expression in B16 melanomas of varying metastatic potential. *Biochem. Biophys. Res. Commun.* **202**: 429–436.

1806. **Rantakokko, J., H. T. Aro, M. Savontaus, and E. Vuorio.** 1996. Mouse cathepsin K: cDNA cloning and predominant expression of the gene in osteoclasts, and in some hypertrophying chondrocytes during mouse development. *FEBS Lett.* **393**: 307–313.

1807. **Rescheleit, D. K., W. J. Rommerskirch, and B. Wiederanders.** 1996. Sequence analysis and distribution of two new human cathepsin L splice variants. *FEBS Lett.* **394**: 345–348.

1808. **Ritonja, A., T. H. T. Coetzer, R. N. Pike, and C. Dennison.** 1996. The amino acid sequences, structure comparisons and inhibition kinetics of sheep cathepsin L and sheep stefin B. *Comp. Biochem. Physiol. B Biochem. Mol. Biol.* **114**: 193–198.

1810. **Sakai, D., H. S. Tong, and C. Minkin.** 1995. Osteoclast molecular phenotyping by random cDNA sequencing. *Bone* **17**: 111–119.

1811. **Shi, G. P., H. A. Chapman, S. M. Bhairi, C. DeLeeuw, V. Y. Reddy, and S. J. Weiss.** 1995. Molecular cloning of human cathepsin O, a novel endoproteinase and homologue of rabbit OC2. *FEBS Lett.* **357**: 129–134.

1812. **Smith, A. M., J. P. Dalton, K. A. Clough, C. L. Kilbane, S. A. Harrop, N. Hole, and P. J. Brindley.** 1994. Adult *Schistosoma mansoni* express cathepsin L proteinase activity. *Mol. Biochem. Parasitol.* **67**: 11–19.

1813. **Tam, S. W., L. R. Cote Paulino, D. A. Peak, K. Sheahan, and M. J. Murnane.** 1994. Human cathepsin B-encoding cDNAs: sequence variations in the 3′-untranslated region. *Gene* **139**: 171–176.

1814. **Tanaka, T., J. Inazawa, and Y. Nakamura.** 1996. Molecular cloning of a human cDNA encoding putative cysteine protease (Prsc1) and its chromosome assignment to 14Q32.1. *Cytogenet. Cell Genet.* **74**: 120–123.

1815. **Volkel, H., U. Kurz, J. Linder, S. Klumpp, V. Gnau, G. Jung, and J. E. Schultz.** 1996. Cathepsin L is an intracellular and extracellular protease in *Paramecium tetraurelia* – purification, cloning, sequencing and specific inhibition by its expressed propeptide. *Eur. J. Biochem.* **238**: 198–206.

1816. **Wijffels, G. L., M. Panaccio, L. Salvatore, L. Wilson, I. D. Walker, and T. W. Spithill.** 1994. The secreted cathepsin L-like proteinases of the trematode, *Fasciola hepatica*, contain 3-hydroxyproline residues. *Biochem. J.* **299**: 781–790.

1817. **Yamamoto, Y., K. Takimoto, S. Izumi, M. Toriyama Sakurai, T. Kageyama, and S. Y. Takahashi.** 1994. Molecular cloning and sequencing of cDNA that encodes cysteine proteinase in the eggs of the silkmoth, *Bombyx mori. J. Biochem. (Tokyo)* **116**: 1330–1335.

Structure–function relationship

1818. **Coulombe, R., P. Grochulski, J. Sivaraman, R. Ménard, J. S. Mort, and M. Cygler.** 1996. Structure of human procathepsin L reveals the molecular basis of inhibition by the prosegment. *EMBO J.* **15**: 5492–5503.

1819. **Coulombe, R., Y. G. Li, S. Takebe, R. Menard, P. Mason, J. S. Mort, and M. Cygler.** 1996. Crystallization and preliminary X ray diffraction studies of human procathepsin L. *Protein Struct. Funct. Genet.* **25**: 398–400.

1820. **Feng, M. H., S. L. Chan, Y. F. Xiang, C. P. Huber, and C. Lim.** 1996. The binding mode of an E-64 analog to the active site of cathepsin B. *Protein Eng.* **9**: 977–986.

1821. **Jia, Z., S. Hasnain, T. Hirama, X. Lee, J. S. Mort, R. To, and C. P. Huber.** 1995. Crystal structures of recombinant rat cathepsin B and a cathepsin B-inhibitor complex. Implications for structure based inhibitor design. *J. Biol. Chem.* **270**: 5527–5533.

1822. **Ménard, R., C. Plouffe, P. Laflamme, T. Vernet, D. C. Tessier, D. Y. Thomas, and A. C. Storer.** 1995. Modification of the electrostatic environment is tolerated in the oxyanion hole of the cysteine protease papain. *Biochemistry* **34**: 464–471.

1823. **Nomura, T., A. Fujishima, and Y. Fujisawa.** 1996. Characterization and crystallization of recombinant human cathepsin L. *Biochem. Biophys. Res. Commun.* **228**: 792–796.

1824. **Sivaraman, J., R. Coloumbe, M. C. Magny, P. Mason, J. S. Mort, and M. Cygler.** 1996. Crystallization of rat procathepsin B. *Acta Crystallogr. D* **52**: 874–875.

1825. **Topham, C. M., N. Srinivasan, C. J. Thorpe, J. P. Overington, and N. A. Kalsheker.** 1994. Comparative modelling of major house dust mite allergen Der p I: structure validation using an extended environmental amino acid propensity table. *Protein Eng.* **7**: 869–894.

1826. **Turk, B., I. Dolenc, E. Zerovnik, D. Turk, F. Gubensek, and V. Turk.** 1994. Human cathepsin B is a metastable enzyme stabilized by specific ionic interactions associated with the active site. *Biochemistry* **33**: 14800–14806.

1827. **Turk, D., M. Podobnik, R. Kuhelj, M. Dolinar, and V. Turk.** 1996. Crystal structures of human procathepsin B at 3.2 and 3.3 Å resolution reveal an interaction motif between a papain-like cysteine protease and its propeptide. *FEBS Lett.* **384**: 211–214.

1828. **Turk, D., M. Podobnik, T. Popovic, N. Katunuma, W. Bode, R. Huber, and V. Turk.** 1995. Crystal structure of cathepsin B inhibited with CA030 at 2.0-Å resolution: a basis for the design of specific epoxysuccinyl inhibitors. *Biochemistry* **34**: 4791–4797.

1829. **Vernet, T., D. C. Tessier, J. Chatellier, C. Plouffe, T. S. Lee, D. Y. Thomas, A. C. Storer, and R. Ménard.** 1995. Structural and functional roles of asparagine 175 in the cysteine protease papain. *J. Biol. Chem.* **270**: 16645–16652.

1830. **Yonezawa, S., T. Takahashi, X. J. Wang, R. N. S. Wong, J. A. Hartsuck, and J. Tang.** 1988. Structures at the proteolytic processing region of cathepsin D. *J. Biol. Chem.* **263**: 16504–16511.